Studies in Classification, Data Analysis, and Knowledge Organization

More information about this series at http://www.springer.com/series/1564

Krzysztof Jajuga · Krzysztof Najman ·
Marek Walesiak
Editors

Data Analysis and Classification

Methods and Applications

Springer

Editors
Krzysztof Jajuga
Department of Financial Investments and Risk Management
Wroclaw University of Economics and Business
Wroclaw, Poland

Krzysztof Najman
Department of Statistics
University of Gdańsk
Sopot, Poland

Marek Walesiak
Department of Econometrics and Computer Science
Wroclaw University of Economics and Business
Jelenia Góra, Poland

ISSN 1431-8814 ISSN 2198-3321 (electronic)
Studies in Classification, Data Analysis, and Knowledge Organization
ISBN 978-3-030-75189-0 ISBN 978-3-030-75190-6 (eBook)
https://doi.org/10.1007/978-3-030-75190-6

Mathematics Subject Classification: 62H25, 62H30, 62H86, 62-09, 68U20, 62P12, 62P20, 62P25

This Springer imprint is published by the registered company Springer Nature Switzerland AG
The registered company address is: Gewerbestrasse 11, 6330 Cham, Switzerland

Preface

This volume presents the papers from the 29th Conference of Section of Classification and Data Analysis of Polish Statistical Society held at the University of Gdansk on September 7–9, 2020. The papers presented refer to a set of studies addressing a wide range of recent methodological aspects and applications of classification and data analysis tools in micro and macroeconomic problems. In the final selection, we accepted 19 of the papers that were presented at the conference. Each of the submissions has been reviewed by two anonymous referees, and the authors have subsequently revised their original manuscripts and incorporated the comments and suggestions of the referees. The selection criteria were based on the contribution of the papers to the theory and applications of modern classification and data analysis.

The chapters have been organized along with the major fields and themes in classification and data analysis: Methodology, Application in Finance, Application in Economics, Application in Social Issues, and Application with COVID-19 Data.

The part on Methodology contains five papers. The paper by Dudek focuses on the new algorithm from spectral clustering family and its applications in large data sets analysis. The author conducted a comparative analysis with other approaches. Rozmus article focuses on the analysis of the number of clusters and stability indicators. The aim of the article is to compare the results in terms of the indicated correct number of groups by classical indexes and stability measures. The paper by Majkowska, Migdał-Najman, Najman, and Raca attempts to characterize words commonly used in the messages published by Twitter users. Text mining methods and techniques were used to carry out the research, which was mainly focused on the analysis of individual words and collocations occurring in the users' tweets. Bryś in his paper conducts research of 1446 selected publications provides insights on classification algorithms applied to information security tasks, their popularity, and the algorithm selection challenges. The paper by Najman and Zieliński investigates the issue of the usefulness of isolation forests in outlier detection. The results of simulations and empirical studies on selected data sets are presented. The assessment takes into account the impact of individual characteristics of big data sets on the effectiveness of the analyzed methods.

The part on Application in Finance contains two papers. Batóg and Wawrzyniak's study was carried out on the basis of selected financial ratios, which in the literature are considered to be nominants with the recommended range of values, with the assumption that the better situation of the examined object is when the values of the indicator-nominant are above the upper limit of the recommended range of values (right-handed asymmetrical nominant) or below the lower limit of this range (left-handed asymmetrical nominant). Trzpiot in her article considers whether the standard risk estimation procedures are in line with investors' expectations. Article is concerned on presenting the assumptions of Gini regression, the selected estimation method, and its application to the systematic risk assessment. The application part is modeling assets listed on the Warsaw Stock Exchange.

The part on Application in Economics contains five papers. The paper by Raca presents an overview of the definitions of the term dark data, a proposal of its interpretation, and a classification of data in a company with regard to: usability, availability, and quality. As part of the research, four universal features of dark data sets have been indicated (unavailability, unawareness, uselessness, and costliness). Cieraszewska, Hamerska, Lula, and Zembura present the results of research including the analysis of abstracts of scientific articles in the field of economics, prepared in English by authors from 36 European countries and registered in the Scopus database in the years 2011–2020. The ontology-based approach is used for identification of concepts related to medical science and economics. The paper also presents the results of research on the relationship between the interdisciplinary nature of research in the field of economics and the number and 'degree of internationalization of authors' teams. The aim of the Putek Szeląg's and Gdakowicz article is to present selected methods of duration analysis to assess the probability of exit from the real estate sale offer system, taking into account various types of competing risk (the year of submitting the property for sale). In the survey, the calculation of the offer duration takes into account the properties that have been sold and are still current (on the day of the end of the survey). Słupik and Trzęsiok's work aims to identify and characterize electricity users in terms of their attitudes toward energy saving. The authors of the article based their analysis on the results of the proprietary research conducted among households in the Silesian Province in Poland, in 2018, and on a review of the literature on profiling individual energy consumers. In the article, the authors also characterize the obtained segments and identify fundamental factors influencing the respondents' behavior toward save energy.

Wolak in the paper presents a study of selected linear ordering algorithms to build a ranking of districts in the Lesser Poland Province in terms of tourist attractiveness using techniques considering potential spatial relationships.

The part on Application in Social Issues contains four papers. Bieszk-Stolorz in her paper assesses the impact of gender of unemployed people on the duration of registered unemployment and on the duration of staying out of the office's register, taking into account different reasons for de-registration. Due to censored observations, i.e., observations not completed with an event in the analyzed period, author decided to use selected methods of survival analysis. The purpose of Grzenda's paper is to indicate the possibility of using Cox regression model to determine direct adjusted probabilities of finding a job by the unemployed depending on their individual characteristics in the context of long-term unemployment risk. The study is based on LFS data from 2017 and 2018 for Poland. Przybysz, Stanimir, and Wasiak proposed to use the methods of multidimensional comparative analysis to assess the level of implementation of the Europe 2020 strategy, indicating areas important for the quality of life of seniors and identifying changes in the assessment of the implementation of this strategy by this generation. The study showed the existence of a very large diversity of seniors in terms of their life quality and their assessment of the strategy. Kos-Łabędowicz and Trzęsiok present two classifications of the elderly in Poland in terms of their preferences regarding means of transport: one prepared on the basis of literature research and expert knowledge, the other with the use of a selected taxonomic method. The aim of the article is to test the agreement between the obtained classifications and thus to verify the validity of the proposed expert segmentation which reflects Polish society specifically.

The part on Application with COVID-19 Data contains three papers. Nojszewska and Sielska analyze the similarities of European countries during COVID-19 pandemic in terms of the following indicators: Economic sentiment indicator (ESI), employment expectations indicator (EEI) from the beginning of 2020. The research shows that after the collapse in March/April 2020, the values of variables reflecting the condition of economies started to increase in most of the identified groups of countries. Salamaga studied a question regarding the influence of the corona crisis on global foreign investment in the near future, especially in the investment market of the Visegrad Group countries. The main purpose of the Landmesser's paper is to analyze the patterns of COVID-19 evolution in a group of 27 EU countries. First, author applies the concept of dynamic time warping (DTW) to identify groups of EU countries affected to varying degrees by the COVID-19 pandemic. Further, within the selected groups, the structure of the time series for infected and deceased COVID-19 patients using ARIMA models was analyzed.

We wish to thank all the authors for making their studies available for our volume. Their scholarly efforts and research inquiries made this volume possible. We are also indebted to the anonymous referees for providing insightful reviews with many useful comments and suggestions.

In spite of our intention to address a wide range of problems pertaining to classification and data analysis theory, there are issues that still need to be researched. We hope that the studies included in our volume will encourage further research and analyses in modern data science.

Wroclaw, Poland
Sopot, Poland
Jelenia Góra, Poland
January 2021

Krzysztof Jajuga
Krzysztof Najman
Marek Walesiak

Contents

About the Editors

Krzysztof Jajuga is a professor of finance at Wroclaw University of Economics and Business, Poland. He holds master, doctoral, and habilitation degree from Wroclaw University of Economics and Business, Poland, title of professor given by the President of Poland, honorary doctorate from Cracow University of Economics and honorary professorship from Warsaw University of Technology. He carries out research within financial markets, risk management, household finance, and multivariate statistics.

Krzysztof Najman is an associate professor at the University of Gdansk, Deputy Dean for Student Affairs and Education at Faculty of Management. He obtained doctoral degree and habilitation degree from University of Gdansk in Poland. He is a member of Main Council of Polish Statistical Association and Section of Classification and Data Analysis SKAD. His field of scientific interests covers cluster analysis and classification methods, artificial intelligence models, self-learning neural networks, multivariate statistical analysis, data mining.

Marek Walesiak is a professor of economics at Wroclaw University of Economics and Business in Department of Econometrics and Computer Science. He holds master, doctoral, and habilitation degree from Wroclaw University of Economics and Business, Poland, title of professor given by the President of Poland. He is a member of the Methodological Commission and Scientific Statistical Council in Statistics Poland (GUS) and an active member of many scientific professional bodies (i.e., Section of Classification and Data Analysis SKAD). His main areas of interest include: classification and data analysis, composite indicators, multivariate statistical analysis, marketing research, computational techniques in R.

Methodology

Evaluation of Two-Step Spectral Clustering Algorithm for Large Untypical Data Sets

Andrzej Dudek

Abstract Researchers analyzing large (>100,000 objects) data sets with the methods of cluster analysis often face the problem of computational complexity of algorithms that sometimes makes it impossible to analyze in an acceptable time. Common solution of this problem is to use less computationally complex algorithms (like k-means), which in turn can in many cases give much worse results than for example algorithms using eigenvalues decomposition. In the article, the new algorithm from spectral clustering family is proposed and compared with other approaches.

Keywords Clustering · Classification · Large data sets · Spectral clustering

1 Introduction

Researchers analyzing large (>100,000 objects) data sets with the methods of cluster analysis often face the number of problems that make analysis very hard or even impossible. Computational complexity of algorithms, sometimes, makes it impossible to analyze in an acceptable time. The other limitation is memory size of standard PC-like computers, which in many cases may be too small for necessary calculations on such data sets. Thus, not all clustering algorithms may be used for that kind of data.

The article is divided into four parts with introduction. First part presents which clustering algorithms can or cannot be used for large data sets in popular statistical **R** framework. The second part is a proposal of modification of spectral clustering procedure. Third part present computational simulation results on over 100,000 objects data matrices with known cluster structure for untypical cluster shapes against the proposed algorithm. The final part contains remarks and conclusions.

A. Dudek (✉)
Wrocław University of Economics and Business, Wrocław, Poland
e-mail: andrzej.dudek@ue.wroc.pl

K. Jajuga et al. (eds.), *Data Analysis and Classification*, Studies in Classification, Data Analysis, and Knowledge Organization,
https://doi.org/10.1007/978-3-030-75190-6_1

2 Limitations of Large Data Sets Classification

Dudek (2013) has examined the following clustering algorithms on one million object multivariate normal distribution data set:

- hierarchical agglomerative methods,
- hierarchical divisive method (diana),
- k-means algorithm,
- partition around medoids (pam, k-medoids algorithm),
- spectral clustering approach (von Luxburg 2006),
- ensemble approach (Dimitriadou et al. 2001).

Only one algorithm (k-means) has passed the following requirements in **R** environment:

- method execution should not report any lack of memory error,
- method should not run longer than five hours.

But in further analysis for untypical cluster shapes, k-means has given the results that not meet the actual structure of clusters.

3 Proposal of New Algorithm

Spectral decomposition algorithm according to von Luxburg (2006) and Ng et al. (2002) can be stated in its general form in the following way:

Let X means data matrix with n rows and m columns, u—number of cluster to divide X (given by researcher before start of decomposition). Sample input data is presented on Fig. 1. Next figures will be showing the same data in transformed space.

Let **A** be similarity matrix of objects from **X**. **A** can be calculated in many ways but most often its elements a_{ij} are defied according to Eq. 1:

$$a_{ij} = e^{-\frac{\sum_{k=1}^{m}\left(x_{ik}-x_{jk}\right)^2}{\sigma}} \tag{1}$$

where: σ—scaling parameter. Most often it is calculated according to Ng et al. (2002) algorithm of iterative choosing of σ, minimalizing the with-class distances of random subset (random rows selected) of **X**: **X′** (this method requires processing of approximately few hundreds clustering procedures of objects in **X′**),

n—number of rows,

m—number of columns,

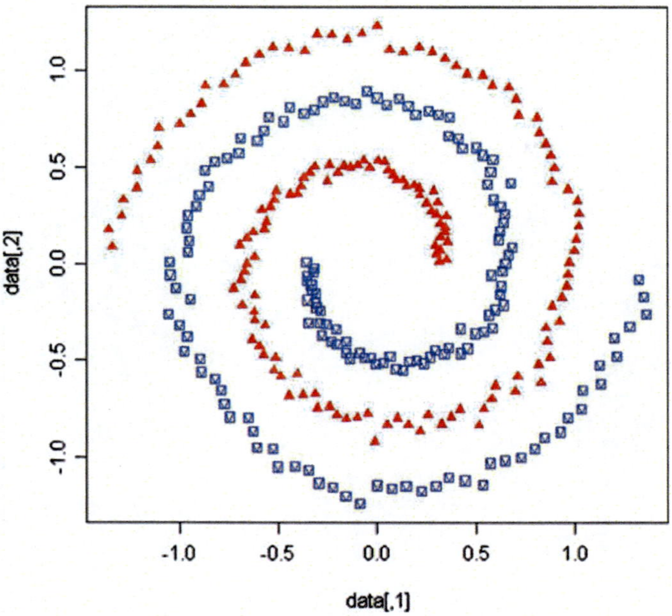

Fig. 1 Input data before spectral decomposition. *Source* Own elaboration with use of *mlbench* **R** library

$$i,j = 1,\ldots,n; k = 1,\ldots,m.$$

For **A** weights matrix **W** is constructed due to Eq. 2:

$$w_{ij} = \begin{cases} \sum_{j=1}^{n} a_{ij} & if \quad i = j \\ 0 & if \quad i \neq j \end{cases} \tag{2}$$

where: **W** = $[w_{ij}]$—weights matrix.

Laplacian **L** is calculated next according to Eq. 3:

$$\boldsymbol{L} = \mathbf{W}^{-\frac{1}{2}} \times \mathbf{A} \times \mathbf{W}^{-\frac{1}{2}} \tag{3}$$

L can be treated as algebraic representation of graph created from objects of **X**.

First *u* eigenvectors of Laplacian **L** creates **E** matrix. Each eigenvector is treated as column of **E** (thus **E** ma has dimensions $n \times u$). The main aim of this step is to widen data in transformed space (see Fig. 2).

Optional matrix **E′** is a result of normalization of **E** due to Eq. 4. This step is narrowing data in transformed space (it can be observed on Fig. 3).

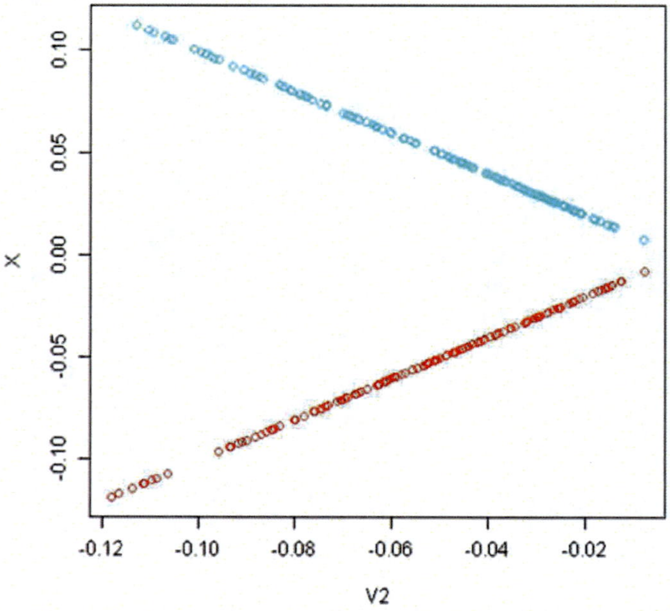

Fig. 2 Data in transformed space. *Source* Own elaboration with use of *mlbench* **R** library

$$E'_{ij} = \frac{E_{ij}}{\sqrt{\sum_{k=1}^{n} E_{kj}^2}} \tag{4}$$

In last stage, **E**′ (or **E** if normalization step is omitted) is clustered with one of "standard" algorithms. Most often *k*-means is used for this purpose.

The two-step spectral clustering algorithm (TSSC) is a try-out of avoiding computational limitations of spectral clustering algorithm. The approach behind it is similar to that used by Shinnou and Sasaki (2008) and Kong et al. (2011).

It can be stated in four steps:

1. Pre-cluster the data set with k-means into given number (500 or 1000) of small clusters (inter-clusters).
2. Calculate the medoids of each cluster.
3. Run normal clustering procedure on cluster medoids from first step.
4. Finally assign object from original data set to the cluster to which belongs the medoid of inter-cluster.

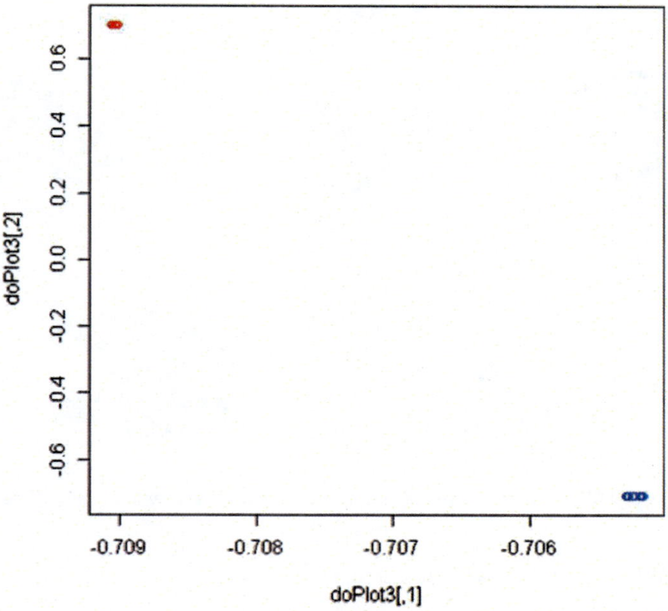

Fig. 3 Data in transformed space after normalization step. *Source* Own elaboration with use of *mlbench* **R** library

4 Simulation Experiment Results

For measuring the quality of methods based on spectral decomposition, an experiment has been carried out. In all simulations, the adjusted Rand (Hubert and Arabie 1985) index has been used for measuring the quality of clustering. The experiment examines this technique in case of non-standard cluster shapes.

In the experiments, the results of clustering with use of spectral decomposition, two-step spectral clustering (TSSC), k-means algorithm, k-medoids algorithm, Ward clustering, and complete link clustering have been compared on five data models: *Spirals, Worms, WWW, Smiley, Cassini*. For each model, fifty random realizations have been generated with use of *mlbench* **R** library. Each data set consists of more than 100,000 objects and is untypical in sense that they are not generated from any distribution mixture. Figure 4 shows sample data sets generated from each model.

Table 1 shows the results of simulation. In all cases, clustering based on spectral decomposition has found the more accurate class structure.

Only two from compared methods have not crushed due to memory limits. From those two, the newly proposed TSSC algorithm gave better results in every case. In two cases (spirals and w3), the difference of performance is at very high level, while for three other, the results measured in average adjusted Rand are comparable, but against with significant TCSS advantage.

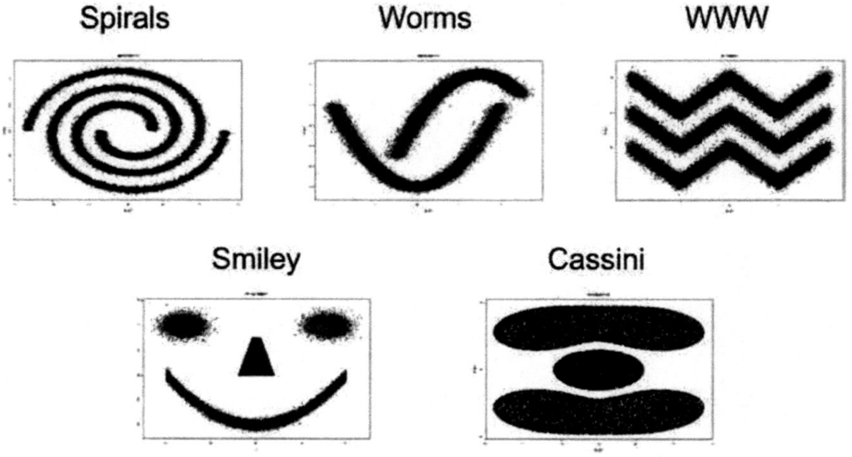

Fig. 4 Data sets used in first experiment. *Source* Own elaboration with use of *mlbench* **R** library

Table 1 Average adjusted Rand values from 50 simulations

	Nr of objects	k-means	k-medoids	Ward	Complete link	Spectral clust	TSSC
Spirals	200,000	0.0350	*	*	*	*	1
Worms	200,000	0.5116	*	*	*	*	0.9999
w3	300,000	0.0049	*	*	*	*	0.9359
Smiley	120,000	0.7981	*	*	*	*	0.8534
Cassini	300,000	0.8157	*	*	*	*	0.9917

*—Calculational/memory complexity exceeded

5 Final Remarks and Conclusions

The problem of classification of large data may be divided into two groups. Classification of large data sets with typical, given from normal distribution, shapes and classification of large data sets with untypical non ellipsoid-like clusters. While in first case, "standard" clustering algorithms give satisfying results in acceptable time, the second type of classification needs further development. In the paper, new algorithm is evaluated for such data sets, and the results of experimental analysis are very promising.

Author is aware that "there is no free lunch" and is far for acclaiming this method as "the best clustering algorithm" but sometimes it can behave better that standard well-known methods of cluster analysis.

References

Dimitriadou E, Weingessel A, Hornik K (2001) Voting-merging: an ensemble method for clustering. In: Dorffner G, Bischop H, Hornik K (eds) Artificial neural networks—ICANN 2001. Lecture notes in computer science, vol 2130. Springer, Heidelberg, pp 217–224

Dudek A (2013) Classification of large data sets. Comparison of performance of chosen algorithms. Acta Universitatis Lodziensis. Folia Oeconomica 285:71–78

Hubert LJ, Arabie P (1985) Comparing partitions. J Classif 2:193–218

Kong T, Tian Y, Shen H (2011) A fast incremental spectral clustering for large data sets, pp 1–5. https://doi.org/10.1109/PDCAT.2011.4

Ng A, Jordan M, Weiss Y (2002) On spectral clustering: analysis and an algorithm. In: Dietterich T, Becker S, Ghahramani Z (eds) Advances in neural information processing systems 14. MIT Press, pp 849–856

Shinnou H, Sasaki M (2008) Spectral clustering for a large data set by reducing the similarity matrix size. In: Proceedings of the sixth international conference on language resources and evaluation (LREC), pp 201–2014

von Luxburg U (2006) A tutorial on spectral clustering. Max planck institute for biological cybernetics, Technical Report TR-149

Determining the Number of Groups in Cluster Analysis Using Classical Indexes and Stability Measures—Comparison of Results

Dorota Rozmus

Abstract In the context of taxonomic methods, in recent years, much attention has been paid to the issue of the stability of these methods, i.e., the answer to the question: to what extent the structure discovered by a given method is actually present in the data. This criterion examines whether the groups that were created as a result of using clustering method to a set of objects are real (the structure is stable), or whether they appeared accidentally. Most often this criterion is used when selecting the number of groups (k), for which should be clustered a set of data. The aim of the article is to compare the results in terms of the indicated correct number of groups by classical indexes and stability measures.

Keywords Clustering · Stability measures · Internal indexes

1 Introduction

The main problem in taxonomy is to determine whether the groups that we received reflect the actual structure of general population (which generated the data). This involves the problem of "clustering model" identification, e.g., the number of groups k, a distance metric, the control parameters of an algorithm. Recently, the stability criterion increasingly gains in popularity in response to these problems.

Informally, this criterion states that if a cluster algorithm is repeatedly used for independent samples of objects (with unchanged parameters of the algorithm), resulting in similar grouping results, it can be considered as stable and reflecting the actual structure of the groups (Shamir and Tishby 2008). Volkovich et al. (2010) even state that the number of groups that maximizes the stability of clustering can serve as an estimate of the "true" number of groups.

The literature proposes a number of different ways for measuring stability (e.g., Ben-Hur and Guyon 2003; Brock et al. 2008; Henning 2007; Fang and Wang 2012;

D. Rozmus (✉)
University of Economics in Katowice, Katowice, Poland
e-mail: dorota.rozmus@ue.katowice.pl

K. Jajuga et al. (eds.), *Data Analysis and Classification*, Studies in Classification, Data Analysis, and Knowledge Organization,
https://doi.org/10.1007/978-3-030-75190-6_2

Lord et al. 2017; Marino and Presti 2019; Suzuki and Shimodaira 2006). Theoretical considerations have also led to the development of computer tools for the practical implementation of the proposed ways to study stability. The practical tools are available within several **R** packages, for example: `clv`, `clValid`, `ClusterStability`, `fpc`, `pvclust`.

Due to the hypothesis that the stability of clusters may be the answer to the question about the appropriate number of clusters (*k*), the aim of the article will be to compare the results in the context of the indicated value of *k* by classical indexes that so far served this issue (e.g., Hubert and Levin index, Dunn index, Silhouette index) and the cluster stability measures proposed in the literature.

2 Measures of Cluster Stability

This part of the article presents the research methods, i.e., cluster stability measures. In this study, only such stability measures that one can find in the **R** program were used, i.e., measures from packages: `clv`, `clValid` and `fpc`.[1] There are much more packages for stability testing of course, but those mentioned libraries can be used with various clustering methods, e.g., *k*-means, *k*-medoids, hierarchical, and others.

As the classical indexes for determining the number of groups in clutering are well known, they will not be discussed in details. It should be mentioned, however, that only internal measures will be used.

2.1 Ben-Hur and Guyon Stability Measure

The concept of stability by Ben-Hur and Guyon (2003) is based on the finding that if the clustering properly represents the structure in the data, it should be stable with respect to small changes in the data set. They proposed two measures of stability: a measure based on the index of similarity between two partitions[2] and a measure based on the pattern-wise agreement concept[3].

The algorithm of calculating of stability measure based on the index of similarity between two partitions can be described in the following steps:

1. Cluster the original data set in order to obtain the reference partition.
2. Select a random subsample of observations from the original data set and group the objects from this subsample.

[1]Packages `clv`, `clValid` and `fpc`, were also selected because the methods implemented there have been the subject of the author's research for a long time (e.g., Rozmus 2017).

[2]This measure is implemented by the function `cls.stab.sim.ind` in `clv` package in **R**.

[3]This measure is implemented by the function `cls.stab.opt.assign` in `clv` package in **R**.

3. Calculate the stability between the reference partition and the partition of the subsample using the index of similarity between two partitions (e.g., Rand index).
4. Repeat the procedure several times.
5. Repeat the procedure for different values of k (number of groups).

The pattern-wise agreement concept of stability measure is based on the idea of pattern-wise agreement and pattern-wise stability.

Given two groupings L_1 and L_2, pattern-wise agreement can be defined as follows:

$$\delta_\sigma(i) = \begin{cases} 1, & \textit{if } \sigma(L_1(i)) = L_2(i), \\ 0, & \textit{if } \sigma(L_1(i)) \neq L_2(i), \end{cases} \tag{1}$$

where $\sigma : \{1,\ldots,k_1\} \to \{1,\ldots,k_2\}$.

Pattern-wise stability is defined as the fraction of subsampled partitions where the subsampled labeling of observation i agrees with that of the reference labeling, by averaging the pattern-wise agreement:

$$\mathrm{n(i)} = \frac{1}{\mathrm{N_i}} \sum \delta_\sigma(\mathrm{i}) \tag{2}$$

where N_i—number of subsamples where pattern i appears.

The stability of group j in the reference partition is the average of pattern-wise stability:

$$c(j) = \frac{1}{|L_1 = j|} \sum\nolimits_{i \in (L_1 = j)} n(i) \tag{3}$$

where $|-|$ means cardinality of the set.

The stability of the reference partition into k groups is defined as:

$$S_k = \min_j c(j). \tag{4}$$

Finally, the most stable clustering is indicated by the maximum of S_k.

2.2 *Brock, Pihur, Datta, and Datta Stability Measure*

Measures of stability by Brock et al. (2008)[4] are dedicated mainly for validating the results of clustering analysis in biology. There are three main types of cluster validation measures available: "internal," "biological," and "stability."

[4]This measure can be found in `clValid` package in **R**.

The article focuses only on the last group of measures. They evaluate the stability of a clustering result by comparing it with the clusters obtained by removing one column (i.e., variable) at a time (Brock et al. 2008). These measures include: the average proportion of non-overlap (APN), the average distance (AD), the average distance between means (ADM), and the figure of merit (FOM).

Only APN was used in experiments because this is the only measure that is normalized in the interval (0, 1), with values close to zero corresponding with highly consistent clustering results. APN measures the average proportion of observations not placed in the same cluster by clustering based on the full data and clustering based on the data with a single column removed:

$$APN = \frac{1}{M \cdot N} \sum_{i=1}^{N} \sum_{j=1}^{M} \left(1 - \frac{n(C^{i,j} \cap C^{i,0})}{n(C^{i,0})}\right), \tag{5}$$

where

$C^{i,0}$ represents the cluster containing observation i using the original clustering (based on all available data),
$C^{i,j}$ represents the cluster containing observation i where the clustering is based on the data set with j column removed,
$n(\cdot)$ is the cardinality of a cluster,
N denotes the total number of observations (rows) in a data set,
M denotes the total number of variables (columns) in a data set.

2.3 Fang and Wang Stability Measure

Fang and Wang stability measures (2012)[5] focus on the concept of stability as robustness to randomness present in the sample. Drawing on the work of Wang (2010), they formulate the concept of stability in the following way: if one draws samples from the population and applies a selected clustering algorithm, the results of grouping should not be very different.

Presented Fang and Wang measure is based on the following general idea: Several times two bootstrap samples are drawn from the data, and the number of clusters is chosen by optimizing an instability estimation from these pairs.

Denoting a cluster algorithm with $k \geq 2$ groups by $\Psi(\cdot, k)$, when we use it to sample X^n, we get the clustering $\Psi_{X^n,k}(x)$; the algorithm can be presented according to the following procedure. For the assumed value of $k = 2, \ldots, K$:

[5]This measure can be found in fpc package in **R**. It includes two functions for measuring stability: clusterboot and nselectboot. In the experiments only the nselectboot function was used.

1. Construct B independent pairs of bootstrap samples $\left(X_b^{n*}, \tilde{X}_b^{n*}\right)$, $b = 1, \ldots, B$.
2. Make groupings $\Psi_{X_b^{n*},k}$ and $\Psi_{\tilde{X}_b^{n*},k}$ on $\left(X_b^{n*}, \tilde{X}_b^{n*}\right)$, $b = 1, \ldots, B$.
3. For each pair, $\Psi_{X_b^{n*},k}$ and $\Psi_{\tilde{X}_b^{n*},k}$ calculate the empirical clustering distance:

$$d\left(\Psi_{X_b^{n*},k}, \Psi_{\tilde{X}_b^{n*},k}\right) = \\ = \frac{1}{n^2}\sum\nolimits_{i=1}^{n}\sum\nolimits_{j=1}^{n}\left|I\left\{\Psi_{X_b^{n*},k}(x_i) = \Psi_{X_b^{n*},k}(x_j)\right\} - I\left\{\Psi_{\tilde{X}_b^{n*},k}(x_i) = \Psi_{\tilde{X}_b^{n*},k}(x_j)\right\}\right|. \tag{6}$$

4. Instability of clustering is calculated as:

$$\hat{s}_B = \frac{1}{B}\sum\nolimits_{b=1}^{B} d\left(\Psi_{X_b^{n*},k}, \Psi_{\tilde{X}_b^{n*},k}\right). \tag{7}$$

3 A Data Set and the Scheme of Research

A data set was built on the data obtained from the sustainable development indicators application developed by the Central Statistical Office in Poland. This application monitors the implementation of the sustainable development policy in the EU countries. The data are divided into four groups, monitoring the implementation of the sustainable development policy within the following domains:

- social,
- economic,
- environmental,
- institutional and political.

The study used the data from 2015, comprising 19 metric variables in the social domain, 18 variables in the economic domain, 11 in the environmental domain, and 15 in the institutional and political domain (only complete data were used).

Clustering was carried out within each domain separately. This is related to the idea of weak and strong sustainability (Borys 2005, 2014; Lorek 2011). In accordance with weak sustainability, it is permissible to consider all domains together, because the resources from these domains are considered substitutable. According to strong sustainability, resources within each domain are considered to be complementary, and therefore, every order should be considered separately because it is not possible to develop one domain at the expense of the other. And in this spirit, the analysis presented in the paper was carried out.

As a clustering methods, two partitioning algorithms were used, i.e., k-means and k-medoids and two hierarchical algorithms, i.e., group average linkage and Ward method.

Among the classical indexes used to determine the number of clusters, only internal measures was chosen, i.e., Hubert and Levin index, Davies and Bouldin index, CalińskiHarabasz index, and Silhouette index. As these are commonly known and recognized measures, they will not be discussed in detail in the paper[6].

In Ben-Hur and Guyon measure of stability, similarity between two partitions were tested with all available in `clv` package indices, i.e., Rand, dot product, similarity index, and Jaccard. For creating subsamples, the subset ratio was equal 0.8.

In stability measure proposed by Fang and Wang (`fpc` package), 100 bootstrap samples were created. Each new data set (of the same size as the original) is created by resampling the original data set with replacement.

4 Empirical Results

As it was mentioned before, the study was carried within each domain separately. The results are discussed below.

In Tables 2, 3, 4, 5, there are presented values of classical indexes and different stability measures used in the experiments. The measures were calculated only for $k = 2, \ldots, 5$. Abbreviations used on those figures are explained in Table 1. In last column, there is presented the information about optimization direction of the criterion.

4.1 Results for the Social Domain

Looking at the values for different stability measures (Table 2), it can be seen that classical indexes and stability measures suggest different value of k (number of groups). Indexes in most cases point $k = 2$ as the optimal value, whereas stability measures the most often indicate $k = 3$. It is also worth paying attention to the Hubert and Levin and Dunn index, which, like Fang and Wang stability measure (FW), suggest the maximum number of groups under consideration. This scheme will also appear in other domains (especially for Dunn index and Fang and Wang stability measure).

[6]The indexes were calculated using the functions from the `clusterSim` and `clusterCrit` packages.

Table 1 Description of abbreviations for the stability measures used in the presentation of the results and the direction of their optimization

Abbreviation	Description	Optimization direction
BH-G rand	Ben-Hur and Guyon measure of stability, with Rand similarity index (implemented by the function `cls.stab.sim.ind`)	max
BH-G dot	Ben-Hur and Guyon measure of stability, with dot product similarity index (implemented by the function `cls.stab.sim.ind`)	max
BH-G sim	Ben-Hur and Guyon measure of stability, with similarity index (implemented by the function `cls.stab.sim.ind`)	max
BH-G jaccard	Ben-Hur and Guyon measure of stability, with Jaccard similarity index (implemented by the function `cls.stab.sim.ind`)	max
BH-G1	Ben-Hur and Guyon measure of stability implemented by the function `cls.stab.sim.opt.assigned`	max
B et al.	Measures of stability by Brock et al. indicated by the average proportion of non-overlap	min
FW	Fang and Wang stability measure (implemented by `nselectboot` function)	min

Source Own computations

4.2 Results for the Economic Domain

Looking at the results for the economic domain, it can be generally stated that the classical indexes and measures of stability are, in most cases, compatible, suggesting $k = 2$. The only exceptions are the average method, where the indexes indicate $k = 5$ as correct, while the stability measures indicate $k = 2$. Moreover, it can be observed that for most of the considered clustering methods, Dunn index and Fang and Wang stability measure (FW) suggest the maximum considered number of groups.

4.3 Results for the Environmental Domain

A huge divergence of results as to the indications of the actual number of groups can be observed for this domain. For example, for the *k*-means method, the indexes most often suggest $k = 4$, while the stability measures indicate very different values (from $k = 2$ to $-k = 5$). A very large variation in the suggested value of the k parameter can be noticed for the *k*-medoids method, both in the context of classical indexes and cluster stability measures. For hierarchical methods, the indices and

Table 2 Results of clustering for partitioning method (social domain)

Index					Stability measures				
k-means	**#2**	**#3**	**#4**	**#5**	**k-means**	**#2**	**#3**	**#4**	**#5**
Hubert and Levin	0.167	0.151	0.112	**0.102**	BH-G rand	0.856	**0.954**	0.907	0.843
Davies and Bouldin	**0.848**	1.526	1.526	1.526	BH-G dot	0.862	**0.928**	0.816	0.651
Caliński-Harabasz	**36.170**	20.356	9.820	9.658	BH-G sim	0.845	**0.940**	0.806	0.629
Silhouette	**0.247**	0.200	0.186	0.183	BH-G jaccard	0.659	**0.897**	0.673	0.519
Dunn	0.388	0.437	0.456	**0.509**	BH-G1	0.855	**0.962**	0.806	0.429
					B et al.	**0.021**	0.025	0.045	0.055
					FW	0.110	0.065	0.067	**0.062**
k-medoids	**#2**	**#3**	**#4**	**#5**	**k-medoids**	**#2**	**#3**	**#4**	**#5**
Hubert and Levin	0.190	0.195	0.178	**0.158**	BH-G rand	0.749	**0.975**	0.914	0.958
Davies and Bouldin	**0.692**	1.787	4.116	4.261	BH-G dot	0.760	**0.959**	0.816	0.884
Caliński-Harabasz	23.754	**31.808**	23.437	17.629	BH-G sim	0.727	**0.974**	0.791	0.894
Silhouette	**0.229**	0.149	0.133	0.109	BH-G jaccard	0.814	**0.883**	0.588	0.84
Dunn	**0.415**	0.318	0.328	0.328	BH-G1	0.918	**0.958**	0.913	0.765
					B et al.	0.037	0.034	**0.033**	0.077
					FW	0.147	0.145	0.135	**0.117**
Average	**#2**	**#3**	**#4**	**#5**	**Average**	**#2**	**#3**	**#4**	**#5**
Hubert and Levin	0.237	0.163	0.146	**0.145**	BH-G rand	0.809	0.789	**0.899**	0.891
Davies and Bouldin	**0.827**	1.142	1.987	1.270	BH-G dot	**0.836**	0.747	0.828	0.765
Caliński-Harabasz	1.358	14.523	9.927	8.654	BH-G sim	0.805	0.673	**0.875**	0.742
Silhouette	**0.236**	0.177	0.150	0.129	BH-G jaccard	0.595	0.682	**0.706**	0.674
Dunn	**0.445**	0.408	0.408	0.408	BH-G1	0.387	0.651	**0.884**	0.437

(continued)

Table 2 (continued)

Index					Stability measures				
					B et al.	0.052	**0.025**	0.027	2,000
					FW	0.212	0.214	0.199	**0.174**
Ward	**#2**	**#3**	**#4**	**#5**	**Ward**	**#2**	**#3**	**#4**	**#5**
Hubert and Levin	0.199	0.174	0.149	**0.100**	BH-G rand	0.772	0.902	**0.914**	0.909
Davies and Bouldin	**0.933**	3.362	4.639	3.814	BH-G dot	0.787	**0.845**	0.821	0.768
Caliński-Harabasz	**28.842**	14.638	9.638	8.150	BH-G sim	0.685	**0.92**	0.881	0.868
Silhouette	**0.229**	0.144	0.165	0.178	BH-G jaccard	0.583	**0.783**	0.600	0.582
Dunn	0.366	0.429	0.429	**0.509**	BH-G1	0.784	**0.847**	0.763	0.708
					B et al.	0.044	**0.034**	0.041	0.047
					FW	0.128	0.130	0.131	**0.119**

# denotes number of clusters

Source Own computations

Table 3 Results of clustering for partitioning method (economic domain)

Index					Stability measures				
k-means	**#2**	**#3**	**#4**	**#5**	**k-means**	**#2**	**#3**	**#4**	**#5**
Hubert and Levin	0.201	0.184	**0.123**	0.128	BH-G rand	0.871	0.883	**0.920**	0.894
Davies and Bouldin	**1.079**	1.248	1.084	1.413	BH-G dot	**0.886**	0.845	0.850	0.743
Caliński-Harabasz	**22.207**	18.320	20.346	15.946	BH-G sim	**0.886**	0.833	0.820	0.724
Silhouette	**0.227**	0.192	0.207	0.173	BH-G jaccard	**0.811**	0.753	0.765	0.618
Dunn	0.190	0.336	0.382	**0.409**	BH-G1	**0.847**	0.817	0.678	0.472
					B et al.	0.018	**0.017**	0.037	0.039
					FW	**0.051**	0.106	0.092	0.079
k-medoids	**#2**	**#3**	**#4**	**#5**	**k-medoids**	**#2**	**#3**	**#4**	**#5**
Hubert and Levin	0.201	0.176	0.133	**0.124**	BH-G rand	**0.942**	0.879	0.865	0.898
Davies and Bouldin	**0.857**	1.628	1.464	1.292	BH-G dot	**0.949**	0.832	0.776	0.816
Caliński-Harabasz	**21.065**	13.811	9.638	10.969	BH-G sim	**0.949**	0.858	0.739	0.851
Silhouette	**0.227**	0.200	0.202	0.184	BH-G jaccard	**0.9136**	0.742	0.641	0.702
Dunn	0.190	0.278	0.362	**0.372**	BH-G1	**0.986**	0.838	0.542	0.821
					B et al.	**0.029**	0.033	0.035	0.042
					FW	0.144	0.145	0.118	**0.102**
Average	**#2**	**#3**	**#4**	**#5**	**Average**	**#2**	**#3**	**#4**	**#5**
Hubert and Levin	0.210	0.165	0.165	**0.095**	BH-G rand	**0.942**	0.878	0.896	0.880
Davies and Bouldin	1.305	1.509	1.251	**0.961**	BH-G dot	**0.953**	0.894	0.886	0.852
Caliński-Harabasz	2.426	1.915	1.417	**8.281**	BH-G sim	**0.947**	0.873	0.837	0.838
Silhouette	**0.232**	0.174	0.141	0.204	BH-G jaccard	**0.922**	0.812	0.806	0.752
Dunn	**0.489**	0.469	0.469	0.481	BH-G1	**0.989**	0.982	0.757	0.764

(continued)

Table 3 (continued)

Index					Stability measures				
					B et al.	**0.004**	0.011	0.024	0.039
					FW	0.203	0.181	0.137	**0.105**
Ward	**#2**	**#3**	**#4**	**#5**	**Ward**	**#2**	**#3**	**#4**	**#5**
Hubert and Levin	0.221	0.189	0.145	**0.107**	BH-G rand	0.914	0.877	0.890	**0.944**
Davies and Bouldin	**0.961**	1.276	1.077	1.231	BH-G dot	**0.930**	0.835	0.802	0.877
Caliński-Harabasz	**24.834**	18.338	24.184	19.687	BH-G sim	**0.925**	0.82	0.737	0.838
Silhouette	**0.211**	0.174	0.193	0.185	BH-G jaccard	**0.882**	0.751	0.676	0.787
Dunn	0.313	0.321	0.321	**0.385**	BH-G1	**0.993**	0.846	0.723	0.815
					B et al.	**0.022**	0.035	0.048	0.058
					FW	0.135	0.126	0.119	**0.101**

Source Own computations

Table 4 Results of clustering for partitioning method (environmental domain)

Index					Stability measures				
k-means	**#2**	**#3**	**#4**	**#5**	**k-means**	**#2**	**#3**	**#4**	**#5**
Hubert and Levin	0.292	0.205	0.117	**0.105**	BH-G rand	0.786	0.849	0.806	**0.891**
Davies and Bouldin	1.586	1.529	**1.236**	1.424	BH-G dot	**0.808**	0.801	0.671	0.788
Caliński-Harabasz	10.336	6.631	**14.146**	11.594	BH-G sim	0.767	0.773	0.666	0.787
Silhouette	0.185	0.228	**0.259**	0.226	BH-G jaccard	**0.712**	0.677	0.529	0.661
Dunn	0.274	0.317	**0.400**	**0.400**	BH-G1	0.711	**0.886**	0.333	0.632
					B et al.	0.155	0.060	**0.054**	0.076
					FW	0.157	0.103	0.102	**0.095**
k-medoids	**#2**	**#3**	**#4**	**#5**	**k-medoids**	**#2**	**#3**	**#4**	**#5**
Hubert and Levin	0.332	0.193	0.194	**0.124**	BH-G rand	0.715	0.798	**0.833**	0.821
Davies and Bouldin	**1.527**	1.785	2.016	1.603	BH-G dot	**0.758**	0.748	0.701	0.597
Caliński-Harabasz	**8.565**	4.553	4.893	6.406	BH-G sim	0.662	**0.714**	0.709	0.642
Silhouette	0.181	**0.238**	0.191	0.197	BH-G jaccard	**0.637**	0.630	0.571	0.443
Dunn	0.291	0.392	0.392	**0.494**	BH-G1	0.768	**0.903**	0.846	0.738
					B et al.	**0.055**	0.056	0.119	0.120
					FW	0.179	0.155	0.137	**0.126**
Average	**#2**	**#3**	**#4**	**#5**	**Average**	**#2**	**#3**	**#4**	**#5**
Hubert and Levin	0.181	0.153	0.130	**0.086**	BH-G rand	0.865	**0.937**	0.886	0.703
Davies and Bouldin	1.621	1.355	**1.018**	1.037	BH-G dot	0.917	**0.959**	0.913	0.727
Caliński-Harabasz	1.091	5.291	5.802	**9.753**	BH-G sim	0.857	**0.927**	0.875	0.652
Silhouette	**0.303**	0.234	0.204	0.242	BH-G jaccard	0.861	**0.928**	0.850	0.558
Dunn	0.425	0.400	0.452	**0.557**	BH-G1	**0.915**	0.847	0.729	0.563
					B et al.	**0.009**	0.035	0.063	0.119
					FW	0.189	0.216	0.184	**0.148**
Ward	**#2**	**#3**	**#4**	**#5**	**Ward**	**#2**	**#3**	**#4**	**#5**
Hubert and Levin	0.257	0.193	0.114	**0.086**	BH-G rand	0.823	**0.957**	0.869	0.893
Davies and Bouldin	1.698	1.439	1.262	**1.037**	BH-G dot	0.854	**0.943**	0.770	0.781
Caliński-Harabasz	7.933	5.587	9.683	**9.753**	BH-G sim	0.782	**0.947**	0.725	0.796
Silhouette	0.206	0.238	**0.255**	0.242	BH-G jaccard	0.775	**0.909**	0.642	0.656
Dunn	0.362	0.392	0.494	**0.557**	BH-G1	0.900	**0.948**	0.733	0.760
					B et al.	**0.039**	0.052	0.116	0.153
					FW	0.152	0.147	0.130	**0.117**

Source Own computations

Table 5 Results of clustering for partitioning method (institutional and political domain)

Index					Stability measures				
k-means	**#2**	**#3**	**#4**	**#5**	**k-means**	**#2**	**#3**	**#4**	**#5**
Hubert and Levin	0.117	0.118	**0.077**	0.084	BH-G rand	0.895	0.949	0.831	**1.000**
Davies and Bouldin	**2.043**	2.337	2.659	2.640	BH-G dot	0.897	0.919	0.643	**1.000**
Caliński-Harabasz	**6.223**	4.186	3.490	2.575	BH-G sim	0.878	0.891	0.674	**1.000**
Silhouette	**0.318**	0.221	0.216	0.180	BH-G jaccard	0.819	0.856	0.482	**0.950**
Dunn	0.393	0.438	**0.461**	0.409	BH-G1	**1.000**	0.928	0.702	0.435
					B et al.	**0.030**	0.043	0.041	0.043
					FW	0.162	0.169	0.140	**0.120**
k-medoids	**#2**	**#3**	**#4**	**#5**	**k-medoids**	**#2**	**#3**	**#4**	**#5**
Hubert and Levin	0.127	0.154	0.109	**0.108**	BH-G rand	0.916	0.900	0.863	**0.962**
Davies and Bouldin	**2.166**	3.746	3.901	4.310	BH-G dot	0.902	0.846	0.733	**0.960**
Caliński-Harabasz	**4.813**	2.358	2.811	1.937	BH-G sim	0.904	0.838	0.763	**0.929**
Silhouette	**0.316**	0.170	0.178	0.148	BH-G jaccard	0.833	0.743	0.606	**0.972**
Dunn	0.349	0.355	0.371	**0.413**	BH-G1	**0.989**	0.891	0.904	0.575
					B et al.	**0.029**	0.069	0.054	0.159
					FW	**0.020**	0.052	0.085	0.088
Average	**#2**	**#3**	**#4**	**#5**	**Average**	**#2**	**#3**	**#4**	**#5**
Hubert and Levin	0.137	0.118	0.077	0.069	BH-G rand	**0.960**	0.897	0.926	0.959
Davies and Bouldin	**2.200**	2.747	2.659	2.485	BH-G dot	**0.970**	0.899	0.889	0.955
Caliński-Harabasz	**5.370**	3.547	3.490	2.634	BH-G sim	**0.947**	0.850	0.898	0.923
Silhouette	**0.300**	0.211	0.216	0.192	BH-G jaccard	0.943	0.823	0.807	**0.962**
Dunn	0.458	0.458	**0.461**	**0.461**	BH-G1	0.875	0.844	0.875	**0.925**

(continued)

Table 5 (continued)

Index					Stability measures				
					B et al.	**0.032**	0.034	0.035	0.044
					FW	0.200	0.189	0.170	**0.149**
Ward	**#2**	**#3**	**#4**	**#5**	**Ward**	**#2**	**#3**	**#4**	**#5**
Hubert and Levin	0.107	0.138	0.097 0	**0.782027**	BH-G rand	0.938	0.916	0.873	**0.969**
Davies and Bouldin	2.301	2.082	3.389	**1.639384**	BH-G dot	0.916	0.878	0.782	**0.970**
Caliński-Harabasz	2.934	4.180	3.044	**7.86463**	BH-G sim	0.927	0.862	0.742	**0.942**
Silhouette	**0.356**	0. 217	0.207	0.2108244	BH-G jaccard	0.862	0.801	0.651	**0.876**
Dunn	0. 386	0.375	0.442	**0.461217**	BH-G1	**0.972**	0.657	0.760	0.675
					B et al.	**0.071**	0.078	0.091	0.086
					FW	0.144	0.160	0.143	**0.113**

Source Own computations

measures of stability are more consistent, suggesting that $k = 5$ by index, and $k = 3$ by stability measures.

Nevertheless, despite the large variation in the selected value of the k, one can again observe very similar behavior of Hubert and Levin, Dunn indexes and Fang and Wang stability measure (often suggest the largest number of groups considered).

4.4 Results for the Institutional and Political Domain

In terms of the institutional and political domain, very large discrepancies in the suggested value of the k parameter can be observed for k-means and k-medoids. For k-means, the indexes most often suggest $k = 2$, while the stability measures most often indicate $k = 5$. Similar conclusions can be drawn for k-medoids (although for stability measures, the clustering into two groups is also often correct). For hierarchical methods, the same conclusions as for the k-means can be seen for the average method. On the other hand, for the Ward method, the indexes and stability measures are quite unanimous, showing in most cases the correctness of the clustering into five groups.

Again, it is also worth paying attention to the Dunn index which behaves similarly to Fang and Wang stability measure, suggesting the largest considered number of clusters.

5 Conclusions

Summing up this research, the results of which are presented in this paper, it should be stated that the final result depends on the chosen method. As a rule, very often classical indexes show a different value of the k parameter than the stability measures. Moreover, it can be seen that this value is lower for indexes than for stability measures.

Another summary conclusion resulting from the conducted research is that the Dunn index (and also the Hubert and Levin index in some cases) very often behaves like Fang and Wang stability measure, favoring the clustering into the largest number of groups under consideration. In additional studies (the results of which are not presented here), the value of the parameter k was increased to 10, and this principle was still revealed.

It seems that an interesting topic of future research will be to compare the results using external indexes (e.g., Rand index) and stability measures on benchmark sets with the known cluster structure.

References

Ben-Hur A, Guyon I (2003) Detecting stable clusters using principal component analysis. Methods Mol Biol 224:159–182

Borys T (ed) (2005) Wskaźniki zrównoważonego rozwoju. Wydawnictwo Ekonomia i Środowisko, Warszawa-Białystok

Borys T (2014) Wybrane problemy metodologii pomiaru nowego paradygmatu rozwoju—polskie doświadczenia. Optimum. Studia Ekonomiczne 3(69):3–21

Brock G, Pihur V, Datta S, Datta S (2008) `clValid`: an R package for cluster validation. J Stat Softw 25(4)

Fang Y, Wang J (2012) Selection of the number of clusters via the bootstrap method. Comput Stat Data Anal 56:468–477

Henning C (2007) Cluster-wise assessment of cluster stability. Comput Stat Data Anal 52:258–271

Lord E, Willems M, Lapointe FJ, Makarenkov V (2017) Using the stability of objects to determine the number of clusters in datasets. Inf Sci 393:29–46

Lorek E (2011) Ekonomia zrównoważonego rozwoju w badaniach polskich i niemieckich. In: Kos B (ed) Transformacja gospodarki—poziom krajowy i międzynarodowy. Zeszyty Naukowe Uniwersytetu Ekonomicznego w Katowicach 90:103–112

Marino V, Presti LL (2019) Stay in touch! New insights into end-user attitudes towards engagement platforms. J Consum Mark 36:772–783

Rozmus D (2017) Using R packages for comparison of cluster stability. Acta Universitatis Lodziensis Folia Oeconomica 330(4):77–86

Shamir O, Tishby N (2008) Cluster stability for finite samples. Adv Neural Inf Process Syst 20:1297–1304

Suzuki R, Shimodaira H (2006) `Pvclust`: an R package for assessing the uncertainty in hierarchical clustering. Bioinformatics 22(12):1540–1542

Volkovich Z, Barzily Z, Toledano-Kitai D, Avros R (2010) The Hotteling's metric as a cluster stability index. Comput Model New Technol 14(4):65–72

Wang J (2010) Consistent selection of the number of clusters via cross-validation, "Biometrika", 97:893–904

Identification of the Words Most Frequently Used by Different Generations of Twitter Users

Agata Majkowska, Kamila Migdał-Najman, Krzysztof Najman, and Katarzyna Raca

Abstract Text data constitutes a significant part of all data generated on the Internet, including the social network users' comments and posts. Each website offers its users different functionalities. LinkedIn mainly focuses on the labor market as well as professional and business contacts, and Facebook offers the possibility of creating groups as well as photo and message sharing with friends, while Twitter allows short text message posting and tracking. One type of information researchers would like to obtain about the users of these portals is their age. Such information is crucial from the perspective of marketing, social and economic research. Each of the social networks, however, has different rules regarding the privacy policy and the publishing of information about the date of birth. This poses a problem for the researchers who would like to obtain such information. The aim of the research presented is to attempt characterization of the words typically used in the messages published by Twitter users. This social networking site was chosen due to the possibility of downloading data without additional user consent. Text mining methods and techniques were used to carry out the research, which was mainly focused on the analysis of individual words and collocations occurring in the users' tweets.

Keywords Text mining · Generation classification · Cluster analysis

A. Majkowska · K. Migdał-Najman · K. Najman (✉) · K. Raca
University of Gdańsk, Gdańsk, Poland
e-mail: krzysztof.najman@ug.edu.pl

A. Majkowska
e-mail: agata.majkowska@phdstud.ug.edu.pl

K. Migdał-Najman
e-mail: kamila.migdal-najman@ug.edu.pl

K. Raca
e-mail: katarzyna.raca@ug.edu.pl

K. Jajuga et al. (eds.), *Data Analysis and Classification*, Studies in Classification, Data Analysis, and Knowledge Organization,
https://doi.org/10.1007/978-3-030-75190-6_3

1 Theory of Generations

The concept of "generation" currently is experiencing a peculiar renaissance. Increasingly often, terms such as X, Y, S, Z, JP2, Ikea, BB, YouTube's kids, multitasking, generation @, Jones, echo boomers, millennials and others, can be encountered not only on the Internet, under the keyword "generations", but also in political, psychological, biological, historiographic, literary and economic discussions. Generation is a distinguishable group of people who share a similar time of birth and experience significant events that take place at critical stages in the life of these groups (Ryder 1965; Macky et al. 2008; Brosdahl and Carpenter 2011; Ruth et al. 2013). Generations are made by history (Strauss and Howe 1991).

In 1924, Dilthy stated that a generation is all those who, in a sense, grew up next to each other, that is, who had shared childhood, had common adolescence, and whose male maturity falls at the same period of time. It makes the persons become interrelated by a deeper communality. Those who experienced the same leadership influence in their adolescence constitute a generation. Generation understood in this way creates a tighter circle of individuals who, as a result of their dependence on the same great events and changes that took place in the period of their excitability, despite the different factors that joined later, are bound into a unified whole (Dilthy 1924). A generation, constituting a social group, establishes a specific bond, co-creates, has analogous expectations of itself and shares a similar way of thinking. It therefore seems that people who were born and brought up in different periods of time may have different experiences and approach to life, may exhibit significant differences in knowledge, values, morality, customs, patterns, behavior and even use different words and phrases during communication. Recognition of this difference in the process of intragenerational and intergenerational communication allows development of new forms of dialog and standards for interpreting the reality that surrounds us. Within one generation, teenagers communicate similarly with the teenagers living on the other side of the globe, as opposed to communicating with someone from another generation whom they live with under the same roof.

M.M. Wallis suggests four criteria that allow differentiation of generations within a given group, i.e., a genealogical criterion (the succession of children), paragenealogical (e.g., Herodotus divides every century into three generations), metrical (groups of peers) and cultural (Wallis 1959). Taking into account the metrical criterion, in Poland, it is now possible to distinguish members of the silent generation[1], Baby Boomers, X, Y, Z and Alpha generation. The silent generation is people born between 1928 and 1945. They are disciplined, risk averse and loyal traditionalists. They quietly share traditional values and are not very loud in supporting radical cultural changes. The baby boomers (BB) generation is people born between 1946 and 1964. It is the generation of the post-war boom, who value all

[1]The term "Silent Generation" first appeared on Novemebr 5, 1951 in the "Time" magazine in the article *The younger generation*.

collective activity. They are egocentric individualists, often accused of narcissism and de-individualization of the old age that has, so far, been characterized by loneliness (Wątroba 2017). Generation X is people born between 1965 and 1980. They strive for independence and self-sufficiency. They are skeptical individualists. They value professional and life achievements, high economic status as well as their own and their relatives' safety. They lead the life of a chameleon (Costanza et al. 2012). Generation Y (Millennials) is people born in the years 1981–1995. They are socially conscious, highly cynical and narcissistic. They were brought up in the era of globalization and universal Internet access. It is a generation that accepts diversity in all areas. They are team players looking for employment in organizations that respect the environment (Fisher and Crabtree 2009). Generation Z, also known as C (connect) or iGen, is people immersed in the world of the Internet (Wątroba 2019). These are people born between 1996 and 2010. The virtual and the real world are the same reality for them. It is a generation that cannot function without electronic and social media. They desire an effortless, stunning career. These people have difficulties in focusing on one activity. They are free to communicate internationally and meet new people (Costanza et al. 2012). Although the timeframe for generation Z has not yet been tightly established, attempts are being made to define the next Alpha (Glass) generation. These are people born after 2010. It is a generation to be the most technologically advanced, compared to previous generations. This means that after the generation Alpha, generation Beta will emerge, etc. The generational time frames specified should be treated indicatively. The literature on the subject proposes more detailed divisions within the groups distinguished and indicates differences in the time intervals characterizing the boundaries of transition between successive generations.

2 Analysis of the Textual Data from the Social Network

From year to year, the amount of digital data has been growing at an increasingly faster pace. According to EMC, the number of digital bites at the end of 2020 will be 44 trillion gigabytes, which is comparable to the number of stars in the universe. Since 2010, there has been a 50-fold increase in the amount of data. The largest percentage is unstructured data, which results from the rapidly developing Internet of Things (IoT) technology, but also from the increasing popularity of the Internet and the online social networks. More than half of the world's population are Internet users, almost half of which are active social network users. Social networks offer a great potential for businesses when it comes to getting closer to customers and thus increase the revenues (Watanabe et al. 2021). Social networks have become the channels of communication between enterprises (B2B), between enterprises and consumers (B2C), and between the consumers themselves (C2C). News or product advertisements are disseminated very quickly, owing to the interconnection of the portal users as well as the possibility of real-time access to the portal at any place, via various devices. Enterprises can participate in conversations, which their potential and current customers engage in (Mills and Plangger 2015).

The most popular social networks include: Facebook, YouTube, WhatsApp, Instagram, Twitter, LinkedIn, Snapchat and TikTok. In 2020, there were over 2.5 billion Facebook users, almost 1 billion instagram users, more than half million LinkedIn users, almost half million snapchat users, and almost half million Twitter users. The dynamics of the data growth on social networks is very fast.

Given the significant business benefits resulting from the information posted by users on social networks, tools have been created that allow for the structuring of data and its subsequent analysis.

2.1 Preparation of Text Data

The content of tweets, the text in documents, or the comments on the Internet are sources of unstructured data. This content does not meet the criteria that are required in traditional databases; therefore, its recording and the subsequent analysis differ. Specially dedicated text mining methods are used to analyze this type of data. Important areas of text mining include: information extraction, information retrieval, data categorization, cluster analysis and summarization; Aggarwal and Zhai (2013).

The process of data preparation, regardless of the method selected, consists of the following stages:

1. Text formatting, including:
    - conversion of capital letters into lowercase,
    - removal of special characters (e.g., # or/),
    - removal of emoticons (e.g., ☺ or ☹),
    - removal of punctuation marks,
    - removal of hyperlinks,
    - removal of numbers.
2. Text tokenization—extraction of words from the text.
3. Removal of non-content words, including prepositions, pronouns and conjunctions (stopwords).
4. Text normalization:
    - stemming—removal of inflectional endings from words, to obtain word roots (e.g., running–run, magically–magic) (Lovins 1968),
    - lemming—transforming words into their basic form (e.g., consultant–consult, is–be) (Hellberg 1972).

In the first stage of text data preparation, it is important to standardize the format. Depending on the research objective, this stage may vary. Usually, all non-letter characters, including emoticons, are removed from the text. If a researcher wants to analyze the emotions contained in the text, however, its removal may be undesirable.

Extraction of words from text is another process in text data preparation. It is not always limited to the separation of text with spaces. Some words can contain multiple words, and tokenization allows their separation. For instance, in the word "wanna," the words "want" and "to" will be separated.

In the next step, words that are not important from the perspective of the research conducted are removed. For this purpose, dictionaries are used, which contain a library of such stop words in a specific language, e.g., English, Polish or Spanish. Often, these dictionaries are industry dictionaries. Researchers can modify the content of these dictionaries by supplementing them with specialized expressions.

The last stage is text normalization, which allows elimination of words with different meanings and identical spelling. Failure to do so, e.g., in Polish, may generate errors in the analysis. Stemming and lemmatization are used here, depending on the complexity of the language analyzed. In the case of stemming, the text normalization process is much faster, due to the lesser complexity of the method.

2.2 *Word Frequency Analysis*

The most basic statistic, which usually begins text analysis, is word frequency. When several text documents are analyzed, term frequency matrix (TF) is performed instead, which indicates the number of times a given word occurs in the document (Luhn 1957):

$$TF_{i,j} = \frac{n_{ij}}{n_{.j}} \quad (1)$$

n_{ij}—number of occurrences of the *i*-th word in the *j*-th document,
$n_{.j}$—the number of words contained in the *j*-th document.

Another common tool is the inverse document frequency (IDF) matrix, which takes into account the frequency of the word in all documents (Spärck Jones 1972):

$$IDF = \log\left(\frac{N}{d_i}\right) \quad (2)$$

N—number of all documents,
d_i—number of documents containing the *i*-th word.

In this way, a word that appears in all text documents will get a lower score than a word that appears only once in one document. By inserting the values calculated into the matrix and then visualizing the results using a heatmap, words with low meaning can be easily distinguished. The words appearing in all documents may turn out to be common enough to qualify as stop words. In order to determine the importance of a word, with respect to all documents, both measures should be multiplied. In this way, the TF−IDF index is obtained (Salton and Yang 1973):

$$\mathrm{TF} - \mathrm{IDF} = \mathrm{TF} \cdot \mathrm{IDF} \tag{3}$$

The words appearing in almost all documents will have an IDF index close to zero; therefore, TF-IDF index will also have a low value. The words that appear in several documents will get a higher index value. These word frequency-based indices are simple and quick to calculate and can constitute an important criterion in the performance of subsequent, in-depth analyzes of textual data.

2.3 N-Gram Analysis

More advanced text analyses focus on the connections of words with other words, which allows familiarization with the context of the entire text document. This is particularly important when expanded or subordinate sentences occur. To identify such relations or word collocations, n-grams can be used, i.e., lists of words or words that have been created by breaking the entire character string into n (Shannon 1951). Depending on the method used by the researcher, the division may be carried out on words or expressions. An exemplary text with the results of the n-gram analysis is presented below (Table 1).

The n-grams in the above table derive from the text splitting used. The words obtained from via the splitting of the text every one word are called unigrams, every two—bigrams, every three—trigrams.

The process of obtaining the n-grams is time-consuming, while its operating time primarily depends on the length of the text document. Additionally, the fact that the whole set of n-grams takes up a large part of computer memory has to be taken into account. If the set of n-grams is so large that it is not possible to illustrate it on one graph, then the n-gram frequency statistics should be determined. This will allow reduction of repetitive phrases and a ranking of the most common relationships between the words. Certain words can collocate with many words (e.g., like – "like swimming," "like eating"), which best can be represented by a graph. This is the most appropriate form of representing the data that indicate existence of relations between the objects. A graph is a graphic form that consists of the vertices representing objects and the edges showing the relationships between them (Diestel 2017). An exemplary graph is presented below, which is based on the bigrams created from the following quote from Ayn Rand (Fig. 1):

Table 1 Exemplary use of the n-gram analysis for the expression “text one text”

n-gram	“text one text”	“text one text”
Unigram	“text,” “one,” “text”	“t,” “e,” “x,” “t,” “o,” “n,” “e”…
Bigram	“text one,” “one text”	“te,” “ex,” “xt,” “to,” “on”…
Trigram	“text one text”	“tex,” “ext,” “xto,” “ton”…

Source Own elaboration

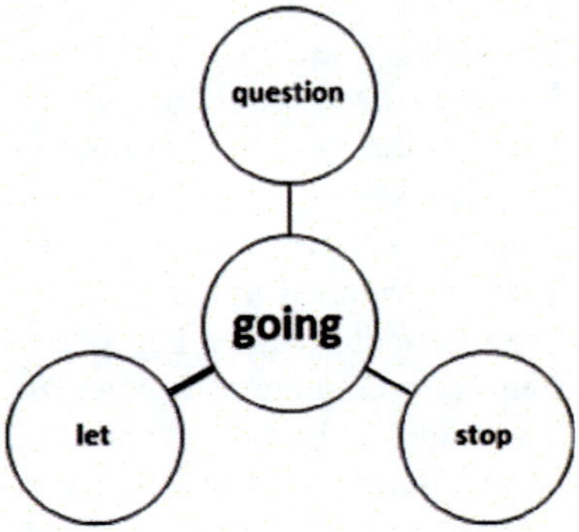

Fig. 1 Web of words created based on a quote from Ayn Rand. *Source* Own elaboration

The question is not who is going to let me; it's who is going to stop me. (Ayn Rand)

In this case, after removing unnecessary words, it turns out that the word most frequently occurring in the bigrams is the word “going,” which plays a significant role in the entire sentence. This is an example of a simple and legible bigram visualization that was created on the basis of a short sentence. The size and the visibility of the graph mainly depends on the amount of the data used for analysis. In the case of an unreadable visualization, the number of the observations can be reduced, or the edge weights can be used.

2.4 Agglomeration Methods of Hierarchical Clustering and Quality Assessment of Group Structure

One method of textual data analysis is classification. In the literature on the subject, different meanings of the term *classification* are propounded (Hull 1970; Pociecha et al. 1988). It is believed that classification entails, inter alia, a division of units in the population examined into classes, in order to select groups of units that are more similar to the units making up the group than to the units outside the group. Mirkin (1996) points to three main areas of classification method development. The possibilities of classification method application are wide and concern various scientific fields and disciplines. Practical application has been described in the works, in which classification methods were used to, e.g., filter spam from relevant e-mail messages (Migdał-Najman and Najman 2013; Tuteja and Bogiri 2017) and analyze Twitter entries in order to anticipate the personalities of social networking users (Pratama and Sarno 2016). The clustering methods can be broken down in many ways. Specific ways of method implementation are written in the form of algorithms. One popular clustering method is hierarchical agglomeration. Commonly used agglomeration methods of hierarchical clustering include: the single connection method (Florek et al. 1951; Sneath 1957), the full connection method (McQuitty 1960; Sneath and Sokal 1963), the centroid method (Sokal and Michener 1958; Gower 1967), the group weighted average (McQuitty 1966, McQuitty 1967), the median (Lance and Williams 1966; Gower 1967), the group mean (Sokal and Michener 1958) and the Ward method (Ward 1963). Hierarchical clustering is a non-uniform procedure, using different principles (methods, criteria) for calculating the distances between units and clusters. In 1967, M. Lance and W. Williams proposed a universal scheme generalizing these principles, modified by Jambu in 1978. In hierarchical methods, the choice of the method for measuring the distance between units plays a key role. There are many measures of distance; the ones frequently used are as follows: the Euclidian metric, the square of Euclidean distance, the Manhattan metric or the Chebyshev distance. These measures are used in typical situations, when the values of variables are measured on typical measurement scales. It often happens that a complex issue is considered which consists of additive elements that together constitute a certain totality. Each element of this totality can be written in the form of some countable or measurable quantity. The measure often used to determine the similarity between texts is the cosine similarity. If the components under analysis are also expressed in the form of proportions that add up to one, a proportion-based distance must be used to measure the distance between the measurement units that are based on this proportion. One such measure of distance, i.e., the half of the sum of the absolute differences in the proportions, is the measure expressed by following Eq. 4 (Balicki 2009):

$$d_{ik} = \frac{1}{2}\sum_{j=1}^{p} |p_{ij} - p_{kj}| \tag{4}$$

$$i, k = 1, \ldots, n \; i \neq k$$

p_{ij}—the share of the *j*-th category (component) in the total phenomenon characterizing the *i*-th unit, where $\sum_{j=1}^{p} p_{ij} = 1$ for each *i*.

The above measure assumes a value equal to zero when the share of the *j*-th category in the total phenomenon for the two units being compared is identical. If mutual disjuncture of occurrence emerges and for each category of the two objects compared the difference is zero, the measure takes the value of one.

The grouping should result in clusters that are as characterized by the largest separability possible, i.e., they should be internally coherent, and be as diverse as possible with regard to each other. Empirical research should check whether these conditions are met. Various types of assessment procedures for the grouping results obtained have been proposed in the literature on the subject. One of those is the

MacQueen's (1967) concept, distinguishing three criteria for assessing the quality of group structure. It is an external, an internal and a relative criterion. In the study, an internal criterion was used to assess the quality of the group structure obtained. One way to assess the degree of matching between the distance matrix and the cophenetic distance matrix (levels of collocation, dendrograms) is the cophenetic correlation coefficient proposed by Sokal and Rohlf in (1962). The coefficient is defined as follows:

$$\mathrm{CCC} = \frac{\sum_{i=1}^{n-1}\sum_{k=i+1}^{n}(d_{ik}-\bar{d})(c_{ik}-\bar{c})}{\sqrt{\sum_{i=1}^{n-1}\sum_{k=i+1}^{n}(d_{ik}-\bar{d})^2\sum_{i=1}^{n-1}\sum_{k=i+1}^{n}(c_{ik}-\bar{c})^2}} \tag{5}$$

where n is the number of units, d_{ik}, c_{ik} is the distance between the i-th and the k-th unit $(i, k = 1, \ldots, n\ i \neq k)$ of the matrices being compared, where the summation extends over all pairs of the set under consideration, while $\bar{d}$ and $\bar{c}$ are the means expressed in the following form (Eq. 6):

$$\bar{d} = \frac{2}{n(n-1)}\sum_{i=1}^{n-1}\sum_{k=i+1}^{n} d_{ik} \tag{6}$$

The coefficient of cophenetic correlation assumes values in the range [−1.1]. A coefficient value close to one means high fit between the distance matrix and the cophenetic distance matrix. The literature on the subject also indicates other measures allowing assessment of the degree of the cophenetic distance matrix and the distance matrix matching. These are the Goodman-Kruskal γ coefficient (Goodman-Kruskal gamma coefficient) (Goodman and Kruskal 1954), the STRESS coefficient (standardized residual sum of square) (Kruskal 1964), the gamma index (γ index) (Hubert 1974; Baker and Hubert 1975), the silhouette coefficient index (Rousseeuw 1987) and the Dunn's index (Dunn 1974) (Dunn's index).

An important element of the grouping process entails determination of the number of clusters which the set examined should be divided into. The simplest method is the dendrogram method, i.e., a particular graphic effect of the hierarchical clustering strategy and the similarity or distance measure applied. It is an approach entailing identification of a large difference between adjacent linkage levels. This method is subjective, however. The literature on the subject indicates various analytical methods for determining the optimal number of classes in the set under analysis. An interesting summary of the basic knowledge in this area was made by Milligan and Cooper (1985). The multitude of the methods mentioned and proposed, however, does not always facilitate the calculation of the correct number of clusters (Migdał-Najman and Najman 2013). One such method suggestion is that of Mojena (1977), which is based on the relative height of the different linkage levels. The number of the classes corresponds to the linkage levels for which the Mojena inequality is satisfied (Mojena 1977; Balicki 2009). For more information on the classification of the methods used to determine the number of classes (see Migdał-Najman and Najman 2013).

3 Applications and Results

Twitter is a social network that allows users to post short text messages (tweets) of 280 characters or less on their timelines. Apart from the text, the messages may contain user tags, locations and topics. Each message is described with a set of metadata, such as: the user's location data, the number of friends or information on the date of account creation. The timeline allows all the user's tweets to be organized according to the date of creation.

To analyze the messages posted on Twitter, it was necessary to download a set of tweets from the Twitter database. This can be done via the Twitter application programming interface (API), which only the users have access to. It limits the number of queries per minute. Depending on the type of information a given researcher wants to obtain, these limits differ.

The Twitter API was used three times in this study, but the limitations regarding data retrieval were the same. One of such limitations was the option of downloading the tweets from the most recent week only. Another limitation was the speed of tweet download (maximum 300 tweets per 15 min).

In the first stage of the study, the Twitter API was used to search for tweets with such phrases as: "happy birthday to me," "happy birthday to you" and "I'm x years old."[2] This allowed identification of the users who publish their age on the social network. Two programming languages were used for this purpose: R (rtweet library) and Python (tweepy library). The algorithms were designed to search for users aged 13–79. In this way, 67 age cohorts were obtained. This was to prevent the errors resulting, among others, from the happy birthday wishes addressed to companies or children who do not have Twitter accounts. Out of the tweets collected in this way, those that did not contain any information regarding user age or usernames were removed. In this way, in the last two weeks of July 2020, a database of 14 224 users, with the age assigned, was collected. The user age distribution is presented in Fig. 2. The largest number of users in the sample examined fell between 13 and 27 years of age, where most commonly they were 20-year-old users. In order to find more users aged 30–80, a different search algorithm for the number of years should probably be used, e.g., an algorithm identifying the age based on what or whom the Twitter users examined follow (Chamberlain et al. 2017).

In the next stage of the study, the Twitter API was used to collect each user's tweets (in the amount of 300 tweets at most for each user). If the user had a greater number of text messages on his/her timeline, they were omitted; if the number of messages was less, all of them were collected. The tweets downloaded in this way were merged according to a given user and then a given age group. This resulted in a database with 67 lines corresponding to the tweets of users from each age cohort. Then, the data preparation stage was carried out, including: text formatting, stop-words deletion and extraction of words from the text.

[2]X is the number of years that has been declared by the users.

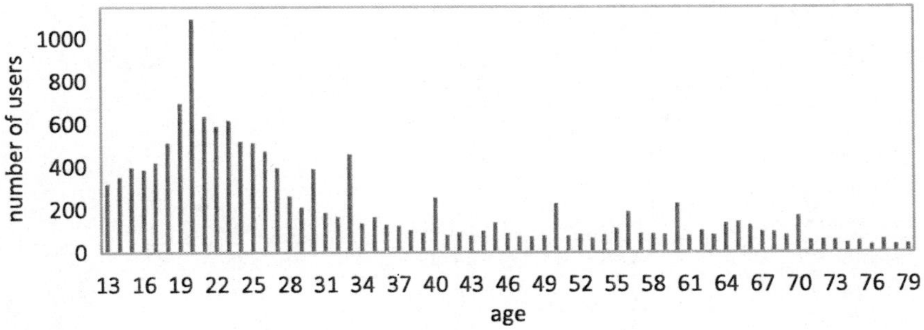

Fig. 2 Age of the Twitter users under analysis. *Source* Own elaboration

3.1 Twitter User Analysis

Before analyzing the text data, the Twitter API was once again used to collect metadata on the users who had been searched out by the age published in the tweets, in an earlier step of the analysis. In this way, a database with metadata of 12,891 users was created and the analyzed[3].

Information was obtained regarding: the number of friends, the geolocation, the profile description, the number of tweets published, the communication platform used and the date of Twitter account creation.

The Twitter users included in the sample differ in terms of the number of followers, the number of friends, the number of likes and the number of tweets (Table 2). The sample includes both the persons who are not active on Twitter as well as those who have posted 2.8 million posts (tweets and retweets).

Another feature contained in the metadata is user location. Users can fill in this information on their own, which causes many data gaps and appearance of names that do not reflect the true locations. In consequence, the geolocation results are not published. Users can also enter a description of their profile in the metadata. After its initial processing, a ranking of the 10 most common words was determined, i.e., "im," "love," "fan," "like," "dont," "sheher," "life," "account," "old," "lover." The last statistic obtained from the metadata was the date of Twitter account creation (Figs. 3, 4). It shows that the user accounts analyzed most often were created in 2020, but several hundred users with a Twitter experience of over 10 years were fund in the sample as well.

Most often, users had set up their accounts in July, June and May. The day-of-the-week distribution is similar to uniform distribution. This means that the

[3]The difference in the number of users results from the interval between the tweet download and the metadata. During that time, the usernames could be changed, user accounts could be deleted or blocked, which resulted in the smaller number of users in the database.

Table 2 Characteristics of the users under examination (quantitative data)

Statistic	Number of followers	Number of friends	Number of likes	Number of tweets
Average	2541.75	958.79	20,820.08	19,285.22
Standard deviation	42,527.53	4242.14	40,754.07	48,427.65
Minimum value	0.00	0.00	0.00	0.00
Quartile 1	64.00	141.00	1082.25	1050.75
Median	280.00	361.50	6434.50	5436.00
Quartile 3	913.00	849.00	22,550.75	19,002.50
Maximum value	3,493,033.00	342,876.00	735,056.00	2,818,872.00

Source Own elaboration

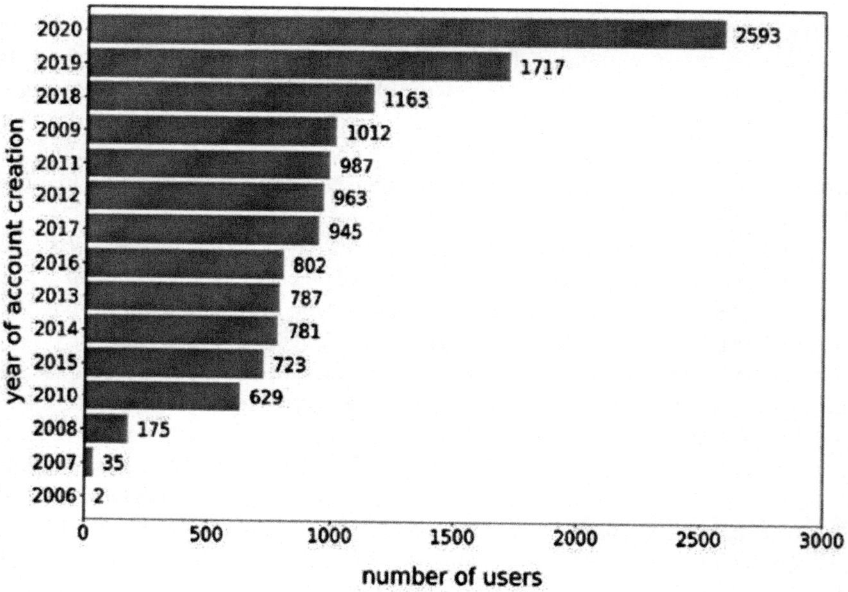

Fig. 3 Year of Twitter account creation. *Source* Own elaboration

day of the week has no impact on these user characteristics. The combination of the month and the year in which Twitter accounts had been created indicates that most users had set up their accounts in July 2020 (743), June 2020 (472) and May 2020 (358).

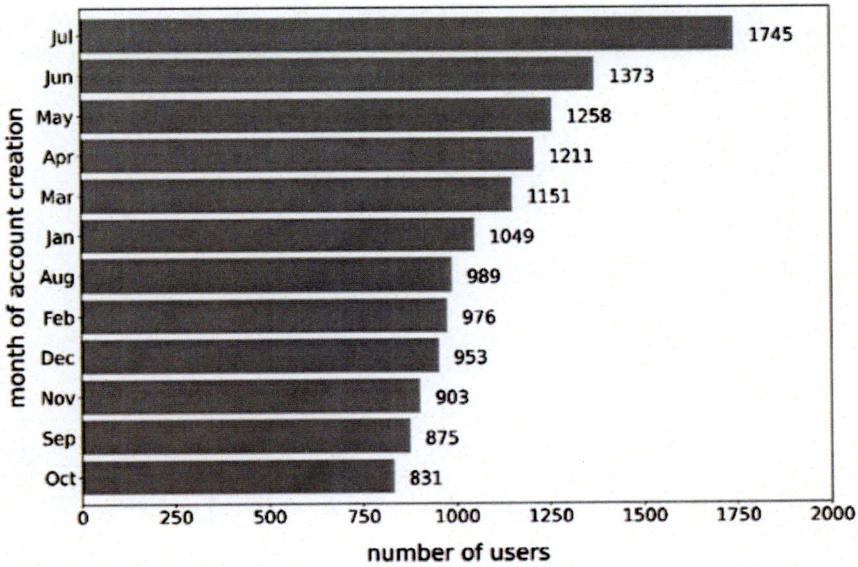

Fig. 4 Month of Twitter account creation. *Source* Own elaboration

3.2 Analysis of the Words Occurring Most Commonly

The data matrix preparation for the cluster analysis began with extraction of words from the tweets posted by the users of specific age groups within the sample examined. On this basis, a matrix with word counts for age groups within 13–79 years of age was created. Then, 100 most frequently used words were extracted in each age group. The repetition of such words as "love," in each of the groups under examination, caused the duplicates to be removed. As a result, 535 unique words were obtained. Based on the occurrence count of a given word, word frequency (TF-Term Frequency) was calculated for each given age group. A matrix with 67 cases and 535 variables reflecting the group profiles was obtained.

The literature indicates various measures of the distance between objects. Nevertheless, the profile data in the research presented forced the use of a distance measure based on proportions (see Eq. 4). To select the clustering method, the coefficient of cophenetic correlation was used (Table 3). The highest value of the coefficient was obtained for the group mean method.

Since some age groups formed single-element clusters and did not connect with others, they were removed from further analysis. These are the users aged: 38, 45, 59, 61, 71, 74, 76, 78, 79.

The dendrogram showing the clustering result after the removal of the above groups is shown in Fig. 5. Based on the Mojena's criterion, four clusters were distinguished.

Table 3 Values of the cophenetic correlation coefficient

Agglomeration method of hierarchical clustering	Cophenetic correlation coefficient	Cophenetic correlation coefficient after removal of single-element groups
The closest neighbor	0.5383	0.6222
The furthest neighbor	0.6389	0.7089
Group average	0.7383	0.7408
Median	0.5646	0.6212
Centroidal	0.6896	0.7015

Source Own elaboration

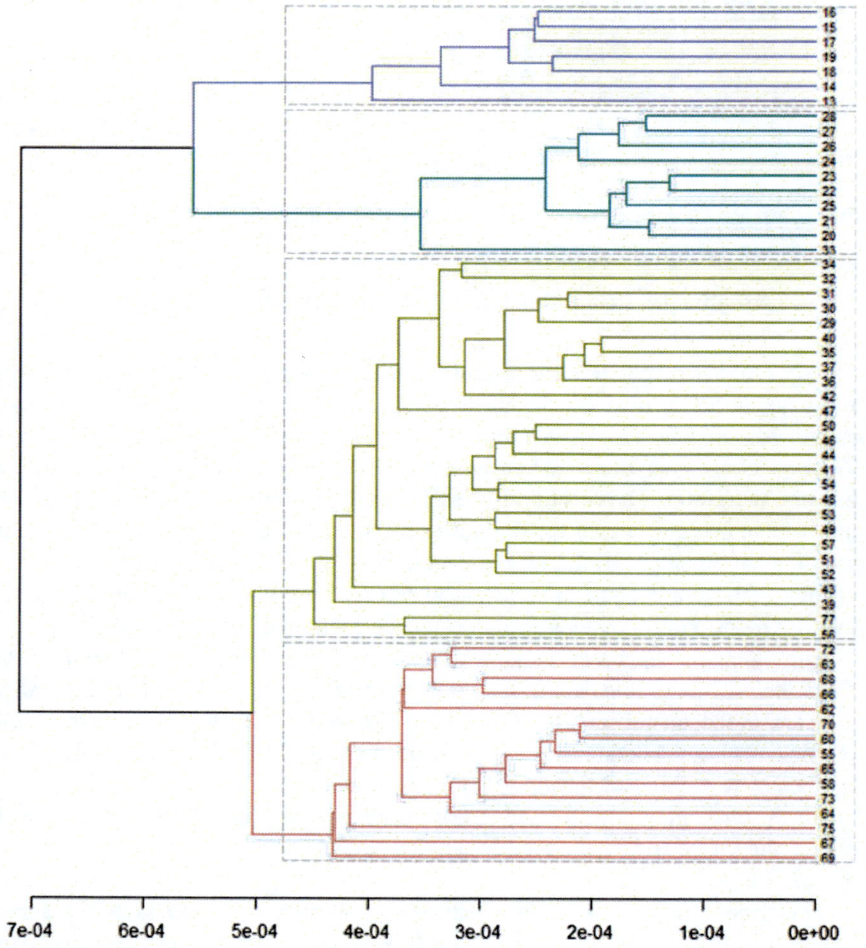

Fig. 5 Dendrogram of the similarity structure in the age groups examined. *Source* Own elaboration

Table 4 Share of users from individual generations in the cluster

Cluster	Name	Baby boomers	Generation X	Generation Y	Generation Z
1	Generation Z	0%	0%	0%	**100%**
2	Generation Z and Y	0%	0%	**29%**	**71%**
3	Generation Y and X	10%	**35%**	**55%**	0%
4	Baby boomers	**93%**	**7%**	0%	0%

Source Own elaboration

While analyzing the share of users from a given generation in the clusters selected, the following groups were assigned: generation Z, generation Z and Y, generation Y and X, baby boomers (Table 4).

Cluster 1 (generation Z) is represented by the youngest users. The user age range is 13–19, with the average age of 16. The words used most frequently include: "im," "like," "dont" or "love" (Table 5). It is worth paying attention to the fact that among the 20 most frequently used words, words characterized by emotions prevail: "like," "love," "good." The next cluster 2 (Generations Z and Y) consists of 24-year-old Twitter users, on average. This group includes the persons from both generation Z (71%) and generation Y (29%). Among the most frequently used words in this group, similarly to cluster 1, the following words can be distinguished: "im," "like," "dont" and "love." The difference between clusters 1 and 2 is small and concerns the frequency of word occurrence only. Cluster 3 (generation Y and X) represents people who are on average 42.4 years old. 55% are generation Y users, 35% generation X and 10% baby boomers. The words "im," "like" and "love," similarly to clusters 1 and 2, appear here most often.

The last cluster is the baby boomers generation, since 93% of the users are from this generation. The remaining 7% are generation X users. The average age is 61.3 years. Here, the words most commonly used in the tweets differ from those listed above. These users most commonly use the word "realdonaldtrump," which

Table 5 Words most commonly occurring in the clusters

Baby boomers	Generation	Generation Z and Y	Generation Z
realdonaldtrump	im	im	im
people	like	like	like
dont	dont	dont	dont
like	people	get	love
im	get	one	one
trump	one	love	u
get	know	people	people
one	good	know	get
know	would	time	na
would	time	good	know

Source Own elaboration

refers to the profile of the US President Donald Trump. Other words are as follows: "people," "dont," "like," while the word "im," which ranked first in the previous clusters, here ranks 5th. Political themes are noticeable in the content of the tweets, as evidenced by the next word "trump," also used quite frequently.

The Twitter users from the X, Y and Z generations most often use the words "im," "like," "dont." Generation Z often writes about their feelings, as evidenced by the frequent use of the words "like" and "love." The most contrasting to the rest is the baby boomers. In the baby boomers group, the tweets are focused on politics, while the tone is much more serious.

3.3 Bigrams and Trigrams

The analysis of single words carried out displays a preliminary picture of the situation. For example, a single word "like" has a completely different meaning than when combined with the word "dont." As such, some words, in juxtaposition with other words, form a completely different meaning of a given utterance. To explore the deeper meanings of the Twitter posts examined, bigrams and trigrams were used.

In the Z generation group, among the 20 most common word pairs, "gon na" or "years old" are most frequently used (Table 6). It is also noticeable that the utterances are characterized by emotions: "love u," "feel like," "love shopee." The use of disaffirmation, i.e., the word "dont," which often occurs in combination with "think," "want," can be observed as well. When analyzing the 20 most common trigrams (Table 7), such verbal abbreviations as, e.g., "ifb" become apparent.

Among the generations Y and Z group, it can be noticed that just as in the case of generation Z group, the use of the word "dont," in combination with "want" and "think," can be noticed.

No abbreviations were found among the 20 most common bigrams in the "generation Y and X" group. With regard to the trigrams analyzed, it can be noted that users are beginning to post entries on more serious topics, for example, work: "job plss please," "get job plss" or on important media topics: "want debate want," "black lives matter."

The bigrams and the trigrams in the oldest group of "baby boomers" are mainly characterized by political themes (as evidenced by the bigrams: "president trump," "donald trump," "white house"). When examining the trigrams, a similarity to the generations Y and X is noticeable, where media topics were also discussed: "black lives matter."

The bigram and trigram analysis allowed identification of the word collocations used by the generational groups distinguished, which would not be possible when analyzing individual words. In particular, differences in the use of the language can be noticed between the youngest and the oldest generation. Generations Z and Y often use abbreviations, which may be understood by the baby boomers generation. The differences between the groups obtained also result from the users' different life

Table 6 Most common bigrams in individual clusters

Baby boomers	Generation Y and X	Generation Z and Y	Generation Z
years old	gon na	gon na	gon na
dont know	years old	wan na	years old
gon na	dont know	years old	wan na
im years	im years	im years	im years
president trump	wan na	dont know	dont know
happy birthday	happy birthday	feel like	happy birthday
dont want	feel like	got ta	bts bts_twt
dont think	im going	im gon	mtvhottest bts
looks like	dont want	happy birthday	im gon
god bless	dont think	i'm going	feel like
via youtube	got ta	im sorry	im sorry
joe biden	looks like	cant wait	im going
im sure	im sure	dont want	thank u
donald trump	cant wait	dont think	dont think
years ago	years ago	dont even	dont want
im sorry	im sorry	first time	good morning
dont care	year old	look like	love u
go back	good morning	last night	got ta
united states	dont like	thank much	thank much
white house	first time	good morning	love shopee

Source Own elaboration

Table 7 Most commonly occurring trigrams in individual clusters

Baby boomers	Generation Y and X	Generation Z and Y	Generation Z
im years old	im years old	im years old	im years old
borrowed time equine	im gon na	im gon na	mtvhottest bts bts_twt
time equine rescue	mtvhottest bts bts_twt	mtvhottest bts bts_twt	im gon na
equine rescue inc	want debate want	dont wan na	love shopee shopeexcarat
black lives matter	debate want debate	bus bus bus	shopee shopeexcarat love
shop httpstcokayazmxar igivedoyou	black lives matter	fahrenheit fahrenheit fahrenheit	shopeexcarat love shopee
im gon na	dethfaktor dumbelon ap	wan na go	boobs boobs boobs
get bonus donation	baldheadman jonfitztv dethfaktor	unfollowed automatically checked	unfollowed automatically checked

(continued)

Table 7 (continued)

Baby boomers	Generation Y and X	Generation Z and Y	Generation Z
bonus donation borrowed	jonfitztv dethfaktor dumbelon	gon na get	exabff exacarat name
donation borrowed time	dont even know	dont even know	say exabff exacarat
rescue inc join	im pretty sure	feel like im	aku mau kejutandarishopee
inc join shop	dont wan na	animalcrossing acnh nintendoswitch	exacarat name pledis
join shop earn	unfollowed automatically checked	아스트로 astro 윤산하	name pledis seventeen
shop earn borrowed	gon na get	sanha ユンサナ 尹産賀	dont wan na
earn borrowed time	help get job	astro 윤산하 sanha	mau kejutandarishopee aku
rescue inc everytime	cant wait see	윤산하 sanha ユンサナ	kejutandarishopee aku mau
inc everytime shop	ive ever seen	ユンサナ 尹産賀 ซานฮา	followed automatically checked
everytime shop httpstcokayazmxar	get job plss	尹産賀 ซานฮา アストロ	ifb ifb ifb
ha ha ha	job plss please	years old im	one person followed
igivedoyou get bonus	plss please help	years old still	years old im

Source Own elaboration

experiences and interests, which is mainly indicated by the trigrams obtained. The youngest people (generation Z) write about love, shopping, entertainment or music, while generation Y and the baby boomers are more involved in social topics, such as politics, justice or employment. It turns out that as the age increases, the Twitter users examined focus less on themselves and devote more tweets to other people.

4 Conclusion

When analyzing data acquired from Internet sources, researchers relatively often face the problem of identifying the age of the authors of the texts published. Information regarding a given author's age, with minimum accuracy of specific age groups, is very useful from the perspective of marketing research, for instance, because it allows familiarization with the differences in purchasing preferences. In many cases, the authors of the content posted on the Internet do not provide their age, which for the above reasons significantly limits the possibility of, e.g., profiling. Language research shows that representatives of different age groups use different vocabulary and grammatical forms.

The research presented entailed an attempt to identify the words characteristic for different age groups of Twitter users. As part of the study, dictionaries of typical words for age groups were developed based on the entries posted by users whose age is known. The data prepared in this way allowed identification of significant differences in the profiles of the words used and determination of four clusters of Twitter users. These clusters have been appropriately named: baby boomers, generation Y and X, generation Z and Y and generation Z. The words used by Twitter users seem to be appropriate for their generation groups.

The research also presents the results of the analysis carried out with regard to the similarity of words and their collocations, based on the representatives of the four clusters distinguished. An in-depth analysis of the bigrams and trigrams extracted from the Twitter entries allowed identification of the users' interests, which differ in terms of the age declared. The younger generation of Twitter uses express their views on entertainment-related topics. The older the Twitter users, the more focused on serious topics, including politics, their tweets are.

The research results can be applied in many fields, including: marketing, psychology, technology. Knowledge of the vocabulary used by different generations allows advertisement profiling with regard to individual age groups, which can help expand the client group. Enterprises developing their own technologies (e.g., computer games) can use, in their applications or user manuals, a language that is understandable for all generations. Knowledge of the words used is an important inter-generation communication factor, which translates into both educational efficiency (teacher–student communication) as well as profit for enterprises (enterprise-client communication).

The results of the study constitute an introduction to more in-depth research in this area. Further research may can entail: an attempt to estimate the age of the respondents based on their vocabulary (in case of missing data) and an attempt to estimate the impact of age on the subjectivism and the emotions expressed by Twitter users. The research could also be extended to analysis of various social groups' interests, based on their retweets and likes.

References

Aggarwal CC, Zhai C (2012) Mining text data. In Springer Science+Business Media, LLC 2012. https://doi.org/10.1007/978-1-4614-3223-4

Baker FB, Hubert LJ (1975) Measuring the power of hierarchical cluster analysis. J Am Statist Assoc 70(349):31–38

Balicki A (2009) Statystyczna analiza wielowymiarowa i jej zastosowania społeczno-ekonomiczne. Wydawnictwo Uniwersytetu Gdańskiego, Gdańsk

Brosdahl DJ, Carpenter JM (2011) Shopping orientations of US males: a generational cohort comparison. J Retail Consum Serv 18(6):548–554. https://doi.org/10.1016/j.jretconser.2011.07.005

Chamberlain BP, Humby C, Deisenroth MP (2017) Probabilistic inference of twitter users age based on what they follow. Lecture notes in computer science (including subseries lecture notes in artificial intelligence and lecture notes in bioinformatics), 10536 LNAI, pp 191–203. https://doi.org/10.1007/978-3-319-71273-4_16

Costanza DP, Badger JM, Fraser RL, Severt JB, Gade PA (2012) Generational differences in work-related attitudes: a meta-analysis. J Bus Psychol 27(4):375–394. https://doi.org/10.1007/s10869-012-9259-4

Diestel R (2017) The basics. In: Graph theory-graduate texts in mathematics, vol 173. Springer, Berlin, Heidelberg. https://doi.org/10.1007/978-3-662-53622-3_1

Dilthy W (1924) Gesammelte Schriften 5: 37. Polish edition: Dilthy W (1924) Rozwój problemu pokolenia (trans: Wyka K). Warszawa

Dunn JC (1974) Well-separated clusters and optimal fuzzy partitions. J Cybern 4(1):95–104. https://doi.org/10.1080/01969727408546059

Fisher TF, Crabtree JL (2009) Generational cohort theory: have we overlooked an important aspect of the entry-level occupational therapy doctorate debate? Am J Occup Ther 63(5):656–660. https://doi.org/10.5014/ajot.63.5.656

Florek K, Łukaszewicz J, Perkal J, Steinhaus H, Zubrzycki S (1951) Taksonomia wrocławska. Przegląd Antropologiczny 17:193–211

Goodman LA, Kruskal WH (1954) Measures of association for cross classifications. J Am Statist Assoc 49(268):732–764

Gower JC (1967) A comparison of some methods of cluster analysis. Biometrics 23(4):623–638

Hellberg S (1972) Computerized iemmatization without the use of a dictionary: a case study from swedish lexicology. Computers and the Humanities, 6(4):209–212. https://doi.org/10.1007/BF02404268

Hubert LJ (1974) Approximate evaluation techniques for the single-link and complete-link hierarchical clustering procedures. J Am Statist Assoc 69(347):698–704

Hull DL (1970) Contemporary systematic philosophies. Annu Rev pf Ecol Systemat 1:19–54. https://doi.org/10.1146/annurev.es.01.110170.000315

Jambu M (1978) Classification automatiqe pour l`analyse des donnees, vol 1. Dunod, Paris

Kruskal JB (1964) Nonmetric multidimensional scaling: a numerical method. Psychometrika 29(2):115–129

Lance GN, Williams WT (1966) A generalized sorting strategy for computer classifications. Nature 212, 218, Letters to Nature

Lovins JB (1968) Development of a stemming algorithm*. Mechanical translation and computational linguistics

Luhn HP (1957) A statistical approach to mechanized encoding and searching of literary information. IBM J Res Dev 1(4):309–317. https://doi.org/10.1147/rd.14.0309

MacQueen JB (1967) Some methods for classification and analysis of multivariate observations. In: Proceedings of 5th Berkeley symposium on mathematical statistics and probability 1. University of California Press, Berkeley, pp 281–297

Macky K, Gardner D, Forsyth S (2008) Generational differences at work: introduction and overview. J Manag Psychol 23(8):857–861. https://doi.org/10.1108/02683940810904358

McQuitty LL (1960) Hierarchical linkage analysis for the isolation of types. Educ Psychol Measur 20(1):55–67

McQuitty LL (1966) Similarity analysis by reciprocal pairs for discrete and continuous data. Educ Psychol Measur 26(4):825–831

McQuitty LL (1967) Expansion of similarity analysis by reciprocal pairs for discrete and continuous data. Educ Psychol Measur 27(2):253–255

Migdał-Najman K, Najman K (2013) Samouczące się sztuczne sieci neuronowe w grupowaniu i klasyfikacji danych. Teoria i zastosowania w ekonomii, Wydawnictwo Uniwersytetu Gdańskiego, Gdańsk

Milligan GW, Cooper MC (1985) An examination of procedures for determining the number of clusters in a data set. Psychometrika 50(2):159–179. https://doi.org/10.1007/BF02294245

Mills AJ, Plangger K (2015) Social media strategy for online service brands. Serv Ind J 35(10): 521–536. https://doi.org/10.1080/02642069.2015.1043277
Mirkin BG (1996) Mathematical classification and clustering. Kluwer Academic Publishers, Dordrecht, The Netherlands
Mojena R (1977) Hierarchical grouping methods and stopping rules: an evaluation. Comput J 20:359–363. https://doi.org/10.1093/comjnl/20.4.359
Pociecha J, Podolec B, Sokołowski A, Zając K (1988) Metody taksonomiczne w badaniach społeczno-ekonomicznych. Wydawnictwo Naukowe PWN, Warszawa
Pratama BY, Sarno R (2016) Personality classification based on Twitter text using Naive Bayes, KNN and SVM. In: Proceedings of 2015 international conference on data and software engineering, ICODSE 2015, pp 170–174. https://doi.org/10.1109/icodse.2015.7436992
Rousseeuw PJ (1987) Silhouettes: a graphical aid to the interpretation and validation of cluster analysis. J Comput Appl Math 20(1):53–65. https://doi.org/10.1016/0377-0427(87)90125-7
Ruth N, Bolton A, Parasuraman A (2013) Understanding generation Y and their use of social media: a review and research agenda. J Serv Manag 24(3):245–267
Ryder NB (1965) The cohort as a concept in the study of social change. Am Sociol Rev 30(6): 843–861. https://doi.org/10.2307/2090964
Salton G, Yang CS (1973) On the specification of term values in automatic indexing. Cornell University
Shannon CE (1951) Prediction and entropy of printed english. Bell System Technical Journal, 30(1):50–64.https://doi.org/10.1002/j.1538-7305.1951.tb01366.x
Sneath PHA (1957) The application of computers to taxonomy. J Gen Microbiol 17(1):201–226
Sneath PH, Sokal RR (1963) Priciples of numerical taxonomy. Freeman, San Fancisco, London
Sokal RR, Michener CD (1958) A statistical method for evaluating systematic relationships. University of Kansas, Scientific Bulletin 38:1409–1438
Sokal RR, Rohlf FJ (1962) The comparison of dendrograms by objective methods. TAXON Wiley 11(2):33–40. https://doi.org/10.2307/1217208
Spärck Jones K (1972) A statistical interpretation of term specificity and its application in retrieval. J Documentation 28(1):11–21. https://doi.org/10.1108/00220410410560573
Strauss W, Howe N (1991) Generations. The history of America's future, 1584 to 2069. William Morrow and Company, Inc., New York
Tuteja SK, Bogiri N (2017) Email Spam filtering using BPNN classification algorithm. In: International conference on automatic control and dynamic optimization techniques, ICACDOT 2016. Institute of Electrical and Electronics Engineers Inc., pp 915–919. https://doi.org/10.1109/icacdot.2016.7877720
Wallis M (1959) Koncepcje biologiczne w humanistyce. In: Kotarbiński T (ed) Fragmenty filozoficzne vol. 2. Warszawa
Ward JH (1963) Hierarchical grouping to optimize an objective function. J Am Statist Assoc 58(301):236–244
Watanabe NM, Kim J, Park J (2021) Social network analysis and domestic and international retailers: an investigation of social media networks of cosmetic brands. J Retail Consum Serv 58:102301. https://doi.org/10.1016/j.jretconser.2020.102301
Wątroba W (2017) Transgresje międzypokoleniowe późnego kapitalizmu. Wydawnictwo Uniwersytetu Ekonomicznego we Wrocławiu, Wrocław
Wątroba W (2019) Transgresywność systemów wartości pokoleń we współczesnym kapitalizmie. Folia Oeconomica, Acta Universitatis Lodziensis 5(344):139–157. https://doi.org/10.18778/0208-6018.344.09

Classification Algorithms Applications for Information Security on the Internet: A Review

Michał Bryś

Abstract The growing use of the Internet in every life area creates an emerging need to provide information security (IS), and numerous classification algorithms approach this problem. This study provides a systematic literature review on the classification algorithms applications for information security on the Internet and cybersecurity. The classification algorithms use cases considered are abusive content, malicious code, information gathering, intrusion attempts, intrusions, availability, information content security, fraud, and vulnerable. As many research papers on that topic were published, this research focuses on recent studies from 2015 to 2020 and includes new areas, like mobile devices and the Internet of things (IoT). The analysis of 1446 selected publications provides insights on classification algorithms applied to IS tasks, their popularity, and the algorithm selection challenges.

Keywords Information security · Classification algorithms · Machine learning

1 Introduction

Information security (IS) importance is growing together with the technology adoption in new areas. Nowadays, most devices we use are connected to the Internet: personal computers, mobile devices, and many others like smart home devices connected to the Internet of Things (IoT). Every byte transferred between devices creates many additional data, like security software logs and operating system logs (Kent and Souppaya 2006). As a result, we have massive datasets of data exchanged between the devices and the additional logs' of frequently more extensive volumes. That datasets can be an input to the classification models supporting the Information security.

The growing adoption of the Internet also has a significant impact on the economy. As the information association report (iA Internet Association 2020)

M. Bryś (✉)
AGH University of Science and Technology, Kraków, Poland

K. Jajuga et al. (eds.), *Data Analysis and Classification*, Studies in Classification, Data Analysis, and Knowledge Organization,
https://doi.org/10.1007/978-3-030-75190-6_4

says, the Internet sector in the US contributed to the equivalent of 10.1% of gross domestic product (GDP) and 4% of national employment. Additionally, according to the European Union data (ENISA 2018), 60% of EU citizens aged 16–74 ordered goods and services over the Internet in 2019 (European Union 2020). It creates a need to invest in the IS to avoid cybersecurity incidents (Brecht and Nowey 2013), which causes financial losses to the companies.

In recent years, also the new law regulations like General Data Protection Regulation (GDPR) (European Parliament and Council of European Union 2016) in Europe, California Consumer Privacy Act (CCPA) (California State Legislature 2018) in the US, The Personal Information Protection and Electronic Documents Act (PIPEDA) (Office of the Privacy Commissioner of Canada 2000) in Canada or Australian Privacy Principles (APPs) (Office of the Australian Information Commissioner 1988) extended the scope of Information security with the user privacy. Companies that are not compliant with the privacy regulations may get financial penalties, which causes IS's additional costs.

The volume of available datasets and the new technologies of big data processing (Ryzko 2020) allows us to apply various classification algorithms to the IS challenges. As Chio and Freeman (2018) discuss the model selection criteria, we can select a classification model family that fits the computational and mathematical complexity requirements. For example, some models (logistic regression, SVMs) are less expensive and faster to train, which is a significant advantage in the fast-changing IS field where attackers are continually modifying their methods. Also, explainability is critical in real-life applications, where we will make business decisions based on the model output. If the model explainability is required, they recommend the decision tree, logistic regression, and Naïve Bayes classifiers.

Recently, some other reviews of classification algorithms application to IS were published (see Haibo and Garcia 2009; Masood et al. 2019; Jing et al. 2018; Apruzzese et al. 2018; Ferrag et al. 2020). However, these studies were narrowed to applications on particular use cases or industries or analyzed a smaller set of publications. This study's contribution is a comprehensive publications dataset (1446 studies), published in recent years, and the use of the industry-approved cyber threats taxonomy in the analysis.

With the extensive landscape of classification algorithms applications on the IS, this study aims to perform a systematic literature review and to answer the following research questions:

- Which classification algorithms are used to address the IS tasks in recent years (2015–2020)?
- In the selected studies, how many classification algorithms are applied? Are they focused only on one algorithm or comparing a few of them?
- What cybersecurity threats were addressed by the classification algorithms?
- What algorithms and cybersecurity threats were included in the highly cited studies?
- What are the challenges in applying classification methods to IS tasks?

The remainder of this paper is divided into four sections. Section 2 contains a cyber threats taxonomy used to map the selected publications to the cybersecurity incidents class. Section 3 describes the systematic literature review method with the literature selection criteria used in this study. Section 4 provides a result of the literature analysis, and finally, Sect. 5 presents the conclusion and possible future research directions.

2 Information Security

A NIST SP 800-12 defines Information Security as "*Protecting information and information systems from unauthorized access, use, disclosure, disruption, modification, or destruction to ensure confidentiality, integrity, and availability*" (Nieles et al. 2017). To provide the IS, we can implement the: authorization, authentication, and authorization verification processes (Liderman 2017). The presented study focuses on the classification algorithms applications on these processes, and this section describes the taxonomy of cybersecurity incidents used to map studies to the cybersecurity incidents.

2.1 *Cybersecurity Incidents Classification Taxonomy*

To have a systematic view on the IS tasks, this study uses Reference Incident Classification Taxonomy created by European Union Agency for Network and Information Security (ENISA). Table 1 presents the cybersecurity incidents classification with incident examples and mapping criteria (regular expressions). To include the study in the analysis and assign it to the cybersecurity incident class, the title, abstract, or keywords should match the mapping criteria column pattern.

2.2 *Application on the Real Data*

There is a big gap between the real data and the data required by the classification algorithms. The actual data, including network traffic logs, metadata, network packets, or user-created content, is frequently available as a structured or unstructured text (see: Fig. 1). However, the majority of classification algorithms described in this study require a structured dataset with numeric data.

Numerous heuristic systems are using raw, unprocessed data as input. For example, for the Intrusion detection task, there is well-established open-source software like tcpdump (1987) (The Regents of the University of California 1987), Zeek (formerly Bro) (1994) (Paxson 1998), or SNORT (1998) (Roesch 1999). Those rule-based anomaly detection systems' main advantages are operating speed,

Table 1 Reference incident classification taxonomy and the literature mapping criteria

<table>
<tr><th>Incident classification</th><th>Incident examples</th><th>Mapping criteria</th></tr>
<tr><td>Abusive content</td><td>Spam, harmful speech, violence</td><td>abusive content|content|spam|harmful speech|porn|violence|fake news|propaganda</td></tr>
<tr><td>Malicious code</td><td>Virus, worm, trojan, spyware, dialer, rootkit</td><td>malicious|malware|worm|trojan|spyware|dialer|rootkit|virus|adware|ransomware|exploit</td></tr>
<tr><td>Information gathering</td><td>Scanning, sniffing, social engineering</td><td>information gathering|scanning|sniffing|social engineering</td></tr>
<tr><td>Intrusion attempts</td><td>Exploiting known vulnerabilities, login attempts, new attack signature</td><td>intrusion attempts</td></tr>
<tr><td>Intrusions</td><td>Privileged account compromise, unprivileged account compromise</td><td>intrusion|intrusion detection|bot|botnet|network traffic|network security</td></tr>
<tr><td>Availability</td><td>DoS, DDoS, sabotage, Ooutage (no malice)</td><td>availability|dos|ddos|sabotage|outage</td></tr>
<tr><td>Information content security</td><td>Unauthorized access to information, unauthorized modification of information</td><td>information content security|unauthorised access|unauthorised modification|keylogger|login|takeover|authentication|authori.ation|privacy</td></tr>
<tr><td>Fraud</td><td>Unauthorized use of resources, copyright, masquerade, phishing</td><td>fraud|unauthorized use|copyright|masquerade|phishing|spoof</td></tr>
<tr><td>Vulnerable</td><td>Open for abuse</td><td>vulnerable|abuse|backdoor|zero day</td></tr>
<tr><td>Other</td><td></td><td></td></tr>
</table>

```
18.58.99.102 - - [26/Jan/2020:00:39:48 +0100] "GET /blog/ HTTP/1.1" 200 8000 "-"
"Mozilla/5.0 (Macintosh; Intel Mac OS X 10_10_1) AppleWebKit/600.2.5 (KHTML,
like Gecko) Version/8.0.2 Safari/600.2.5 (Applebot/0.1;
+http://www.apple.com/go/applebot)" US –
```

Fig. 1 Example of web server logs

simple architecture, easy implementation, and the community's rules database. Beyond all benefits of using real raw data for information security challenges, many questions arise. How often should rules be updated to keep the high system efficiency? How to set and update the thresholds? Are the general rules set good enough, or creating a new set of specific rules is needed (Chio and Freeman 2018)?

We can use classification algorithms to build more complex information security systems using preprocessed data to address those challenges. In this scenario, we need to perform an extensive feature engineering process. The publicly available datasets KDD-CUP-99 (UCI KDD 1999), NSL-KDD (University of New Brunswick 2009), or UNSW-NB15 (UNSW Canberra 2015) contain example features set for the information security task.

To create numeric features from the raw text data, the natural language processing (NLP) and information retrieval (IR) techniques apply. Usually, to extract the information from text, the initial step is to convert text to the vector, where one word corresponds to a feature, with a number of its occurrences in the text as a value (Joachims 1998). Then, algorithms like TF-IDF may apply to calculate the word importance. Another technique is to create a binary feature of the presence (1) or not (0) a word in the text using one-hot encoding. On top of the extracted information, the typical practice is to generate numeric features by dataset aggregations, i.e., the number of events in time per user. The feature engineering part is critical for the future classification model accuracy.

3 Methodology

In this section, we present the literature selection criteria to perform the systematic literature review (Snyder 2019). This study analyzes the publications from January 2015 to September 2020, indexed in the scientific publications databases, and including the classification algorithms application on the Information Security. Table 2 shows the detailed literature selection criteria and the assessment methods.

We queried the scientific publications databases (*Web of Science, Scopus, IEEE Xplore*) using keywords: *cybersecurity, computer crime, information security, classification* and created an initial dataset for the analysis. In the next steps, we

Table 2 Literature selection criteria

Criteria	Desired value	Assessment
Contextual criteria		
Use of the classification algorithm	Yes	Analyzing the abstract and keywords
Applied field	Information security, cybersecurity, computer crime	Checking the publication visibility under desired search terms
Purpose of study	The use of a classification methods to provide the information security	Analyzing the abstract and keywords; validating the topic relevance
Bibliographical criteria		
Date of publication	January 2015–September 2020	Validating if the publication date is within the desired date range
Type of publication	Article, conference proceedings paper, book, book chapter	Checking the publication type
Source of publication	Indexed in the scientific publications database, described by keywords	Analyzing the source of publication

Table 3 Literature selection process

Criteria	Web of science	Scopus	IEEE Xplore
Keywords: (information security OR cybersecurity OR computer crime) AND classification	662	1312	7471
Publication date: January 2015–September 2020	563	927	4437
Type of publication: article, conference proceedings paper, book, book chapter	532	911	4318
Join Web of Science, Scopus, and IEEE Xplore results, remove duplicates		5150	
Contextual criteria (the use of classification algorithm)		1991	
Contextual criteria (the application to cybersecurity)		1446	
Total publications to analyze		1446	

assessed the results against the publication date and publication type bibliographical criteria.

After the initial selection process, each study was analyzed and labeled by (1) the classification algorithm used and (2) the cybersecurity threat addressed. As a result, we created a dataset of 1446 publications. Table 3 shows the detailed publication selection process. Table 4 contains the final dataset columns grouped in publication metadata, classification algorithm used in the study, and the IS tasks addressed (binary features).

4 Application of Classification Algorithms to Information Security

We performed a series of summaries on the selected publications dataset to address the research questions about the classification algorithms applications to IS popularity in recent years, the combinations of algorithms evaluated, and the cybersecurity incidents examined. We also discuss the main challenges in the classification algorithm selection in the IS field. In this section, we present the study results.

4.1 Popular Classification Algorithms

To check what classification algorithms applied to the IS are popular in the studies published in recent years (2015–2020), we calculated the occurrences in the titles, abstracts, and keywords. Table 5 contains the most popular classification algorithms used for the IS (note that one study could examine more than one classification algorithm). The frequently used algorithms were:

Table 4 Selected publications dataset columns

Study metadata	Classification algorithms used	Cybersecurity threats
Document title	Auto-encoder (AE)	Abusive content
Authors	Artificial neural network (ANN)	Malicious code
Publication year	Recurrent neural network (RNN)	Information gathering
Abstract	Deep neural network (DNN)	Intrusion attempts
DOI	Naïve Bayes (NB)	Intrusion
Article citation count	C4.5	Availability
Database	Convolutional neural network (CNN)	Information content security
Keywords	Decision trees	Fraud
Text (title, abstract, keywords)	Random forest	Vulnerable
	Extreme learning machine (ELM)	Total threats
	J48	
	Long short-term memory (LSTM)	
	Support vector machine (SVM)	
	k-nearest neighbors (kNN)	
	Genetic fuzzy systems	
	Fuzzy C-means	
	Fuzzy pattern tree	
	Fuzzy K-mean	
	Fuzzy neural network	
	Logistic regression (LR)	
	Boltzmann machine	
	Self organizing feature maps	
	Cauchy possibilistic clustering (novel)	
	Genetic programming (GP)	
	k-means	
	Latent Dirichlet Allocation (LDA)	
	Noise-resistant statistical traffic classification (NSTC)	
	Self-normalizing neural network (SNN)	
	Feed-forward neural networks (FNN)	
	Total algorithms	

- Support vector machines (SVMs) (occurred in 39% studies),
- Decision trees (23%),
- Random forest (21%),
- Convolutional neural networks (CNNs) (14%),
- Naïve Bayes (10%).

Table 5 Popularity of classification algorithms applied to IS

Algorithm	Studies	% of total
Support vector machine (SVM)	549	38%
Decision trees	328	23%
Random forest	302	21%
Convolutional neural network (CNN)	200	14%
Naïve Bayes (NB)	151	10%
Deep neural network (DNN)	120	8%
Artificial neural network (ANN)	97	7%
k-means	94	7%
Auto-encoder (AE)	81	6%
Logistic regression (LR)	79	5%
k-nearest neighbors (kNN)	76	5%
J48	64	4%
Recurrent neural network (RNN)	57	4%
C4.5	45	3%
Long short-term memory (LSTM)	44	3%
Extreme learning machine (ELM)	24	2%
Latent Dirichlet Allocation (LDA)	19	1%
Boltzmann machine	17	1%
Other	28	2%

It is worth to notice that the most popular algorithms are not novel. They were discussed in the 80 s [Decision Trees—1986 (Quinlan 1986)] and 90 s [Naïve Bayes—1992 (Langley et al. 1992), SVMs—1995 (Cortes and Vapnik 1995), Random Forest—1995 (Ho 1995), and Convolutional Neural Networks (CNNs)—1999 (LeCun et al. 1999)]. However, growing data and technology availability provides new use cases for these algorithms.

The less popular algorithms (<0.5% for each algorithm, grouped into category Other) were feed-forward neural networks (FNN), fuzzy neural network, genetic programming (GP), self-normalizing neural network (SNN), fuzzy C-means, self-organizing feature maps, Cauchy possibilistic clustering, genetic fuzzy systems, fuzzy pattern tree, fuzzy K-mean, noise-resistant statistical traffic classification (NSTC).

We also identified algorithms with constantly growing popularity in the recent years (2015–2019), excluding the top five algorithms described above:

- Auto-encoders (AE),
- Deep neural networks (DNN),
- Long short-term memory (LSTM),
- K-nearest neighbors (kNN),
- Logistic regression (LR).

We present the popularity trends of each classification algorithms on Fig. 2.

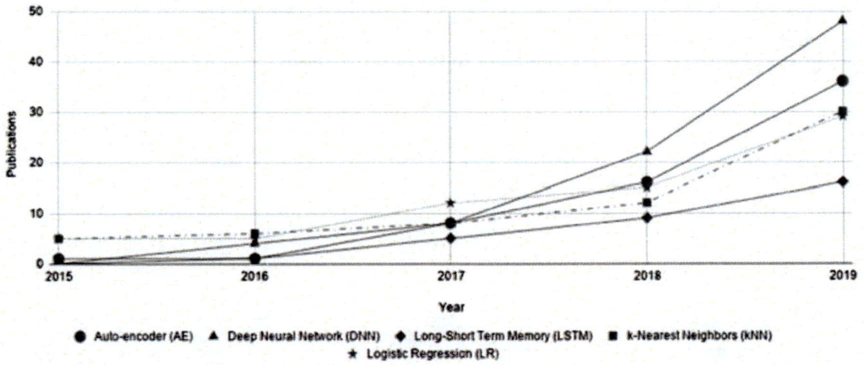

Fig. 2 Classification algorithms with growing popularity

4.2 *Classification Algorithms Used Per Study*

As we discovered the use of multiple classification algorithms per study, we wanted to investigate if other researchers are more focused on applying one algorithm or are comparing a few and compare their performance. The results in Table 6 show that the majority of studies (78%) were focused only on applying one classification algorithm.

To examine the studies using more than one algorithm and find the sets of algorithms used frequently, we used the Apriori algorithm. Table 7 shows that researchers combine random forest with decision trees or SVMs, and SVMs with random forest, decision trees, or Naïve Bayes.

4.3 *Cybersecurity Incidents Examined*

As the dataset contains publications mapped to the cybersecurity incidents, we checked the classes of incidents examined by classification algorithms. The most

Table 6 Number of algorithms used per study

Algorithms per study	Studies	% of total
1	1134	78,4%
2	484	33,5%
3	189	13,1%
4	57	3,9%
5	29	2,0%
6	9	0,6%
7	1	0,1%
Total studies	1446	

Table 7 Classification algorithms examined together—Apriori algorithm

Sets of classification algorithms	Support
(Random forest, decision trees)	0.23
(Random forest, support vector machine (SVM))	0.17
(Support vector machine (SVM), decision trees)	0.17
(Naïve Bayes (NB), support vector machine (SVM))	0.10

Table 8 Cybersecurity incidents addresses by studies

Cybersecurity incident	Studies	% of total
Intrusion	938	64.9%
Malicious code	510	35.3%
Information content security	225	15.6%
Fraud	187	12.9%
Abusive content	164	11.3%
Availability	157	10.9%
Vulnerable	90	6.2%
Information gathering	18	1.2%
Intrusion attempts	1	0.1%
Total studies	1446	

studied cybersecurity incidents are intrusion detection (65% studies) and malicious code detection (35%) (note that one study could analyze multiple cybersecurity incidents). The specific of both use cases—high availability of large datasets from network logs for Intrusions and the possibility of isolating malicious code behavior—makes it suitable to apply the classification algorithms. Table 8 presents the detailed results.

4.4 Highly Cited Studies

We also analyzed the highly cited papers (more than 100 citations) in the selected dataset. In terms of covered incidents, they follow the top cybersecurity incidents examined by the classification algorithms: 6 out of 8 highly cited papers discuss the Intrusion incident. Moreover, highly cited papers fit the most popular classification algorithms on the IS field: 5 out of 8 studies include SVMs, and 2 out of 8 uses random forest. Table 9 presents a full summary of the highly cited papers.

Table 9 Highly cited papers on the classification algoritms applied to Information security field

Document title	Authors	Year	Article citation count	Algorithms	Cybersecurity incidents
Face spoof detection with image distortion analysis	D. Wen; H. Han; A. K. Jain	2015	228	Support vector machine (SVM)	Information content security, fraud
Robust joint graph sparse coding for unsupervised spectral feature selection	X. Zhu; X. Li; S. Zhang; C. Ju; X. Wu	2016	189	k-Nearest neighbors (kNN)	Intrusion
Building an intrusion detection system using a filter-based feature selection algorithm	M.A. Ambusaidi; X. He; P. Nanda; Z. Tan	2016	169	Support vector machine (SVM)	Intrusion
A deep learning approach for intrusion detection using recurrent neural networks	C. Yin; Y. Zhu; J. Fei; X. He	2017	146	Artificial neural network (ANN), recurrent neural network (RNN), random forest, J48, support vector machine (SVM)	Intrusion
Robust network traffic classification	J. Zhang; X. Chen; Y. Xiang; W. Zhou; J. Wu	2015	144	Random forest	Intrusion
Long short-term memory recurrent neural network classifier for intrusion detection	J. Kim; J. Kim; H L. Thi Thu; H. Kim	2016	129	Recurrent neural network (RNN), long short-term memory (LSTM)	Intrusion
Detection of face spoofing using visual dynamics	S. Tirunagari; N. Poh; D. Windridge; A. Iorliam; N. Suki; A.T.S. Ho	2015	116	Support vector machine (SVM)	Abusive content, malicious code, fraud
Decision tree and SVM-based data analytics for theft detection in smart grid	A. Jindal; A. Dua; K. Kaur; M. Singh; N. Kumar; S. Mishra	2016	107	Decision trees, support vector machine (SVM)	Intrusion

4.5 *Challenges in Classification Algorithms Application to the Information Security*

Researchers frequently indicate the challenges in classification algorithms applications to information security. One of them is unbalanced datasets. It is a situation when the model training data has underrepresented target value, i.e., a small number of fraudulent transactions in all transactions dataset. To analyze unbalanced datasets, the researcher needs to oversample or undersample the data with technics like synthetic minority over-sampling technique (Wang et al. 2006).

Another challenge reported is the data quality problem, when the data collected from the production systems has errors, gaps, or are malformed. This challenge also includes the missing data delete due to anonymization or legal requirements. It creates limitations in the data analysis and could make creating accurate classifications model harder.

Third often emphasized challenge is the data volume. Nowadays, it is the common scenario in the information security area, especially for the network traffic logs and analyzing data to detect denial of service attacks. Implementation of the classification algorithm to the large datasets requires careful model family selection, keeping the impact of computational and mathematical complexity on model training process. Table 10 includes the selected challenges with the examples publications.

5 Conclusions and Future Research Directions

This study presents the results of an extensive systematic literature review of classification algorithms applications to information security. Based on the results, we found the following conclusions.

In recent years (2015–2020), the algorithms frequently used in the IS field were: support vector machines (SVMs), decision trees, random forest, convolutional

Table 10 Challenges in classification algorithms applications for the information security

Challenge	Description	Examples
Unbalanced datasets	The training dataset contains underrepresented sample of the target value, i.e., small sample of fraudulent e-commerce transactions in the all transactions set	Liu et al. (2018), Al-Azani and El-Alfy (2018)
Missing data, data quality	Incomplete data with missing labels, gaps in observations or other data errors	Kaur and Bansal (2016), Ahmed et al. (2018)
Data volume	The training dataset volume exceeds the one machine capacity	Hou et al. (2018), Gurulakshmi and Nesarani (2018)

neural networks (CNNs), and Naïve Bayes. Since these algorithms are not novel, the growing data availability and new technologies make IS the new application field. Although, the other classification algorithms are gaining popularity in the IS field:

- Auto-encoders (AE),
- Deep neural networks (DNN),
- Long short-term memory (LSTM),
- K-nearest neighbors (kNN),
- Logistic regression (LR).

The majority of analyzed papers (78%) focus on applying the one classification algorithm. We used the Apriori algorithm and identified a set of algorithms used together for those that examine more than one: Random forest with decision trees or SVMs and SVMs with random forest, decision trees, or Naïve Bayes.

Regarding cybersecurity incidents, classification algorithms are extensively used for intrusion detection (65% studies) and malicious code detection (35%) tasks.

We also analyzed the highly cited papers in the field. They fit the most popular classification algorithms on the IS field pattern: 5 out of 8 studies include SVMs, and 2 out of 8 uses random forest.

The last finding is the list of challenges in applying classification algorithms to the IS. Researchers indicate working with unbalanced datasets, fixing the missing data and overall data quality, and working with large data volume. It requires extensive data preprocessing (i.e., sampling methods) and fewer algorithms fast and less expensive in training.

The future research on this area may include the whole of 2020 for a more comprehensive data range. Future research may include the classification algorithms grouping into the model families and assess them against computational and mathematical complexity. The other researchers could also map the IS applications to the various industries and examine how algorithms fit the industry-specific use cases.

References

Ahmed A, Krishnan V, Foroutan S et al (2018) Cyber physical security analytics for anomalies in transmission protection systems. IEEE Ind Appl Soc Annu Meet (IAS) 2018:1–8

Al-Azani S, El-Alfy EM (2018) Imbalanced sentiment polarity detection using emoji-based features and bagging ensemble. In: 2018 1st International conference on computer applications and information security (ICCAIS), pp 1–5

Ambusaidi MA, He X, Nanda P et al (2016) Building an intrusion detection system using a filter-based feature selection algorithm. IEEE Trans Comput 65:2986–2998

Apruzzese G, Colajanni M, Ferretti L et al. (2018) On the effectiveness of machine and deep learning for cyber security. In: 2018 10th international conference on cyber conflict (CyCon), pp 371–390

Brecht M, Nowey T (2013) A closer look at information security costs. In: Böhme R (ed) The economics of information security and privacy. Springer, Berlin, Heidelberg, pp 3–24

California State Legislature (2018) California consumer privacy act of 2018. http://leginfo.legislature.ca.gov/faces/codes_displayText.xhtml?division=3.&part=4.&lawCode=CIV&title=1.81.5. Accessed on 25 Oct 2020
Chio C, Freeman D (2018) Machine learning and security. O'Reilly Media Inc, Massachusetts
Cortes C, Vapnik V (1995) Support-vector networks. Mach Learn 20:273–297
European Parliament and Council of European Union (2016) Regulation (EU) 2016/679. https://eur-lex.europa.eu/legal-content/EN/TXT/HTML/?uri=CELEX:32016R0679&from=EN. Accessed on 25 Oct 2020
European Union (2020) Ordering or buying goods and services. https://ec.europa.eu/eurostat/statistics-explained/index.php/Digital_economy_and_society_statistics_-_households_and_individuals#Services_ordered_from_other_individuals_via_the_internet. Accessed on 27 Oct 2020
European Union Agency for Network and Information Security (ENISA) (2018) Reference incident classification taxonomy. https://www.enisa.europa.eu/publications/reference-incident-classification-taxonomy/at_download/fullReport. Accessed on 25 Oct 2020
Ferrag MA, Shu L, Yang X et al (2020) Security and privacy for green IoT-based agriculture: review, blockchain solutions, and challenges. IEEE Access 8:32031–32053
Gurulakshmi K, Nesarani A (2018) Analysis of IoT bots against DDOS attack using machine learning algorithm. In: 2018 2nd International conference on trends in electronics and informatics (ICOEI), pp 1052–1057
Haibo H, Garcia EA (2009) Learning from imbalanced data. IEEE Trans Knowl Data Eng 21:1263–1284
Ho TK (1995) Random decision forests. In: Proceedings of the 3rd international conference on document analysis and recognition, pp 278–282
Hou J, Fu P, Cao Z et al (2018) Machine learning based DDos detection through NetFlow analysis. MILCOM 2018—2018 IEEE military communications conference (MILCOM)
iA Internet Association (2020) IA Industry Indicators. Data and analysis for the U.S. internet industry. Q1 2020 Data, Q3 2020 Release. https://internetassociation.org/wp-content/uploads/2020/09/IA_Internet-Industry-Indicators-Report_Q3-2020_digital.pdf. Accessed on 27 Oct 2020
Jindal A, Dua A, Kaur K et al (2016) Decision tree and SVM-based data analytics for theft detection in smart grid. IEEE Trans Industr Inf 12:1005–1016
Jing X, Yan Z, Pedrycz W (2018) Security data collection and data analytics in the internet: a survey. IEEE Commun Surv Tutorials 21:586–618
Joachims T (1998) Text categorization with support vector machines: learning with many relevant features. Eur Conf Mach Learn 1398:137–142
Kaur R, Bansal M (2016) Multidimensional attacks classification based on genetic algorithm and SVM. In: 2016 2nd International conference on next generation computing technologies (NGCT), pp 561–565
Kent K, Souppaya M (2006) NIST SP 800-92. Guide to computer security log management
Kim J, Kim J, Thi Thu HL et al (2016) Long short term memory recurrent neural network classifier for intrusion detection. Int Conf Platform Technol Serv (PlatCon) 2016:1–5
Langley P, Iba W, Thompson K (1992) An analysis of Bayesian classifiers. In: AAAI'92: proceedings of the tenth national conference on artificial intelligence vol 90, pp 223–228
LeCun Y, Haffner P, Bottou L et al (1999) Object recognition with gradient-based learning. Shape, contour and grouping in computer vision lecture notes in computer science vol, 1681, pp 319–345
Liderman K (2017) Bezpieczeństwo informacyjne. Wydawnictwo Naukowe PWN, Warszawa
Liu K, Fan Z, Liu M et al (2018) Hybrid intrusion detection method based on K-Means and CNN for smart home. In: 2018 IEEE 8th annual international conference on CYBER technology in automation, control, and intelligent systems (CYBER), pp 312–317
Masood F, Ammad G, Almogren A et al (2019) Spammer detection and fake user identification on social networks. IEEE Access 7:68140–68152

Nieles M, Dempsey K, Pillitteri VY (2017) Special Publication (NIST SP)—800-12 Rev. 1. An introduction to information security
Office of the Australian Information Commissioner (1988) Australian Privacy Principles. https://www.oaic.gov.au/privacy/australian-privacy-principles/. Accessed on 25 Oct 2020
Office of the Privacy Commissioner of Canada (2000) The personal information protection and electronic documents act (PIPEDA). https://www.priv.gc.ca/en/privacy-topics/privacy-laws-in-canada/the-personal-information-protection-and-electronic-documents-act-pipeda/. Accessed on 25 Oct 2020
Paxson V (1998) Bro: a system for detecting network intruders in real-time. 7th USENIX Secur Symp 31(23–24):2435–2463
Quinlan JR (1986) Induction of decision trees. Mach Learn 1:81–106
Roesch M (1999) Snort—lightweight intrusion detection for networks. LISA '99: 13th Syst Admin Conf 99(1):229
Ryzko D (2020) Modern big data architectures. John Wiley & Sons, New Jersey, New York
Snyder H (2019) Literature review as a research methodology: an overview and guidelines. J Bus Res 104:333–339
The Regents of the University of California (1987) tcpdump. https://opensource.apple.com/source/tcpdump/tcpdump-56/tcpdump/tcpdump.1. Accessed on 17 Dec 2020
Tirunagari S, Poh N, Windridge D et al (2015) Detection of face spoofing using visual dynamics. IEEE Trans Inf Forensics Secur 10:762–777
UCI KDD (1999) KDD-CUP-99 Dataset. http://kdd.ics.uci.edu/databases/kddcup99/kddcup99.html. Accessed on 17 Dec 2020
University of New Brunswick (2009) NSL-KDD Dataset. https://www.unb.ca/cic/datasets/nsl.html . Accessed on 17 Dec 2020
UNSW Canberra (2015) UNSW-NB15 Dataset. https://www.unsw.adfa.edu.au/unsw-canberra-cyber/cybersecurity/ADFA-NB15-Datasets/. Accessed on 18 Dec 2020
Wang J, Xu M, Wang H et al (2006) Classification of imbalanced data by using the SMOTE algorithm and locally linear embedding. In: 2006 8th international conference on signal processing, p 3
Wen D, Han H, Jain AK (2015) Face spoof detection with image distortion analysis. IEEE Trans Inf Forensics Secur 10:746–761
Yin C, Zhu Y, Fei J et al (2017) A deep learning approach for intrusion detection using recurrent neural networks. IEEE Access 5:21954–21961
Zhang J, Chen X, Xiang Y et al (2015) Robust network traffic classification. IEEE/ACM Trans Networking 23:1257–1270
Zhu X, Li X, Zhang S et al (2016) Robust joint graph sparse coding for unsupervised spectral feature selection. IEEE Trans Neural Netw Learn Syst 28:1263–1275

Outlier Detection with the Use of Isolation Forests

Krzysztof Najman and Krystian Zieliński

Abstract Appropriate preparation of data for analysis is a key element in empirical research. Considering the source of data or the nature of the phenomenon studied, some observations may differ significantly from others. Inclusion of such cases in a research may seriously distort the profile of the population under examination. Nevertheless, their omission can be equally disadvantageous. When analyzing dynamically changing phenomena, especially in case of big data, a relatively small amount of outliers may constitute a coherent and internally homogeneous group, which, along with the registration of subsequent observations, may grow into an independent cluster. Whether or not an outlier is removed from the dataset, researcher must be first aware of its existence. For this purpose, an appropriate method of anomaly detection should be used. Identification of such units allows the researcher to make an appropriate decision regarding the further steps in the analysis.

Assessment of the usefulness of outlier value detection methods has been increasingly influenced by the possibility of their application for big data problems. The algorithms should be effective for large volume and diverse sets of data, which are additionally subject to constant changes. For these reasons, apart from high sensitivity, the following are also important: low computational time and the algorithm's adaptability.

The aim of the research presented is to assess the usefulness of Isolation Forests in outlier detection. Properties of the algorithm, with its extensions, will be analyzed. The results of simulation and empirical research on selected datasets will be presented. The algorithm evaluation will take into account the impact of particular features of big datasets on the effectiveness of the methods analyzed.

Keywords Outliers · Anomalies · Isolation forests

K. Najman (✉) · K. Zieliński
University of Gdańsk, Gdańsk, Poland
e-mail: krzysztof.najman@ug.edu.pl

K. Jajuga et al. (eds.), *Data Analysis and Classification*, Studies in Classification, Data Analysis, and Knowledge Organization,
https://doi.org/10.1007/978-3-030-75190-6_5

1 The Essence of Outliers in Cluster Analysis

By analyzing the properties of complex populations, researchers collect data on many of the characteristics describing them. Individual units within a population differ from each other, while the level of the differentiation is one of the significant features characterizing it. It may happen, however, that some, usually a few, units differ so much, that a suspicion can emerge, that a mistake has been made (measurement error) during the data collection process or that the units do not belong to a given population. Since, due to its values, such a unit is on the edge of a given feature's distribution, these values are called outliers (Grubbs 1969). Hawkins formally defined the concept of an outlier as follows:

"An outlier is an observation that deviates so much from the other observations as to arouse suspicions that it was generated by a different mechanism." (Hawkins 1980).

Occurrence of outliers in a dataset possesses a considerable problem, from the perspective of its analysis. In many cases, even a single observation can significantly change the statistics describing a given phenomenon. It is particularly visible in analyzes employing the arithmetic mean as well as in regression analysis (Anscombe 1973). In terms of cluster analysis, this is also an important issue, because even a small number of atypical values can distort the resulting group structure. Some methods of cluster analysis, e.g., the self-learning artificial neural networks of growing neural gas (GNG) type are so sensitive that they will try to create a separate cluster even for a few single outlying observations (Migdał-Najman and Najman 2013). As such, one-element clusters can emerge, which do not contribute to the understanding of the phenomenon under examination. It may happen, however, that deformed clusters including an outlying value are obtained, which in turn may significantly distort the clusters.

From the perspective of the impact on the results of grouping, the nature of anomalies can be diverse. Such a unit may be distant from all the others in the dataset, while its feature values are well outside the range of all other units' variation. Such an unusual value is called an extreme value. It may happen that the values of the features examined with respect to a given unit do not differ in the scope, and yet, such a unit is outside the group structure. This is an unusual value, i.e., an outlier, but not an extreme value (Aggarwal 2015).

Figure 1 illustrates such problems. Point A is considerably distanced from all the others; thus, it is an outlier and an extreme value. Many methods of grouping will create a separate cluster for it, making it relatively easy to be detected and removed as non-contributing to the understanding of the structure of the population under examination. Point B is also an outlier, but the values of its features fall within the variability range of the remaining part of the population. It is also not that considerably distant from the other objects. Yet, the point is controversial, because it distorts, to a large extent the parameters of the cluster which it belongs to. The red line marks the boundary of the cluster that would emerge without this point, while the navy blue line marks the boundary including it. It is easy to see that these

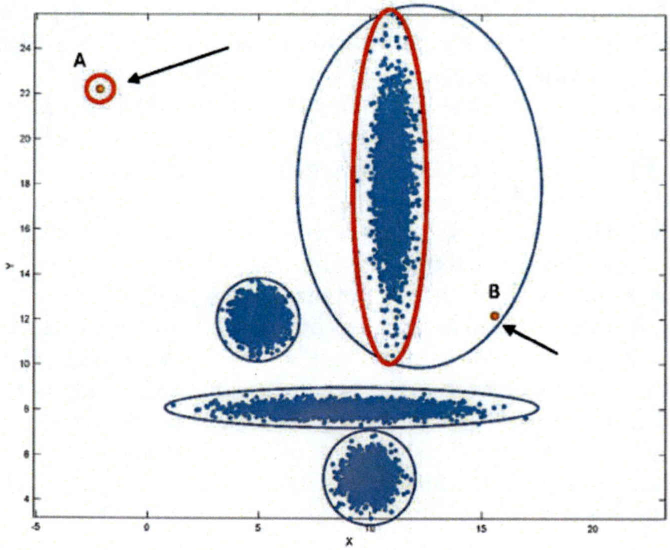

Fig. 1 Outliers and extreme values. *Source* Own elaboration

boundaries differ significantly. The cluster parameters, with and without point B, would be completely different.

For the above reasons, outliers should be detected and removed from the dataset at the stage of data preparation for analysis. A number of methods can be used for this purpose, including Isolation Forests. The purpose of the research presented is to assess the usefulness of this method in outlier detection, in particular in cluster analysis. Properties of a basic algorithm and its extensions will be analyzed. The results of simulation and empirical research on selected datasets will be presented. The algorithm evaluation will take into account the impact of particular features of big datasets on the effectiveness of the methods analyzed.

2 Introduction to Isolation Forests and Extended Isolation Forests

Isolation Forests (iForest) (Liu et al. 2008) are a method of outlier detection, propounded in 2008 and developed by Fei Tony Liu, Kai Ming Ting, and Zhi-Hua Zhou. The algorithm aims to isolate outliers by random partitioning of the dataset, using multiple isolation trees. The mechanism steams from the assumption of anomalies, i.e., their small share in the dataset and their considerable distance from other typical observations. In Isolation Forests, the step of building a given population's profile is omitted, owing to which the impact of the distance of the anomaly from the rest of the data is not biased by inaccurate mapping of the dataset structure. Taking into account the risk of examining inconsistent datasets, the method gains in its universality.

To better understand how the iForest algorithm works, the Isolation Tree algorithm needs to be explained. As the name suggests, their structure is closely related to Classification Trees. The main difference entails the fact that Isolation Trees are unsupervised algorithms; thus, they can be used for outlier detection. Unlike Classification Trees, where the dataset is partitioned into nodes, based on a selected measure that determines the quality of the partition, Isolation Trees divide the dataset according to a random variable and its random value. This change allowed significant acceleration of the algorithm, since selection of the best partition requires many potential variants to be checked. The degree of observation isolation depends on the length of the path between the tree root and the leaf containing the observation. This is equivalent to the number of the dataset partitions needed to extract a given value. Typical observations require more nodes for their isolation, because there are more units within their small area. The probability that outlier at a random data partition will be separated earlier is higher (Fig. 2).

Definition 1 Let X be the set of observations, and y the variables in the set. The *Isolation Tree* is a type of a tree, in which node T is a leaf or a node with one partition and exactly two descendants $T_{i,j}$, so that the random variable $y_k \in y$ and the random value $p \in (min(y_k : y_k \in T), max(y_k : y_k \in T))$, while the elements of the set are divided into:

$$T_i : y_k < p, T_i : y_k \geq p. \tag{1}$$

The degree of isolation of a given observation x is determined by the path length $h(x)$, from the root of the tree to the leaf which point x is located in. Due to the fact that the value of $h(x)$ depends on many factors, such as the sample size or the dimensionality, it cannot in itself be regarded as a universal coefficient of isolation.

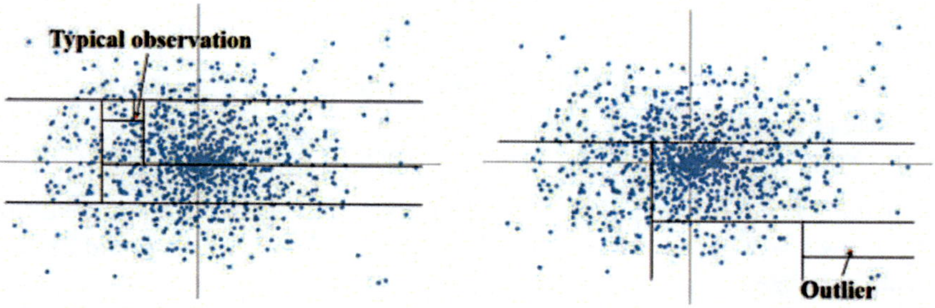

Fig. 2 Isolation of a typical and an outlying observation in Isolation Trees. *Source* Own elaboration

Using the binary search tree (BST) theory, it is possible to estimate the average height of the Isolation Tree in a set of n elements. This value is the same as the average length of a failed BST search, that is:

$$c(n) = 2H(n-1) - 2\frac{n-1}{n} \quad (2)$$

where $H(i) \approx ln(i) + e$ is a harmonic number. The value is used to calculate the anomaly score in the iForest algorithm.

Similarly to random forests, the iForest algorithm is more effective when individual trees have low isolation capabilities; nevertheless, a large number of trees will be used for the prediction. For this reason, to train an Isolation Tree, not all the observations from the set are needed; it is even advisable to sample a small number of observations. In this way, the tree can work on separating the unusually atypical values rather than unnecessarily creating a large number of partitions among the values that are consistent with the population profile.

Definition 2 Let X be the dataset. The *Isolation Forest* is a statistical model consisting of t single, independent Isolation Trees $\{h(\psi,\Theta_k), k = 1,\ldots,t\}$, where ψ is a subset of X, drawn without a return, and $\Theta_k \sim$ *iid* is a random vector.

Using the algorithm, the following equation results from the anomaly score prediction for observations x:

$$s(x,n) = 2^{-\frac{E(h(x))}{c(n)}}, \quad (3)$$

where E(h(x)) is the average path length for observations x in all the Isolation Trees from the iForest. Dividing this value by $c(n)$, a normalized path length is obtained, which can be interpreted universally, regardless of the set of observations. If:

- $s(x_o)$ is significantly higher than 0.5, the observation is an outlier,
- $s(x_o)$ is lower than 0.5, the observation is not treated as an outlier,
- $\forall_i s(x_i) \approx 0.5$, no significant outliers have been observed in the dataset.

The iForest algorithm comprises two steps. In the first, Isolation Trees are trained on the training part of the dataset. Then, new observations are introduced to the model, which calculates their anomaly score. In the process of the model's training, the researcher determines a priori the values of two parameters: the sample size in the drawing and the number of independent Isolation Trees.

One important advantage of the algorithm is the fact that the iForest is distinguished by low computational requirements. The structure of single trees is the same as the BST; thus, the computational complexity for an Isolation Tree is $O(\psi\lceil\log_2\psi\rceil)$. For the iForest, the complexity is $O(t\psi\lceil\log_2\psi\rceil)$. As a result, the increase in complexity is only linear (it is dependent on the sample size and the number of trees), which makes the Isolation Forest perfect for analysis of big datasets. In the process of evaluation, the observations go through each of the t trees, and based on the path length, the value of $s(x, \psi)$ is calculated. The computational complexity, just as in the process of learning, is $O(tn\lceil\log_2 \psi\rceil)$, where n is the number of observations which the anomaly score is calculated for.

In the process of Isolation Tree training, data partition according to random values of individual variables causes the observations from the dataset to partition perpendicularly to the given variable, at the partition point p. The problem is reflected in higher dimensions and increases the risk of including an anomaly in further analysis, if the observation assumes typical values for at least one of the variables. Optimization (Liu et al. 2019) of the partitioning value increases the method's effectiveness; nevertheless, it still assumes partition according to one variable. One idea reducing the impact of the phenomenon on the value of the anomaly score is the Extended Isolation Forest (EIF) (Sahand et al. 2019).

The change in the approach to outlier isolation entails a different definition of dataset partition in the Isolation Tree nodes. In classic Isolation Forests, for a k-dimensional dataset, partition by a random value p from the ith variable is the same as a cross-section of hyperplane R^k by a $k - 1$ dimensional hyperplane that is parallel to $R^k \backslash R^i$ and shifted by the value of p. In this way, each partition is parallel to the rest of the variables in the dataset. Nothing, however, prevents the partition hyperplane from sloping with reference to the other variables. If the slope angle is random, the maps of the anomaly score values should be topologically closer to the dataset under examination.

Selection of a random slope is equivalent to indication of a normal vector $\vec{n}$, at a random point in a k-dimensional sphere. To do so, it is enough to draw each of the vector $\vec{n}$ coordinates as a random value from a normal standardized distribution. Then, to determine the displacement vector $\vec{p}$, values are randomly selected for each of the k coordinates within the range of the subset being divided. The formula for partition of the observations in a node into two consecutive subsets can be written as follows:

$$(\vec{x} - \vec{p}) \cdot \vec{n} \leq 0 \tag{4}$$

A partition taking into account a dimension value greater than one has positive impact on the isolation capabilities of the algorithm, owing to which the activation map better reflects the dataset. At the same time, it does not increase the computational time needed to train the Isolation Tree. It should be added that in the EIF, the number of the variables taken into account when determining the normal vector can be established, whereas for the variables that are not included, vector $\vec{n}$ assumes the value of 0. In the following part of the article, the algorithm's behavior will be tested, taking into account all the variables, as to determine the partition of the dataset (Fig. 3).

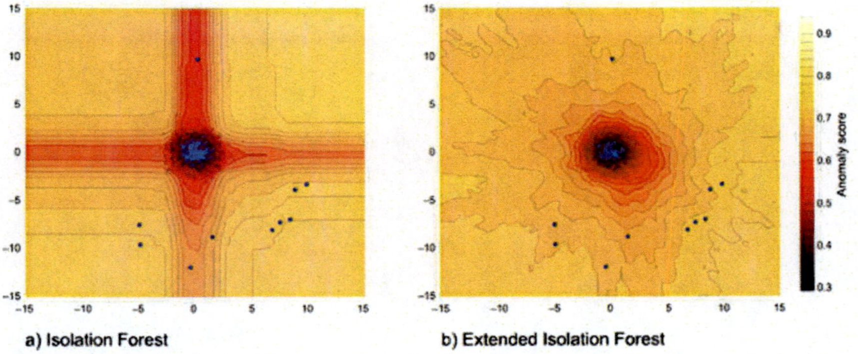

Fig. 3 Map of iForest and EIF anomaly score. *Source* Own elaboration

3 The Impact of Algorithm Parameters on the method's Effectiveness

Despite the small number of parameters, the value of which is determined a priori by the researcher, their impact on the iForest and the EIF is very significant. The number of the trees used in the iForest and in the Extended Isolation Forest algorithms affects the method's effectiveness as well as the computational time needed to train the model and to predict new observations. As in other ensemble learning algorithms (Probst and Boulesteix 2018), a greater number of trees reduces the deviation of the anomaly score values for both typical observations and outliers. The model is better fitted for the input data; therefore, small changes in the values of observations do not drastically affect the value of the anomaly score. Equally, insertion of new trees linearly increases the computational complexity of the algorithm, owing to which the parameter value should be the convergence point for the anomaly index value.

The empirical research conducted on a dataset of 100,000 observations from a normal distribution and on a thousand outliers shows stabilization of the standard deviation of the anomaly index value at as little as 100 trees (Fig. 4).

Another parameter, the value of which is determined by the researcher, is the sample size. The number of the observations used to train individual Isolation Trees, similarly to the number of trees, affects the fit of the model to the input dataset. The difference between the parameters entails the fact that an increase in the size carries the risk of tree overtraining. When sampling a large number of observations, the probability of anomaly inclusion in the process of training is higher, due to which some of these values can be treated as typical. Concurrently, the mean path length for typical observations is extended, which heightens the differences in the values of the anomaly index for inlying and outlying observations (Fig. 5).

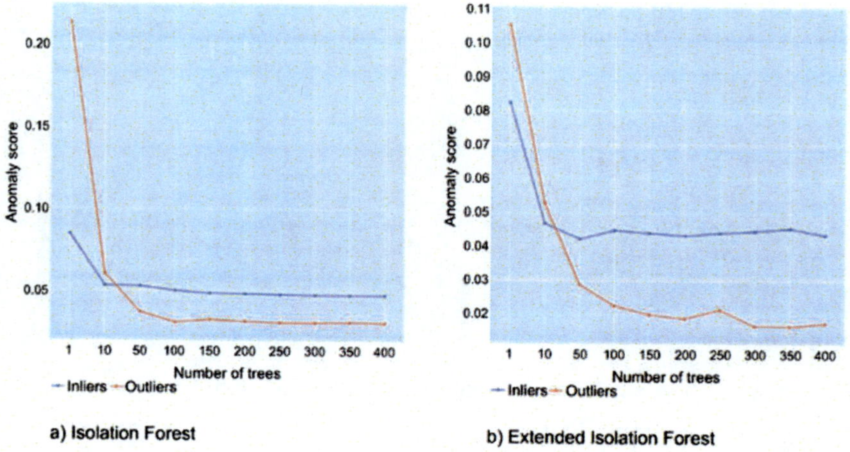

Fig. 4 Impact of the number of trees on the anomaly score deviation. *Source* Own elaboration

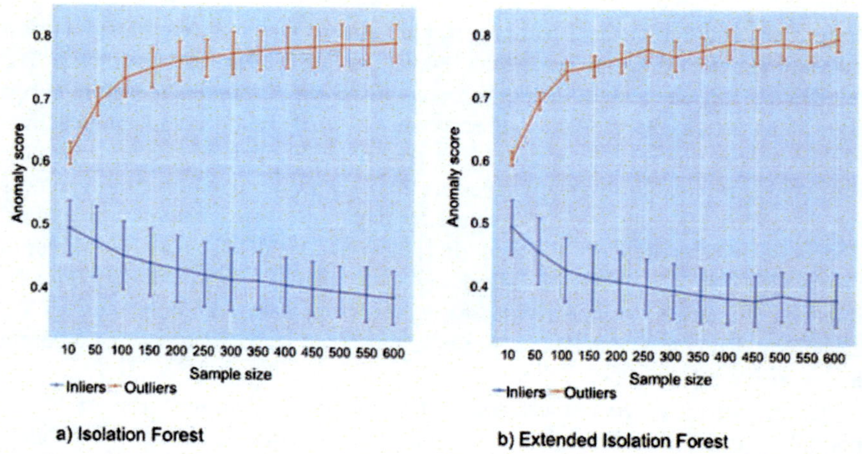

Fig. 5 Impact of the sample size on the anomaly score values. *Source* Own elaboration

4 The Impact of Dataset Characteristics on the Anomaly Score Values

Parameters of the method are not the only factors which influence significantly the anomaly score. One important element that may affect the method's effectiveness is the dimensionality of the dataset. More variables mean a greater number of potential dataset partitions in individual isolation trees, which means that the values of the anomaly index provide poorer reflection of the phenomenon under examination (Fig. 6).

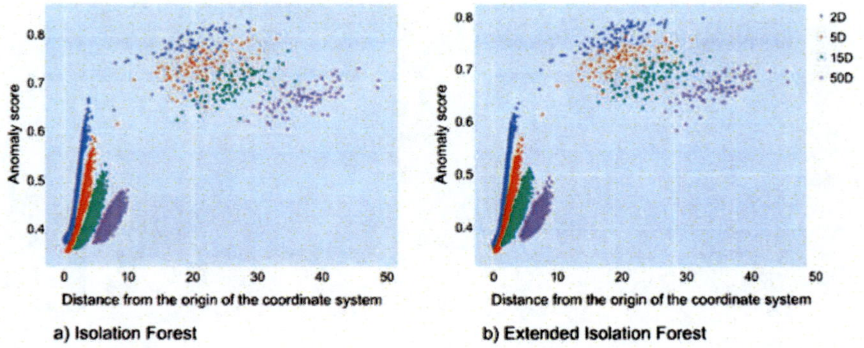

Fig. 6 Impact of the dataset dimensionality on the anomaly score value. *Source* Own elaboration

The Isolation Forest and the Extended Isolation Forest were trained on datasets of various dimensions, where typical observations assume, for each variable, the values from the N(0, 1) distribution, whereas 1% of the outliers is clearly separable from them. As the number of the features describing the data increases, the values of the anomaly index are more dispersed—small distances between the observations result in a smaller difference in the values of the anomaly index. Concomitantly, the values of the anomaly index for outlying observations become lower. The impact of the data's high dimensionality on the distribution of the s(x, n) values can be effectively compensated for by the use of a greater number of individual Isolation Trees in the process of the model training.

Insofar, research involved datasets with one cluster point. The method for calculating the anomaly score in Isolation Forests and Extended Isolation Forests indicates the global nature of the algorithms. Often, in empirical data, researchers encounter situations when various groups of observations with completely different distributions of the variables stand out in a given dataset. In such a case, the algorithm's ability to detect local anomalies in the dataset is significant.

Simulation studies show that the effectiveness of Isolation Forests in detecting local anomalies is most affected by the disproportions in the number and the dispersion of the observations for individual groups. The assumption regarding the anomalies in the algorithms described entails their rare occurrence and large distance from the other values. Observations from smaller clusters have a lower chance of being included in the sample, which makes the path length from the tree root to the observation significantly shorter. Observation dispersion also significantly affects the anomaly score values. The random value of partition p in each node is selected from between the minimum and the maximum value of the variable in the observations drawn. For instance, 100 observations assume values within the range interval [0, 1] and another 100 within [2, 50]. In the best case scenario, the first partition will separate the objects in disparate groups from each other, owing to which the dispersion effect will be reduced. If the value of p is within the range [2, 50], however, the objects in the group with the greater standard deviation will be treated as atypical in the tree branch encompassing the values of the variable that

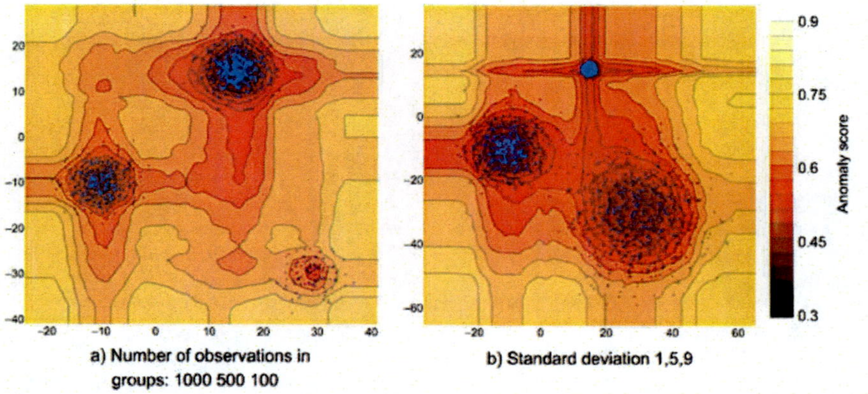

Fig. 7 Impact of the group structures on the anomaly score value. *Source* Own elaboration

are less than p. The values from the dispersed group will be underrepresented in this part of the tree. At the same time, in its remaining part, the partition will take place without a compact group, owing to which the observation dispersion will not affect the path length from the root to the observation. Summing up, both phenomena significantly impact the method's effectiveness. In parallel, the disproportion in the number of observations in a group has greater impact on the increase in the anomaly score (Fig. 7).

The last feature characterizing the datasets constituting the object of simulation research is the share of noise in the data. Data noise occurs when some observations do not form any group structures, and their distribution approximates randomness. Simultaneously, the spatial distribution of these values does not differ significantly from typical units. If the share of such observations constitutes a significant part of the dataset examined, they cannot be referred to as outliers. Data noise should be considered in two contexts when detecting outliers. First of all, a good method of anomaly detection should be able to distinguish anomalies from typical values and noise. Due to the conventional boundary between noise and outliers, it is difficult to determine exactly how the method would distinguish between the two phenomena. From the perspective of data analysis, it seems more important to distinguish typical observations from noise, e.g., by assigning higher values to the anomaly score. Another aspect to be kept in mind is the impact of the noise itself on the effectiveness of the method used. In terms of anomaly detection, it manifests itself by inclusion of typical observations as outliers or outliers as typical (Fig. 8).

An Extended Isolation Forest with parameters $t = 100$ and $\psi = 256$ was trained for three datasets characterized by different spatial structures. Regardless of the percentage share of noise in the data, the value of the anomaly score is noticeably lower near the typical observations. At a thirty percent share of noise in the data, the algorithm is able to distinguish between typical values, anomalies and noise; whereas along with an increase in the distance from the main cluster, the coefficient's values increase significantly. This allows the researcher to make a decision regarding

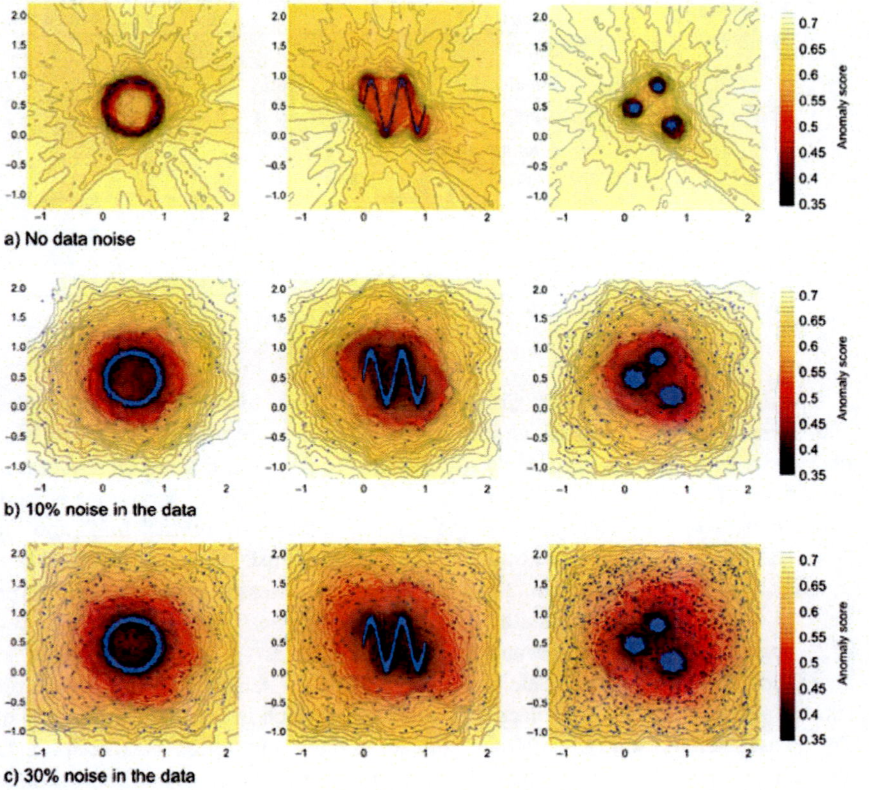

Fig. 8 Impact of the share of data noise on the anomaly score values. *Source* Own elaboration

the cut-off point between the anomalies, the noise, and the typical observations. Due to the fact that a properly trained Isolation Forest does not require large samples in individual trees, data noise does not have any critical impact on the effectiveness of the iForest or the EIF algorithms. Interestingly, the share of any noise in the dataset causes the anomaly score values for typical observations to be lower.

5 Discussion of the Empirical Research Results

The simulation studies show the algorithms' strong outlier detection abilities. The next stage of the research entails the testing of the Isolation Forests and the Extended Isolation Forests on the datasets described in more detail in the literature. The local outlier factor (LOF) method, with a neighborhood parameter $k = 10$, was used as a comparative algorithm. The datasets examined differ in the number of observations, the dimensions, and the share of anomalous values (Table 1).

Table 1 Description of empirical datasets

Dataset	n	d	Anomaly percentage (%)
SMTP	95,156	3	0.03
Statellite*	5100	36	9.7
Shuttle	49,097	9	7
Mammography	11,183	6	2
Wine_red	1599	11	0.6
Wine_white	4898	11	0.4
Lymphography	148	18	4.1
Musk	3062	30	3.2

Source Own elaboration

The dataset "Wine Quality" (Cortez 2009) FF consists of a description of red and white wines, the quality of which has been rated on a scale of 1–10. Observations, the quality of which was extremely low (Quality = 3), were marked as outliers. In the dataset "Lymphography" (Zwitter and Soklic), most of the observations were in the "metastases" and "malignant lymph" classes, while the "normal find" and the "fibrosis" ones were marked as outliers. The dataset "Musk" (Dua and Graff 2019a) describes various molecular structures that have been assessed by experts in a given field. In outlier detection, the dataset is limited to muskless classes (j146, j147 and 252), marked as typical observations, and musk classes (213 and 211), added as anomalies. For the dataset "Satellite" (The Center for remote sensing), the three least numerous classes were selected as anomalies, which in total constitute 32% of the dataset. In the study, only one, the least numerous, class was adopted as atypical, in consistency with the definition of outliers, which should constitute a small part of the observation. "SMTP3" (Dua and Graff 2019b) is a subset of the KDD CUP 99 dataset, in which the "attacks" class is treated as anomalies. In the sets "Shuttle" (Dua and Graff 2019c) and "Mammography" (https://www.openml.org/d/310), the observations assigned to the least numerous classes were marked as outliers (Table 2).

Table 2 Results of outlier detection in empirical data

Comparison of empirical datasets	Area Under Curve (AUC)		
	IForest	EIF	LOF
SMTP	0.879	*0.882*	0.811
Statellite	*0.804*	0.741	0.515
Shuttle	*0.998*	0.996	0.521
Mammography	0.859	*0.862*	0.67
Wine_red	0.828	*0.85*	0.598
Wine_white	0.775	*0.788*	0.698
Lymphography	0.984	*0.993*	0.982
Musk	*1*	*1*	0.392

Source Own elaboration

The values of the area under the receiver operating characteristic (ROC) curve indicate the effectiveness of the iForest and the EIF algorithms, with regard to anomaly detection. In most cases, the area under the ROC curve (AUC) values are significantly higher, compared to the local outlier factor (LOF) algorithm. The results of both tree-based algorithms are comparable. The reason for the significantly lower AUC values, in the case of the LOF method, may be the global nature of the outliers considered, which favors the isolation methods. In some of the datasets, observations from the least numerous classes were marked as atypical. The simulation studies showed that the number of observations in a group is the decisive factor, in the context of the anomaly score value.

6 Final Conclusions

Summing up, the simulation and empirical studies have confirmed the outlier detection capability of the iForest and the EIF algorithms. The main advantages of the methods analyzed can be recapped in the following points:

- Linear computational complexity, which makes the method suitable for the analysis of big datasets. The duration of the forest training stage depends on the sample size and the number of trees; thus, the dataset dimensionality and the total number of observations do not play any role here. Additionally, it is worth remembering about the convergence of the algorithms at a small number of trees and a small sample size.
- Resistance of the algorithms to data noise. The simulation studies have shown that the anomaly score for outliers and for the noise differ significantly, owing to which the researcher can independently decide about the cut-off point between them. At the same time, at a higher share of noise in the data, the anomaly score values for typical observations are lower.
- Ability to analyze multidimensional datasets. A larger number of variables does not negatively affect the computational time or the algorithm's ability to detect anomalies. Even when an observation is considered an outlier only because of one of the variables, the algorithm, with an appropriate number of trees, is able to successfully detect such a unit.
- Effectiveness of the algorithms is independent of the spatial structure of the data.
- The small number of the parameters to be determined by the researcher, which facilitates the adaptation of the method to the dataset examined.

The limitations of the algorithms are less obvious and debatable. Although in certain situations, they may pose a difficulty in outlier analysis; in other cases, they may prove to be helpful. The features that raise most doubts pertain to:

- Inefficiency of datasets with significant differences in the number of observations in each group. The small sample size used in the algorithm means that the values of the smallest clusters are used less frequently to train trees, which

results in higher values of the coefficient of isolation. This relationship, however, does not have to be a disadvantage; because in the case of global anomalies, small clusters of observations are still treated as outliers.
- Lack of exact cut-off point for the anomaly score values between the anomalies, the data noise, and the typical observations. Before marking some observations as anomalies, one should analyze the distribution of the coefficient of isolation in the dataset, which makes full automatization of the algorithm difficult. In parallel, it allows consideration of expert knowledge or of the business needs in the final decision.

Despite the extensive analysis of the iForest and the EIF algorithms, some issues related to the detection of the outlying values that are typical for big datasets remain to be settled. Increasingly, often the datasets analyzed are characterized by a real-time influx of observations (streaming data) (Zhiguo 2013), which means that the methods used should have high adaptability. With regard to the iForest and the EIF, the question of how often the model should be re-trained to adequately reflect the current relations between observations should be answered. Another way to analyze streaming data could entail addition of newly trained trees to the forests, with simultaneous removal of the oldest ones, yet the method could still be ineffective for analysis of periodically changing phenomena. Effectiveness of the algorithm, in regards to local outliers, is another issue. It is most clearly visible in the case of group structures; thus, it could be eliminated by combining iForest or EIF with grouping methods (Rongfang et al. 2019). In the literature, attempts have been made to make such modernization, but the most effective combination of methods, taking into account a larger number of grouping algorithms, has not yet been established. It should be remembered, however, that combination of different methods will affect the computational time of the algorithm. The key determining aspect can therefore be the appropriate use of grouping algorithms with linear computational complexity. At the same time, it should be remembered that incorrect classification of observations into the corresponding clusters, at the stage of grouping, may disturb the effectiveness of the entire method. As such, combination of grouping methods with algorithms that isolate outliers should only take place when it is possible to properly separate the clusters of observations from each other.

References

Anscombe FJ (1973) Graphs in statistical analysis. Am Stat 27(1):17–21

Aggarwal CC (2015) Data mining. Springer International Publishing Switzerland. https://doi.org/10.1007/978-3-319-14142-8

Cortez P (2009) http://www3.dsi.uminho.pt/pcortez. Viticulture Commission of the Vinho Verde Region(CVRVV), Porto, Portugal. https://archive.ics.uci.edu/ml/datasets/wine+quality. Accessed 4 Sept 2020

Dataset was published by a courtesy of Aleksandar Lazarevic. https://www.openml.org/d/310 Accessed 5 Sept 2020

Dua D, Graff C (2019a) UCI machine learning repository [http://archive.ics.uci.edu/ml]. University of California, School of Information and Computer Science, Irvine, CA. https://archive.ics.uci.edu/ml/datasets/Musk+%28Version+2%29. Accessed 4 Sept 2020

Dua D, Graff C (2019b) UCI machine learning repository [http://archive.ics.uci.edu/ml]. University of California, School of Information and Computer Science, Irvine, CA. https://archive.ics.uci.edu/ml/machine-learning-databases/kddcup99-mld/. Accessed 4 Sept 2020

Dua D, Graff C (2019c) UCI machine learning repository [http://archive.ics.uci.edu/ml]. University of California, School of Information and Computer Science, Irvine, CA. https://archive.ics.uci.edu/ml/datasets/Statlog+%28Shuttle%29. Accessed 4 Sept 2020

Grubbs FE (1969) Procedures for detecting outlying observations in samples. Technometrics 11(1):1–21

Hawkins DM (1980) Identification of outliers. Chapman and Hall

Liu FT, Ting KM, Zhou Z (2008). Isolation forest, 2008 Eighth IEEE International Conference on Data Mining, Pisa, pp 413–422. https://doi.org/10.1109/ICDM.2008.17

Liu Z, Liu X, Ma J, Gao H (2019) An optimized computational framework for isolation forest. Mathematical Problems in Engineering, vol 2018, Article ID 2318763

Migdał-Najman K, Najman K (2013) Samouczące się sztuczne sieci neuronowe w grupowaniu i klasyfikacji danych. Teoria i zastosowania w ekonomii, Wydawnictwo Uniwersytetu Gdańskiego, Gdańsk

Probst P, Boulesteix A-L (2018) To tune or not to tune the number of trees in random forest. J Mach Learn Res 18:10–18

Rongfang G, Tiantian Z, Shaohua S, Zhanyu L (2019) Research and improvement of isolation forest in detection of local anomaly points. J Phys Conf Ser 1237:052023. https://doi.org/10.1088/1742-6596/1237/5/052023

Sahand H, Kind MC, Brunner RJ (2019) Extended isolation forest. IEEE Transactions on Knowledge and Data Engineering, pp 1–1. Crossref. Web

The Centre for Remote Sensing, University of New South Wales, Kensington, PO Box 1, NSW 2033. https://archive.ics.uci.edu/ml/datasets/Statlog+%28Landsat+Satellite%29. Accessed 4 Sept 2020

Zhiguo D (2013) An anomaly detection approach based on isolation forest algorithm for streaming data using sliding window. IFAC Proc 46:12–17. https://doi.org/10.3182/20130902-3-CN-3020.00044

Zwitter M, Soklic M University Medical Centre, Institute of Oncology, Ljubljana, Yugoslavia. https://archive.ics.uci.edu/ml/datasets/Lymphography. Accessed 4 Sept 2020

Application in Finance

Propositions of Transformations of Asymmetrical Nominants into Stimulants on the Example of Chosen Financial Ratios

Barbara Batóg and **Katarzyna Wawrzyniak**

Abstract The paper is a continuation of the Authors' research on transformation of nominants with a recommended range of values for stimulants, which should ensure the greatest possible compatibility of the order of examined objects according to the values of variable-nominant before and after their transformation. In earlier studies, the Authors focused on the symmetric nominants. The present paper attempts to propose similar solutions as in the case of the symmetrical nominant, which can be used for left- and right-handed asymmetrical nominants. In the theoretical part, the transformations proposed by other Authors were analyzed and compared with Authors' propositions. The study was carried out on the basis of selected financial ratios, which in the literature are considered to be nominants with the recommended range of values, with the assumption that the better situation of the examined object is when the values of the indicator-nominant are above the upper limit of the recommended range of values (right-handed asymmetrical nominant) or below the lower limit of this range (left-handed asymmetrical nominant). The data on the financial ratios come from Notoria Serwis and concern companies from the *Machinery industry* sector listed on the Warsaw Stock Exchange in 2018.

Keywords Asymmetrical nominants with the recommended range of values · Stimulants · Transformations · Financial ratios

B. Batóg (✉)
University of Szczecin, Szczecin, Poland
e-mail: barbara.batog@usz.edu.pl

K. Wawrzyniak
West Pomeranian University of Technology in Szczecin, Szczecin, Poland
e-mail: katarzyna.wawrzyniak@zut.edu.pl

K. Jajuga et al. (eds.), *Data Analysis and Classification*, Studies in Classification, Data Analysis, and Knowledge Organization,
https://doi.org/10.1007/978-3-030-75190-6_6

1 Introduction

The results presented in the paper are a continuation of the authors' research (Batóg and Wawrzyniak 2020) on the transformation of indicator-nominants with the recommended range of values for stimulants normalized in the range [0; 1]. The terms "stimulant" and "destimulant" were introduced in Polish literature by Hellwig (1968, 1972), and the term "nominant" was introduced by Borys (1978, 1984). When objects are ordered it is important to determine the character of the variables describing these objects, and then unify them and make them comparable by means of an appropriate normalizing transformation. In the above-mentioned studies of Batóg and Wawrzyniak, the focus was on symmetric indicators, where the situation of the examined object with the values of indicator below the lower and above the upper limit of the recommended range of values is evaluated in the same way. The closer the lower and upper limit of the recommended range of values are the indicator-nominant values, the better the situation in the examined object. If we consider two companies for which the current ratio (nominant with the recommended range of values from 1.2 to 2) is 1.1 and 2.1, respectively, then they should be evaluated equally, as the value of the current ratio in both cases is equally distant from the lower and upper limit of the recommended range of values. On the other hand, if one company has a current ratio of 1.1 and the other 2.3, the situation in the first company should be evaluated better than in the second company. In order to transfer this principle also after the transformation of the symmetrical nominant into a stimulant normalized in the range [0; 1], an Author's modification to the known in the literature formulas for the transformation of indicators-nominant into stimulants has been proposed. The aim of the study, the results of which are presented in this paper, was to propose original formulas for the transformation of right and left asymmetrical indicator-nominant to stimulants normalized in the range [0; 1], while maintaining a similar approach as in case of the symmetrical indicator-nominant. The results of the transformations obtained on the basis of the proposed transformations were compared with the results obtained on the basis of the transformations presented in the literature. The study used data on the current ratio (right-handed asymmetrical nominant) and on the debt margin (left-handed asymmetrical nominant) for companies from the *Machinery industry* sector listed on the Warsaw Stock Exchange in 2018.

2 Previous Proposition of Modification of Minimum and Maximum

In the paper of Batóg and Wawrzyniak (2020), it was proposed to modify the formulas for the transformation of the nominant to stimulant normalized in the range [0; 1], in which the symmetry of the ranges of nominant values below and above the recommended range of values was introduced by setting new minimum and maximum values. Moreover, it is assumed that the distance of the new minimum from the lower limit of the recommended range of values is the same as the distance of the new maximum from the upper limit of the recommended range of values. Thanks to this modification, the consistency of the order of the examined objects according to the values of indicator-nominant before and after the transformation was obtained.

Below we compare the results of the transformation of nominants into stimulants normalized in the range [0; 1] obtained according to the transformation proposed by Kukuła (2000) and the results of the transformation using the modification proposed by the Authors. Equation 1 presents a linear transformation of the nominant with the recommended range of values to the stimulants normalized in the range [0; 1] proposed by Kukuła. In turn, Figs. 1 and 2 show the results of this transformation in the case of symmetrical and asymmetrical ranges of nominant values below and above the recommended range of values (in short, symmetrical and asymmetrical minimum and maximum).

$$x_{ij}^S = \begin{cases} \frac{1}{c_{1j}-a_j}\left(x_{ij}^N - a_j\right) & \text{for} \quad x_{ij}^N < c_{1j} \\ 1 & \text{for} \quad c_{1j} \leq x_{ij}^N \leq c_{2j} \\ \frac{1}{c_{2j}-b_j}\left(x_{ij}^N - b_j\right) & \text{for} \quad x_{ij}^N > c_{2j} \end{cases} \tag{1}$$

where

x_{ij}^N value of *j*th indicator-nominant for *i*th object,
x_{ij}^S value of normalized stimulant of *j*th indicator-nominant for *i*th object,
c_{1j} lower limit of the recommended range of values of *j*th indicator-nominant,
c_{2j} upper limit of the recommended range of values of *j*th indicator-nominant,
a_j minimum value of *j*th indicator-nominant,
b_j maximum value of *j*th indicator-nominant.

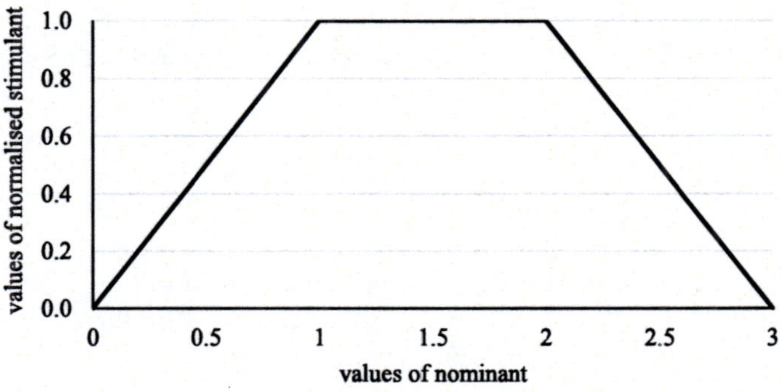

Fig. 1 Linear transformation of a nominant with a recommended range of values into a stimulant with symmetrical minimum and maximum

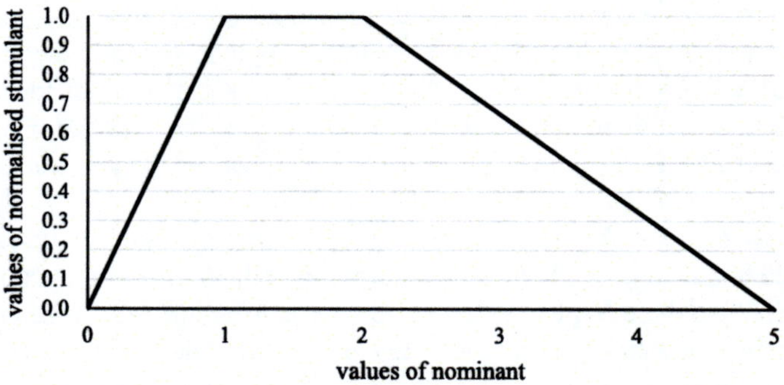

Fig. 2 Linear transformation of a nominant with a recommended range of values into a stimulant with asymmetrical minimum and maximum

Figure 1 shows that in the case of symmetrical minimum and maximum, the property is observed that the same values of the normalized stimulant are assigned to the values of the nominant equally distant from the lower and upper limit of the recommended range, that is

$$x_{ij}^{N} = 0.5 \rightarrow x_{ij}^{S} = 0,5,$$

$$x_{ij}^{N} = 2.5 \rightarrow x_{ij}^{S} = 0,5.$$

However, in the case of asymmetrical minimum and maximum (Fig. 2), this property is not observed, because

$$x_{ij}^{N} = 0.5 \rightarrow x_{ij}^{S} = 0,5,$$

$$x_{ij}^{N} = 2.5 \rightarrow x_{ij}^{S} = 0,83.$$

The modification proposed in the paper of Batóg and Wawrzyniak (2020) allows to remove this defect by setting a new minimum or maximum value according to Eqs. 2a and 2b.

$$a_j^* = \begin{cases} a_j & \text{for} \quad c_{1j} - a_j \geq b_j - c_{2j} \\ c_{1j} - \left(b_j - c_{2j}\right) & \text{for} \quad c_{1j} - a_j < b_j - c_{2j} \end{cases} \tag{2a}$$

$$b_j^* = \begin{cases} c_{2j} + \left(c_{1j} - a_j\right) & \text{for} \quad c_{1j} - a_j \geq b_j - c_{2j} \\ b_j & \text{for} \quad c_{1j} - a_j < b_j - c_{2j} \end{cases} \tag{2b}$$

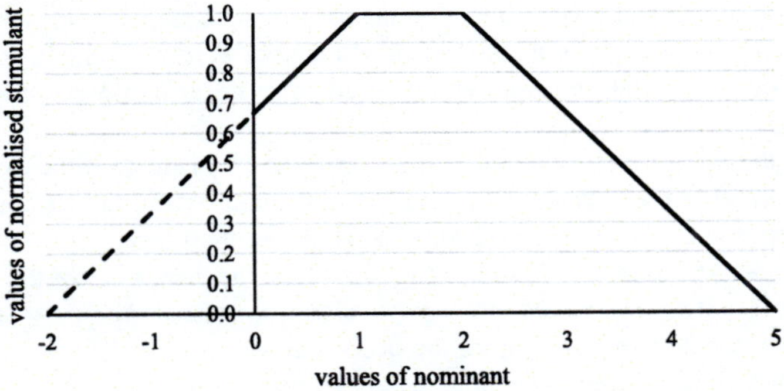

Fig. 3 Linear transformation of a nominant with a recommended range of values into a stimulant with a modified minimum

where

- c_{1j} lower limit of the recommended range of values of *j*th indicator-nominant,
- c_{2j} upper limit of the recommended range of values of *j*th indicator-nominant.
- a_j minimum value of *j*th indicator-nominant before modification,
- b_j maximum value of *j*th indicator-nominant before modification,
- a_j^* minimum value of *j*th indicator-nominant after modification,
- b_j^* maximum value of *j*th indicator-nominant after modification.

Figure 3 presents the linear transformation of a nominant with a recommended range of values into a stimulant with a modified minimum (Eq. 2a). Thanks to this modification we get the same values of the normalized stimulant for the nominant values which are in the same distance from a recommended range of values:

$$x_{ij}^N = 0.5 \rightarrow x_{ij}^S = 0,83,$$

$$x_{ij}^N = 2.5 \rightarrow x_{ij}^S = 0,83.$$

3 Proposals of Nonlinear Transformation of Nominant into Stimulants Normalized in the Range [0; 1]

In the literature, one can find various proposals for the transformation of the nominant into stimulants (e.g., Strahl and Walesiak 1997; Strahl and Dziechciarz 1999; Kukuła 2000; Kowalewski 2002; Wójciak 2003). They use both linear and nonlinear transformations. In this part of the paper, the original proposals of nonlinear transformations of the nominant with the recommended range of values into stimulant normalized in the range [0; 1] are presented. These proposals are based on the exponential and logarithmic functions with the use of minimum or maximum modifications so that the applied transformation is characterized by the symmetry of minimum and maximum to a recommended range of values. This is a reference to the modification which the Authors proposed for a linear transformation (Eqs. 2a and 2b).

Since the proposals for nonlinear transformations will concern the symmetrical and right- and left-handed asymmetrical nominants, the definitions of such nominants proposed by Kowalewski (2002, 2006) are recalled here.

A symmetrical nominant is a nominant for which the values below the lower and upper limits of the recommended range of values are evaluated equally—the closer the limits the better and the further the limits the worse.

A right-handed asymmetrical nominant is a nominant for which the values above the upper limit of the recommended range of values are evaluated better than the values below the lower limit of the recommended range of values. A left-handed asymmetrical nominant is a nominant for which the values below the lower limit of the recommended range of values are evaluated better than the values above the upper limit of the recommended range of values.

Transformations described by Eqs. 3 and 4 refer to the symmetrical nominant. They differ in the rate of decrease of obtained values of normalized stimulant.

Transformation described by Eq. 3 refers to the case when the decrease of values of the normalized stimulant close to the lower and upper limits of a recommended range of values is faster than for values close to minimum and maximum (convex functions).

$$x_{ij}^{S} = F_j\left(x_{ij}^{N}\right) = \begin{cases} e^{\alpha_j x_{ij}^{N} + \beta_j} - e^{a_j^*} & \text{for} \quad x_{ij}^{N} < c_{1j} \\ 1 & \text{for} \quad c_{1j} \le x_{ij}^{N} \le c_{2j} \\ e^{\alpha_j (c_{1j} + c_{2j} - x_{ij}^{N}) + \beta_j} - e^{a_j^*} & \text{for} \quad x_{ij}^{N} > c_{2j} \end{cases} \tag{3}$$

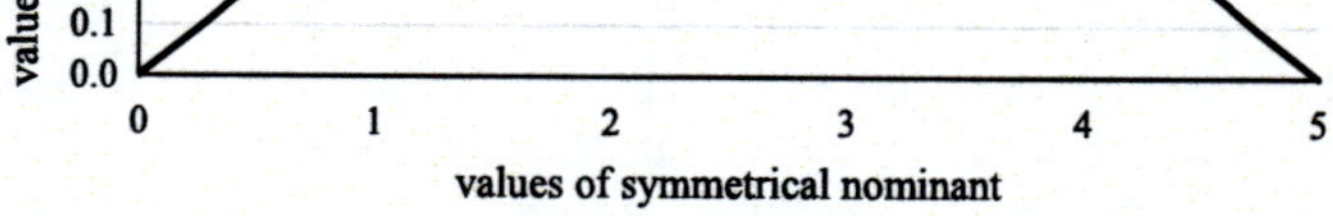

Fig. 4 Nonlinear transformation of a symmetrical nominant into a normalized stimulant according to Eq. 3

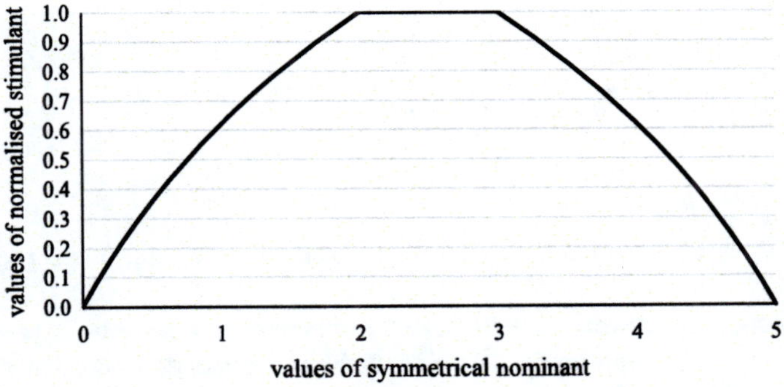

Fig. 5 Nonlinear transformation of a symmetrical nominant into a normalized stimulant according to Eq. 4

where

$\alpha_j = \frac{\ln\left(1+e^{a_j^*}\right)-a_j^*}{c_{1j}-a_j^*}$, $\beta_j = a_j^*(1-\alpha_j)$

The values of a normalized stimulant obtained using Eq. 3 are presented in Fig. 4.

Transformation described by Eq. 4 refers to the case when the decrease of values of the normalized stimulant close to the lower and upper limits of a recommended range of values is slower than for values close to minimum and maximum (concave functions).

$$x_{ij}^S = F_j\left(x_{ij}^N\right) = \begin{cases} \ln\left(\alpha_j x_{ij}^N + \beta_j\right) & \text{for} \quad x_{ij}^N < c_{1j} \\ 1 & \text{for} \quad c_{1j} \le x_{ij}^N \le c_{2j} \\ \ln(\alpha_j(c_{1j} + c_{2j} - x_{ij}^N) + \beta_j) & \text{for} \quad x_{ij}^N > c_{2j} \end{cases} \tag{4}$$

where

$\alpha_j = \frac{e-1}{c_{1j}-a_j^*}$, $\beta_j = 1 - \alpha_j a_j^*$

The values of a normalized stimulant obtained using Eq. 4 are presented in Fig. 5.

Similar transformations can be made for the right-handed and left-handed asymmetrical nominants. The transformation of the right-handed asymmetrical nominant is presented by Eq. 5 (convex function on the left side of the recommended range of values and concave function on the right side of the recommended range of values).

$$x_{ij}^S = F_j\left(x_{ij}^N\right) = \begin{cases} e^{\alpha_{j1} x_{ij}^N + \beta_{j1}} - e^{a_j^*} & \text{for} \quad x_{ij}^N < c_{1j} \\ 1 & \text{for} \quad c_{1j} \le x_{ij}^N \le c_{2j} \\ \ln(\alpha_{j2}(c_{1j} + c_{2j} - x_{ij}^N) + \beta_{j2}) & \text{for} \quad x_{ij}^N > c_{2j} \end{cases} \tag{5}$$

where

$$\alpha_{j1} = \frac{\ln\left(1+e^{a_j^*}\right)-a_j^*}{c_{1j}-a_j^*},\ \beta_{j1} = a_j^*\left(1-\alpha_{j1}\right),$$

$$\alpha_{j2} = \frac{e-1}{c_{1j}-a_j^*},\ \beta_{j2} = 1-\alpha_{j2}a_j^*$$

The values of a normalized stimulant obtained using Eq. 5 are presented in Fig. 6.

The transformation of the left-handed asymmetrical nominant is presented by Eq. 6 (concave function on the left side of the recommended range of values and convex function on the right side of the recommended range of values).

$$x_{ij}^S = F_j\left(x_{ij}^N\right) = \begin{cases} \ln\left(\alpha_{j1}x_{ij}^N + \beta_{j1}\right) & for \quad x_{ij}^N < c_{1j} \\ 1 & for \quad c_{1j} \le x_{ij}^N \le c_{2j} \\ e^{\alpha_{j2}\left(c_{1j}+c_{2j}-x_{ij}^N\right)+\beta_{j2}} - e^{a_j^*} & for \quad x_{ij}^N > c_{2j} \end{cases} \tag{6}$$

where

$$\alpha_{j1} = \frac{e-1}{c_{1j}-a_j^*},\ \beta_{j1} = 1-\alpha_{j1}a_j^*$$

$$\alpha_{j2} = \frac{\ln\left(1+e^{a_j^*}\right)-a_j^*}{c_{1j}-a_j^*},\ \beta_{j2} = a_j^*\left(1-\alpha_{j2}\right)$$

The values of a normalized stimulant obtained using Eq. 6 are presented in Fig. 7.

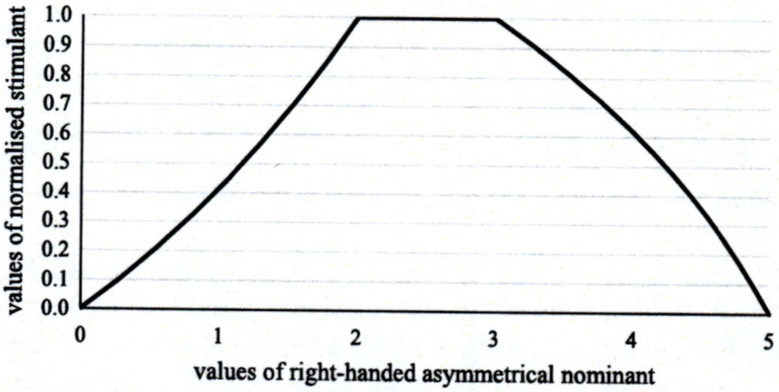

Fig. 6 Nonlinear transformation of a the right-handed asymmetrical nominant into a normalized stimulant according to Eq. 5

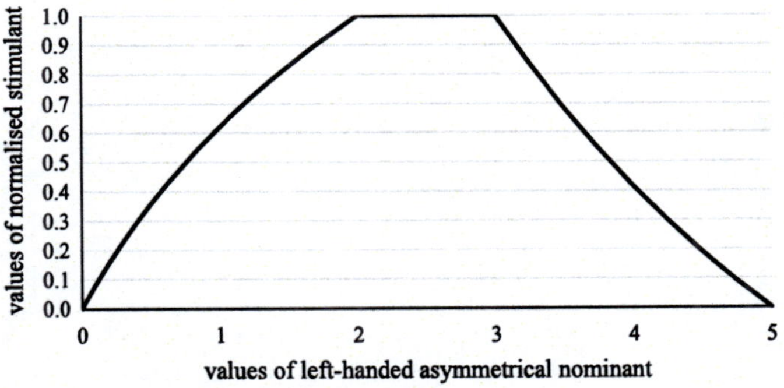

Fig. 7 Nonlinear transformation of a the left-handed asymmetrical nominant into a normalized stimulant according to Eq. 6

4 Data and Empirical Results

The empirical verification of values normalized in the range [0; 1] of the stimulant obtained by means of proposed nonlinear transformations of symmetrical and asymmetrical nominants was performed using two indicators-nominants with the recommended range of values. The theoretical recommended ranges of values can be found, among others, in works of Gabrusewicz (2014), Hozer et al. (1997), Sierpińska and Jachna (2004), Waśniewski and Skoczylas (2004). These indicators-nominants are:

- current ratio—the right-handed asymmetrical nominant (theoretical recommended range of values: [1.2; 2]),
- debt margin—the left-handed asymmetrical nominant (theoretical recommended range of values: [0.57; 0.67]).

In addition to the theoretical recommended ranges of values, the verification also used the empirical recommended ranges of values determined according to the formula:

$$[M - MAD; M + MAD], \tag{7}$$

where

M median,
MAD median absolute deviation (Młodak 2006)

The data on selected indicators refer to companies from the *Machinery industry* sector listed on the Warsaw Stock Exchange in 2018.

Figures 8, 9, 10, 11 show the results of the nonlinear transformation of current ratio assuming that it is a symmetrical nominant. Figures 8 and 10 present the values of normalized stimulants obtained according to nonlinear transformations (downward and upward quadratic functions) proposed by Kukuła (2000) with the

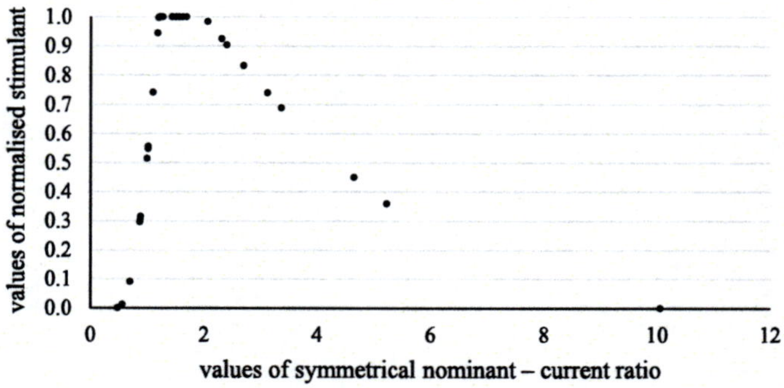

Fig. 8 Current ratio—nonlinear (upward quadratic function) transformation of the symmetrical nominant proposed by Kukuła (2000) with the theoretical recommended range of values

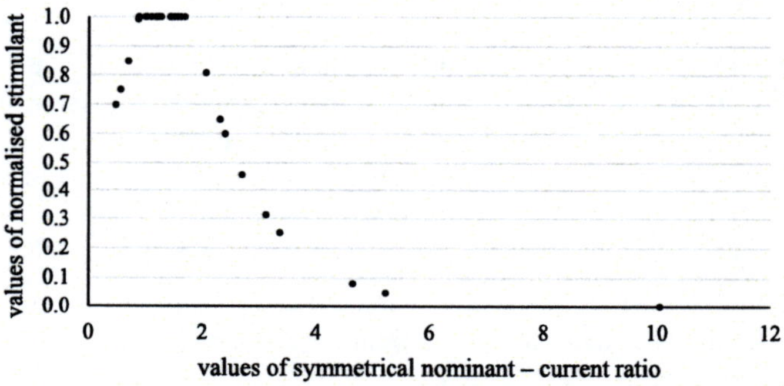

Fig. 9 Current ratio—nonlinear transformation of the symmetrical nominant proposed by Authors (Eq. 3) with the empirical recommended range of values

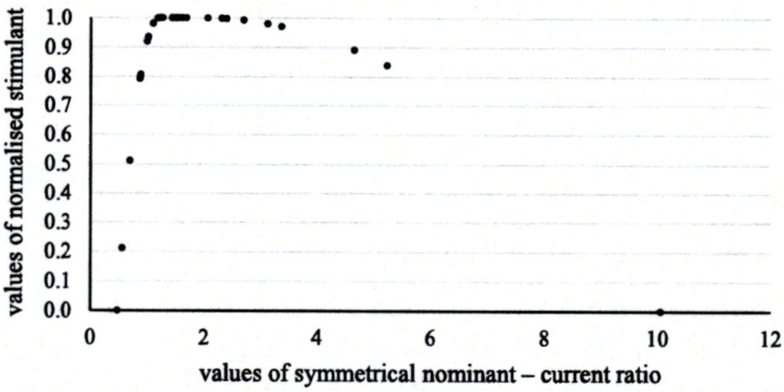

Fig. 10 Current ratio—nonlinear (downward quadratic function) transformation of the symmetrical nominant proposed by Kukuła (2000) with the theoretical recommended range of values

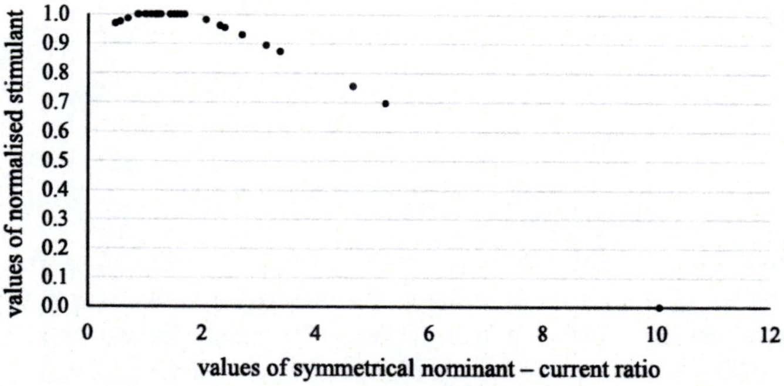

Fig. 11 Current ratio—nonlinear transformation of the symmetrical nominant proposed by Authors (Eq. 4) with the empirical recommended range of values

theoretical recommended range of values [1.2; 2]. Figures 9 and 11 present the values of normalized stimulants obtained according to nonlinear transformations proposed by Authors (Eqs. 3 and 4) with the empirical recommended range of values [0.88; 1.82].

The comparison of the values of normalized stimulants, which are presented in Figs. 8 and 9, shows significant differences. This can be illustrated on the basis of two selected values of the current ratio: 0.69 and 2.39, which lie below the lower limit and above the upper limit of the recommended range of values, respectively. In the case of Fig. 8, the values of normalized stimulant are equal to:

$$x_{ij}^{N} = 0.69 \rightarrow x_{ij}^{S} = 0.09,$$

$$x_{ij}^{N} = 2.39 \rightarrow x_{ij}^{S} = 0.90.$$

But in the case of Fig. 9, the values of normalized stimulant are equal to:

$$x_{ij}^{N} = 0.69 \rightarrow x_{ij}^{S} = 0.85,$$

$$x_{ij}^{N} = 2.39 \rightarrow x_{ij}^{S} = 0.60.$$

The situation is similar when we compare Figs. 10 and 11. In this case, the values of normalized stimulants for selected values of the current ratio are equal to:

- Figure 10.

$$x_{ij}^{N} = 0.69 \rightarrow x_{ij}^{S} = 0.51,$$

$$x_{ij}^{N} = 2.39 \rightarrow x_{ij}^{S} = 0.99,$$

- Figure 11.

$$x_{ij}^{N} = 0.69 \rightarrow x_{ij}^{S} = 0.98,$$

$$x_{ij}^{N} = 2.39 \rightarrow x_{ij}^{S} = 0.95.$$

The following figures (Figs. 12, 13) illustrate the results of the nonlinear transformation of current ratio under the assumption that current ratio is a right-handed asymmetrical nominant. Figure 12 presents the values of normalized stimulant calculated by means of transformation proposed by Kowalewski (2002) with the theoretical recommended range of values [1.2; 2]. In turn, Fig. 13 presents the values of normalized stimulant calculated by means of transformation given by Eq. 5 with the empirical recommended range of values [0.88; 1.82].

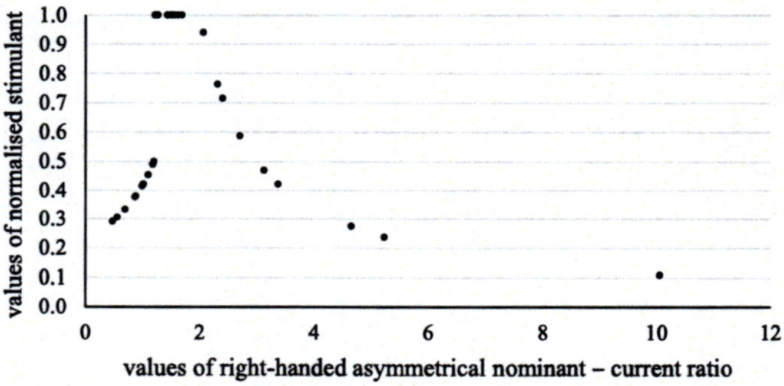

Fig. 12 Current ratio—nonlinear transformation of the right-handed asymmetrical nominant proposed by Kowalewski (2002) with the theoretical recommended range of values ($k_p = 2,\ k_l = 1$)

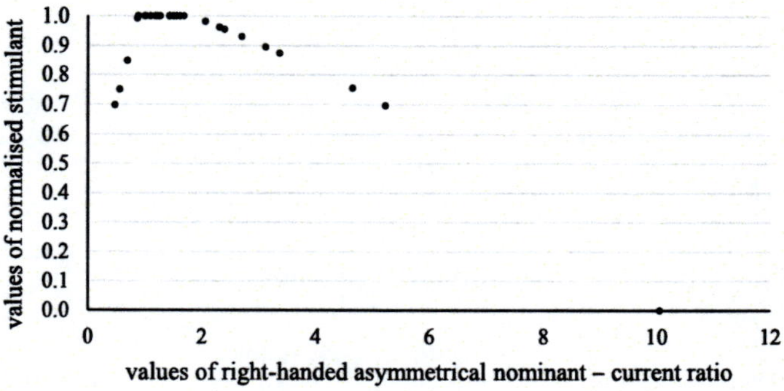

Fig. 13 Current ratio—nonlinear transformation of the right-handed asymmetrical nominant proposed by Authors (Eq. 5) with the empirical recommended range of values

The values of the normalized stimulants obtained according to Kowalewski's proposal and according to Eq. 5 for the current ratio values of 0.69 and 2.39 are, respectively, equal to:

- Figure 12.

$$x_{ij}^{N} = 0.69 \rightarrow x_{ij}^{S} = 0.33,$$

$$x_{ij}^{N} = 2.39 \rightarrow x_{ij}^{S} = 0.71.$$

- Figure 13.

$$x_{ij}^{N} = 0.69 \rightarrow x_{ij}^{S} = 0.85,$$

$$x_{ij}^{N} = 2.39 \rightarrow x_{ij}^{S} = 0.95.$$

Figures 14, 15, 16, 17 show the results of the nonlinear transformation of debt margin assuming that it is a symmetrical nominant. Figures 14 and 16 present the values of normalized stimulants obtained according to nonlinear transformations (downward and upward quadratic functions) proposed by Kukuła (2000) with the theoretical recommended range of values [0.57; 0.67]. Figures 15 and 17 present the values of normalized stimulants obtained according to nonlinear transformations proposed by Authors (Eqs. 3 and 4) with the empirical recommended range of values [0.28; 0.68].

The values of the normalized stimulants obtained according to Kukuła's proposal and according to Eq. 3 for the debt margin values of 0.24 and 0.71 are, respectively, equal to:

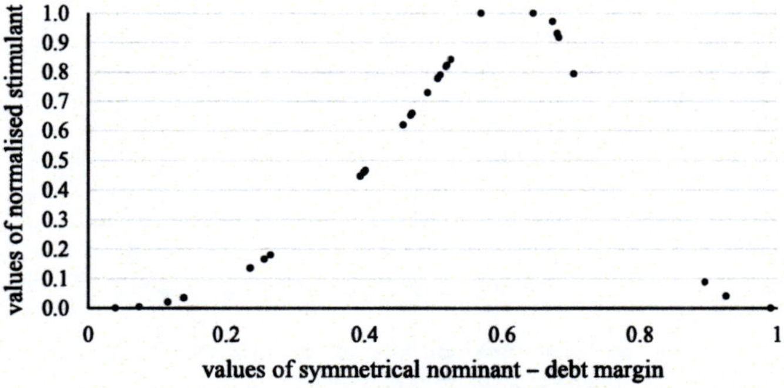

Fig. 14 Debt margin—nonlinear (upward quadratic function) transformation of the symmetrical nominant proposed by Kukuła (2000) with the theoretical recommended range of values

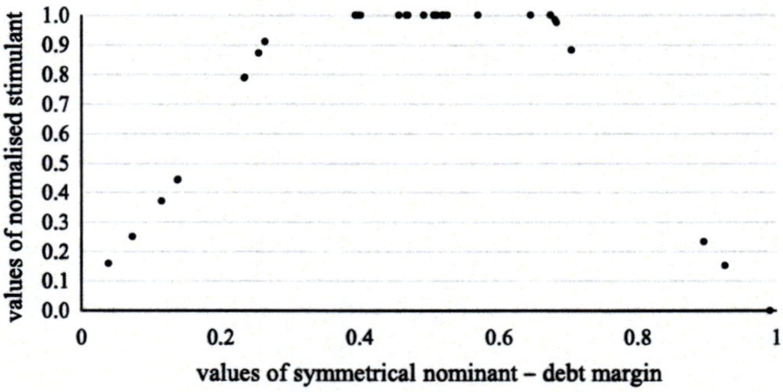

Fig. 15 Debt margin—nonlinear transformation of the symmetrical nominant proposed by Authors (Eq. 3) with the empirical recommended range of values

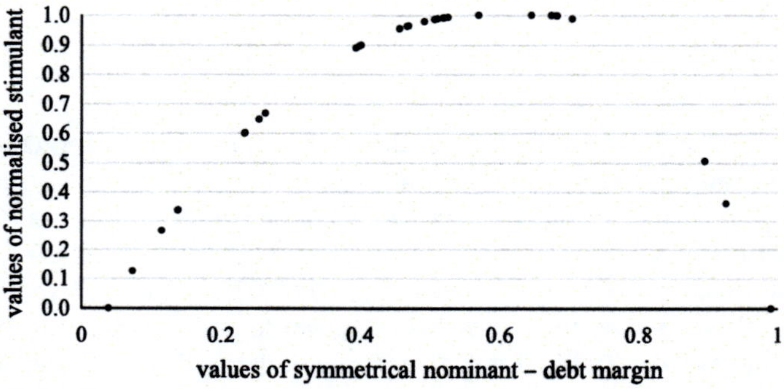

Fig. 16 Debt margin—nonlinear (downward quadratic function) transformation of the symmetrical nominant proposed by Kukuła (2000) with the theoretical recommended range of values

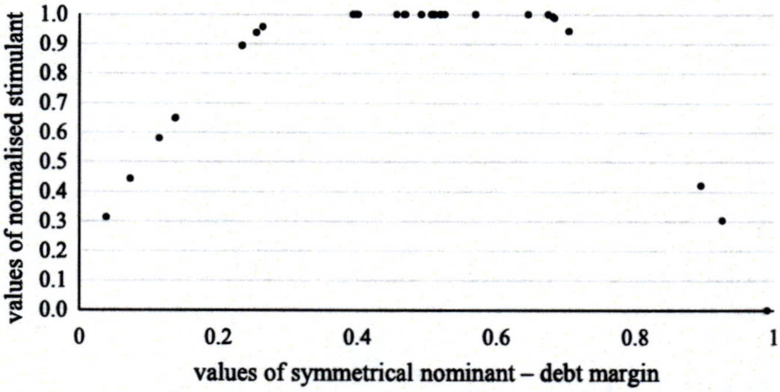

Fig. 17 Debt margin—nonlinear transformation of the symmetrical nominant proposed by Authors (Eq. 4) with the empirical recommended range of values

- Figure 14.

$$x_{ij}^{N} = 0.24 \rightarrow x_{ij}^{S} = 0.14,$$

$$x_{ij}^{N} = 0.71 \rightarrow x_{ij}^{S} = 0.79.$$

- Figure 15.

$$x_{ij}^{N} = 0.24 \rightarrow x_{ij}^{S} = 0.79,$$

$$x_{ij}^{N} = 0.71 \rightarrow x_{ij}^{S} = 0.88.$$

In the case of the transformations shown in Figs. 16 and 17, differences in the obtained values of the normalized stimulants can also be observed. According to the transformation proposed by Kukuła (Fig. 16), these values are equal to:

$$x_{ij}^{N} = 0.24 \rightarrow x_{ij}^{S} = 0.60,$$

$$x_{ij}^{N} = 0.71 \rightarrow x_{ij}^{S} = 0.99.$$

While in the case of transformation given by Eq. 4, we obtain:

$$x_{ij}^{N} = 0.24 \rightarrow x_{ij}^{S} = 0.89,$$

$$x_{ij}^{N} = 0.71 \rightarrow x_{ij}^{S} = 0.94.$$

Figures 18 and 19 illustrate the results of the nonlinear transformation of debt margin under the assumption that debt margin is a left-handed asymmetrical nominant. Figure 18 presents the values of normalized stimulant calculated by means of transformation proposed by Kowalewski (2002) with the theoretical recommended range of values [0.57; 0.67]. In turn, Fig. 19 presents the values of normalized stimulant calculated by means of transformation given by Eq. 6 with the empirical recommended range of values [0.28; 0.68].

The values of the normalized stimulants obtained according to Kowalewski's proposal and according to Eq. 6 for the debt margin values of 0.24 and 0.71 are, respectively, equal to:

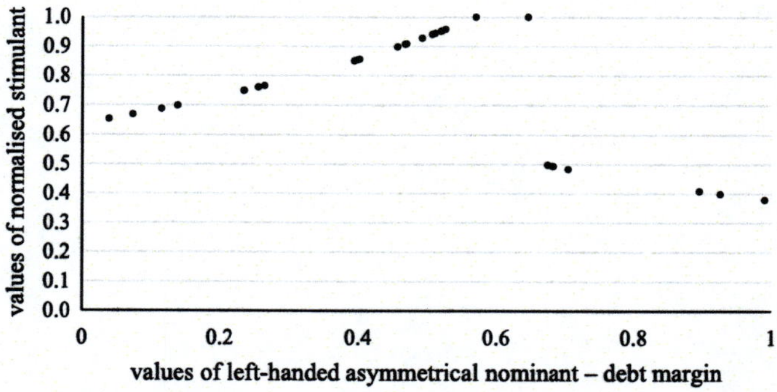

Fig. 18 Debt margin—nonlinear transformation of the left-handed asymmetrical nominant proposed by Kowalewski (2002) with the theoretical recommended range of values ($k_p = 1,\ k_l = 2$)

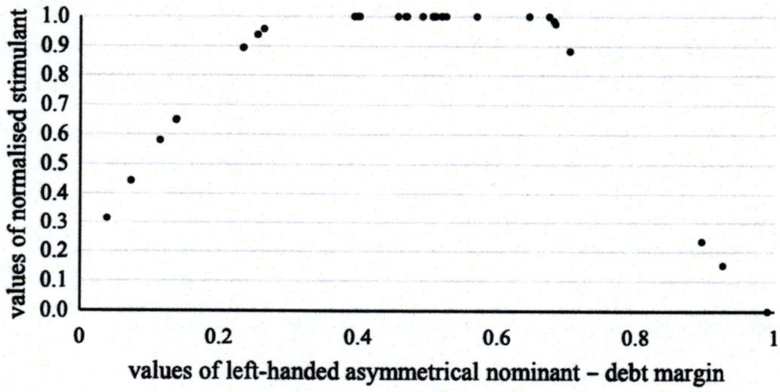

Fig. 19 Debt margin—nonlinear transformation of the left-handed asymmetrical nominant proposed by Authors (Eq. 6) with the empirical recommended range of values

- Figure 18.

$$x_{ij}^N = 0.24 \to x_{ij}^S = 0.75,$$

$$x_{ij}^N = 0.71 \to x_{ij}^S = 0.48.$$

- Figure 19.

$$x_{ij}^N = 0.24 \to x_{ij}^S = 0.89,$$

$$x_{ij}^N = 0.71 \to x_{ij}^S = 0.88.$$

5 Conclusions

The study shows that the proposed by Authors' nonlinear transformations of symmetrical and asymmetrical nominants into stimulants normalized in the range [0; 1] allow to obtain a higher consistency of the order of the examined objects (companies) before and after the transformation. This is evidenced by the values of normalized stimulants obtained by means of those transformations, which, in comparison with the values of normalized stimulant obtained according to Kukuła's and Kowalewski's proposals, reflect much better the original ordering of objects (companies) resulting from the values of indicators-nominants.

The paper focuses on nonlinear transformations that can be applied for both symmetrical and asymmetrical nominants. In the case of symmetrical nominations, two approaches have been proposed which differ in the rate of decrease in the values of the normalized stimulant below the lower limit and above the upper limit of the recommended range of values. Nonlinear transformations have been proposed for asymmetrical nominants, which are a combination of nonlinear transformations for symmetrical nominants. Therefore, for the right-handed asymmetric nominants, the transformation of the values of nominant above the upper limit of the recommended range of values was conducted by nonlinear transformation according to Eq. 4 (slower decrease of values of normalized stimulant, i.e., stimulant values closer to 1), while the transformation of the values of nominant below the lower limit of the recommended range of values was conducted by nonlinear transformation according to Eq. 3 (faster decrease of values of normalized stimulant, i.e., stimulant values closer to 0)—the combination of these two cases is given by Eq. 5. In turn, for the left-handed asymmetric nominants, the transformation of the values of nominant above the upper limit of the recommended range of values was conducted by nonlinear transformation according to Eq. 3 (faster decrease of values of normalized stimulant, i.e., stimulant values closer to 0), while the transformation of the values of nominant below the lower limit of the recommended range of values was conducted by nonlinear transformation according to Eq. 4 (slower decrease of values of normalized stimulant, i.e., stimulant values closer to 1)—the combination of these two cases is given by Eq. 6.

In all proposed formulas for nonlinear transformations, the principle of symmetry of the ranges of the values of nominant below and above the recommended range of values has been kept by calculating new minimum or maximum values, assuming that the distance of the new minimum from the lower limit of the recommended range of values is the same as the distance of the new maximum from the upper limit of the recommended range of values.

In conclusion, it is worth mentioning that the proposed nonlinear transformations make it possible to obtain the values of a stimulant with greater diversity, i.e., there is a possibility of greater diversity in the final evaluation of the examined companies. On the other hand, the use of the empirical recommended range of values instead of the theoretical recommended range of values makes it possible to link the transformation with the specificity of the functioning of companies in a given economic sector—thanks to this, more companies will obtain transformed values closer to 1 than when the theoretical recommended range of values had been used.

References

Batóg B, Wawrzyniak K (2020) Comparison of proposals of transformation of nominants into stimulants on the example of financial ratios of companies listed on the Warsaw Stock Exchange. In: Jajuga K, Batóg J, Walesiak M (eds) Classification and data analysis. Theory and applications, Springer Nature, Switzerland, pp 3–17. https://doi.org/10.1007/978-3-030-52348-0

Borys T (1978) Metody normowania cech w statystycznych badaniach porównawczych [Methods of characteristics normalization in statistical comparative studies]. Przegląd Statystyczny 25 (2):227–239

Borys T (1984) Kategoria jakości w statystycznej analizie porównawczej [Category of Quality in Statistical Comparative Analysis]. Prace Naukowe Akademii Ekonomicznej we Wrocławiu 284, Seria: Monografie i opracowania 23, Wydawnictwo Akademii Ekonomicznej we Wrocławiu, Wrocław

Gabrusewicz W (2014) Podstawy analizy finansowej. PWE, Warszawa

Hellwig Z (1968) Zastosowanie metody taksonomicznej do typologicznego podziału krajów ze względu na poziom ich rozwoju oraz zasoby i strukturę wykwalifikowanych kadr [Procedure of evaluating high level manpower data and typology of countries by means of the taxonomic method]. Przegląd Statystyczny 15(4):307–327

Hellwig Z (1972) Procedure of evaluating high-level manpower data and typology of countries by means of the taxonomic method. In: Gostkowski Z (ed) Towards a system of human resources indicators for less developed countries, Papers prepared for UNESCO research project, Ossolineum, The Polish Academy of Sciences Press, Wrocław, pp 115–134

Hozer J, Tarczyński W, Gazińska M, Wawrzyniak K, Batóg J (1997) Metody ilościowe w analizie finansowej przedsiębiorstwa. Główny Urząd Statystyczny, Warszawa

Kowalewski G (2002) Nominanty niesymetryczne w wielowymiarowej analizie sytuacji finansowej jednostek gospodarczych. Przegląd Statystyczny 2:123–132

Kowalewski G (2006) Jeszcze o nominantach w metodach porządkowania liniowego zbioru obiektów, Taksonomia 13. Klasyfikacja i analiza danych—teoria i zastosowania, Prace Naukowe Akademii Ekonomicznej 1126:519–528

Kukuła K (2000) Metoda unitaryzacji zerowanej. Wydawnictwo Naukowe PWN, Warszawa

Młodak A (2006) Analiza taksonomiczna w statystyce regionalnej. Difin, Warszawa

Sierpińska M, Jachna T (2004) Ocena przedsiębiorstwa według standardów światowych. Wydawnictwo Naukowe PWN, Warszawa

Strahl D, Dziechciarz J (1999) Study major choice—factor preference measurement. In: Gaul W, Locarek-Junge H (eds) Classification in the information age. Springer-Verlag, Berlin, Heidelberg, pp 473–481

Strahl D, Walesiak M (1997) Normalizacja zmiennych w skali przedziałowej i ilorazowej w referencyjnym systemie granicznym. Przegląd Statystyczny 1:69–77

Waśniewski T, Skoczylas W (2004) Teoria i praktyka analizy finansowej w przedsiębiorstwie. Fundacja Rozwoju Rachunkowości w Polsce, Warszawa

Wójciak M (2003) Niesymetryczne metody wartościowania nominant. Taksonomia 10. Klasyfikacja i analiza danych—teoria i zastosowania, Prace Naukowe Akademii Ekonomicznej 988:519–528

Gini Regression in the Capital Investment Risk Assessment—Sensitivity Risk Measures in Portfolio Analysis

Grażyna Trzpiot

Abstract One of the basic market models is the Sharpe model, which is used to position equity investments. The use of linear regression is a classic approach used in modeling, replacing this approach with a Gini regression model is subject of the article. Outliers and extreme values, which we observe in the distribution of rates of return on listed assets, were the motivation to use the Gini regression model. Indication of the advantages of such an approach and verification of suitability for market data was the main goal. The application part is modeling assets from the Warsaw Stock Exchange.

Keywords Gini regression model · Systematic risk · Portfolio analysis

1 Introduction

In this article, we ask whether the standard risk estimation procedures are in line with investors' expectations. The mathematical model of portfolios theory relies on the assumption of market efficiency. As a general rule, investments either move in line with the market or in the opposite direction. According to portfolio theory, these correlations should stay the same for some period of time. So people assume correlations that were accurate some time ago will remain still accurate. Stock prices are determined based on the buying and selling behavior of people. For prices to accurately reflect a stock's true value, we must assume that people act rationally.

Starting from Markowitz model we use the variance as the most popular measure of variability and the risk measures. There are two properties which seem natural and are implicit when dealing with the variance: the symmetry and the decomposition. There are two kinds of symmetric relationships that are imposed in the conventional statistical analysis. The first one is the symmetry of the variability

G. Trzpiot (✉)
University of Economics in Katowice, Katowice, Poland
e-mail: grazyna.trzpiot@ue.katowice.pl

K. Jajuga et al. (eds.), *Data Analysis and Classification*, Studies in Classification, Data Analysis, and Knowledge Organization,
https://doi.org/10.1007/978-3-030-75190-6_7

measure with respect to the underlying distribution and the second one is the symmetry in the relationship between variables. We have two types of decompositions of variance. One is the decomposition of a variability measure of a linear combination of random variables into the contributions of the individual variables and the contributions of the relationships between them. The other decomposition is the one that decomposes the variability of a population that is composed of several subpopulations into the contributions of the subpopulations and some extra terms. The Gini approach deviates from this conventional approach in both cases symmetric relationship and the decomposition of the GMD (Gini's Mean Difference) includes the structure of the decomposition of the variance as a special case. The usefulness of the GMD and its contribution to our statistical analysis is especially important whenever the concepts that are used are not symmetric by definition. Among those concepts are regression in statistics and elasticity in economics. The Gini describes the variability by two attributes: the variate and its rank.

One of the basic market models is the Sharp model, which is used to position equity investments. In practice the OLS (ordinary least squares) method was used for estimation of this model. Empirical studies show that the estimator obtained by means of the OLS is not robust to the observed extreme values. The assumption about the square loss function influences the estimation results. OLS requires linear relationship between conditional expectation of the dependent variable and explanatory variables and errors are identical and independent distributed and uncorrelated with the independent variables. Often monotonic transformations are applied to linearize the model, which can lead to changes of the sign of the estimated coefficients and OLS is sensitive to outliers.

The Gini approach deviates from this conventional approach. The use of linear regression is a classic approach used in modeling, replacing this approach with a Gini regression model is the main goal of the article. Assuming risk aversion, when making an investment, we consider the beta estimation procedure relating to Gini regression. Gini regression has better properties of resistance to extreme values and improves the quality and thus the reality of the estimation results.

The Gini regression proposed by Olkin and Yitzhaki (1992) is concerned with the minimization of a particular target function, instead of the traditional variance. The estimators derived from this criterion are more robust than those of OLS when the regressors deviate from the multinormal distribution. The Gini regression approach also provides a better adjustment in the presence of extreme values (outliers) (Yitzhaki and Schechtman 2013) for an overview of the Gini methodology. Outliers can excessively affect the amplitude and the signs of the coefficient estimates (Choi 2009). Data contamination may also exclude the possibility of obtaining a valid inference since the coefficient estimates exhibit important variances (instability). The Gini regression allows the problems of instability and of inconsistent signs of the coefficient estimates to be solved.

In the paper/work by Schechtman et al. (2011) we can find two approaches to the estimation of the regression model parameters using the mean Gini difference (GMD). The first approach is based on the weighted average of the slope coefficients defined between adjacent observations (semi-parametric approach), and the second approach uses minimization of GMD residuals.

The semi-parametric approach is based on the estimation of the regression coefficient as follows: it is the weighted average of the slopes defined between adjacent cases (or all pairs of cases) of the regression curve. The procedure is partially like OLS, estimators can be accurately written and all expressions used are referenced in linear regression. The derivation of the estimators and their properties are discussed in detail in Schechtman et al. (2008). This regression model does not require specification of the functional form of the model. It can be used when the researcher is interested in estimating the mean slope or studying the elasticity of an arc without having to meet the formal assumptions about the model.

The second approach is based on minimizing the mean Gini distance (GMD) of residuals. This approach requires a linear model. This is similar to the Least Absolute Deviation (LAD) regression (Koenker and Bassett 1978). Instead of minimizing the sum of the absolute residual variations, the residual GMD, which is the mean of the absolute differences between all pairs of residuals, is minimized. As with LAD, the estimators can be derived numerically.

In this article, we present the assumptions of Gini regression, the selected estimation method and its application to the systematic risk assessment. The application part is asset modeling from the Warsaw Stock Exchange.

2 Systematic Risk—Estimation Beta

The systematic risk is reflected by the beta coefficient. It measures average changes of an asset when the market index increases or decreases. The sensitivity of beta as a systematic risk estimation can be related to two factors:

1. Inconsistencies between standard statistical methods and financial theory. In particular, the OLS regression coefficient estimator uses square weights that contradict risk aversion.
2. Probability distributions of market rates of return do not meet the assumption that the distribution is normal, they often have "fat tails".

The beta estimator determined with the OLS is the weighted average of the slope coefficients obtained from two adjacent observations situated along the Security Characteristic Curve. This makes it impossible to check how large weights we assign to the extreme values of the rate of return in the sample.

We consider a market model where the rates of return on investment are random and continuous with the total density function $f(R_k, M)$, where R_k is the rate of return, and k and M—the market portfolio. We will write as f_M, F_M, μ_M, and σ_M^2 respectively the boundary density, boundary distribution, expected value and variance M. We assume that there are first and second moments and define $R_k(m) = E(R_k| M = m)$ as the conditional expected rate return on shares k assuming a market return of $M = m$. $R_k(m)$ marks the securities market line (Sharpe 1981), but we will also use the name Security Characteristic Curve. The following relationship is usually assumed to estimate the beta value of shares:

$$R_k = \alpha_k + \beta_k M + \varepsilon_k \tag{1}$$

with the additional assumption that the random components ε_k, are independent, with the same distribution with the expected value of zero and constant variance. The OLS estimator can be written as follows:

$$\beta_{MNK} = \frac{\mathrm{cov}(R_k, \mathrm{M})}{\mathrm{cov}(\mathrm{M}, \mathrm{M})} \tag{2}$$

where the index k was omitted.

Based on beta coefficient, several types of stocks can be distinguished:

- $\beta > 1$—an aggressive stock, that is more volatile than the market,
- $0 < \beta < 1$—a defensive stock, that is less volatile than the market.
- $\beta = 0$—the rate return of a stock does not follow the market portfolio. According to Sharpe's formula, $\beta = 0$ for risk-free assets,
- $\beta < 0$—stock is negatively correlated with the market index.

It is a regression equation in which the profit rate of the k-th share is the dependent variable, while the stock index return rate acts as an explanatory variable. The random component takes into account the influence of other factors that affect the action. The above relationship has been called the security characteristic line. The beta factor of this equation determines the degree of sensitivity of a given stock to changes in the stock index profit rate. This ratio is also equated with the measure of systematic risk.

Despite the fact that the weaknesses of the beta coefficient have been shown in empirical studies, it is still one of the important theoretical measures of risk. To emphasize the importance of systematic risk in financial theory, we assume that only two parameters, expected value and systematic risk, are sufficient to capture the full effect of the distribution of the share return in relation to the investor's utility function. Systematic risk is expressed as the covariance between the rate of return of a share and the marginal utility of capital. However, the marginal utility of capital is assumed. The discussed procedure of systematic risk estimation requires assuming a certain expected marginal utility of capital. LSM implies a quadratic utility function, which means that the marginal utility of capital is a linear function of capital. This assumption of linearity is responsible for the beta sensitivity to "fat tails" (Trzpiot 2007, 2019).

We consider the task of maximizing the expected utility of an investor who has a portfolio of risky and safe assets. The investor's goals can be written as an optimization of the model:

$$\max E(U(M)) \tag{3}$$

$$M = M_0[y + \sum\nolimits_{i=1}^{n} \alpha_i R_i],$$
$$\sum\nolimits_{i=1}^{n} \alpha_i = 1,$$
$$M_0 \equiv 1,$$

where:

$E[U(M)]$—expected utility,
U—utility function which is continuous, monotone, rising and concave function,
M_0—fixed initial capital taken as 1, with no loss of generality,
α_i and R_i—the share of capital M invested in the share i and the rate of return of the share i, respectively,
y—the rate of return of other income, whether deterministic or stochastic.

Consider an expected utility maximizing investor who holds a mixed portfolio of risky and the investor holds a given portfolio $\{\alpha_0\}$, whose shares are α_i^0, $i = 1, \ldots, n$. Note that the only requirement on α_0 is that it is held by the investor. Assume the investor wants to change the holdings of asset k in the portfolio. The effect of increasing α^0 k on expected utility is given by:

$$\frac{\partial E(U(M)}{\partial \alpha_k} = E(U'(M)R_k) - \lambda \tag{4}$$

where λ is the Lagrange multiplier associated with the portfolio constraint. By adding and subtracting $E[U'(M)]\mu_k$, where μ_k is the expected return on asset k, we can rewrite this equation as:

$$\frac{\partial E(U(M)}{\partial \alpha_k} = E(U'(M))\mu_k + \text{cov}(U'(M), R_k) - \lambda \tag{5}$$

Our purpose here is to compare between assets. Hence, all factors that are equal for all assets can be ignored. Last equation expresses the effect of a marginal increase in asset k on expected utility as a function of the expected return on asset k, and the asset's systematic risk, defined as the covariance between marginal utility of wealth and the return on asset k. Assuming a specific utility function enables us to obtain an explicit expression for systematic risk. If the utility function is defined in terms of the rate of return on the portfolio rather than the level of wealth, the standard expression for systematic risk is produced.

The beta coefficient is still one of the theoretical measures of risk. To emphasize the importance of systematic risk in financial theory, it was assumed that only two parameters, expected value and systematic risk, are sufficient to cover the full effect of the distribution of the share return in relation to the investor's utility function. Systematic risk is expressed as the covariance between the rate of return on shares and the ultimate utility of capital. However, the marginal utility of capital is assumed. The discussed procedure of systematic risk estimation requires assuming a certain expected marginal utility of capital. OLS implies a quadratic utility function, which means that the marginal utility of capital is a linear function of capital. This assumption of linearity is responsible for beta sensitivity to "fat tails" (Trzpiot 2008). If traders identify risk with another volatility index such as a semi-variance or a Gini index (that is, the Gini regression coefficient), an alternative expression could be:

$$\beta_{GINI} = \frac{cov(R_k, F(M))}{cov(M, F(M))} \tag{6}$$

where $F(M)$ is the distribution of the investor's capital.

We will write down this formula (Trzpiot 2008) for the beta calculations for the action k using Gini regression. Since a prerequisite for a small increase in the share of k to increase the expected utility for all risk averse investors with an α^0 portfolio is that $\mu_k - \Gamma_M \, \beta_{GIN} \geq 0$.

3 Gini Regression—Multiple Regressions Model

We analyze $(Y, X_1, \ldots, X_K) - (K + 1)$ dimensional vector of random variables with finite expected values, respectively $(\mu_Y, \mu_1, \ldots, \mu_K)$ and with the covariance variance matrix S. Assume that we have a general regression function defined as:

$$g(x_1, \ldots, x_K) = E(Y|X_1 = x_1, \ldots, X_K = x_K). \tag{7}$$

The resulting vector of the regression coefficients β_N is as follows:

$$\beta_N = [E(V'X)]^{-1} E(V'Y) \tag{8}$$

where:

$\beta_N = (\beta_{N1}, \ldots, \beta_{Nk})$ is $(K \times 1)$ column vector of the (conditional) regression coefficients,

V is $(n \times K)$ matrix of the cumulative distributions distribution of random variables $X_1, \ldots, X_K$,

Y is a $(n \times 1)$ vector of the value of the dependent variable and X is a $(n \times K)$ matrix of the deviations of the explanatory variables from their expected values.

Y is a $(n \times 1)$ vector of the value of the dependent variable and X is a $(n \times K)$ matrix of the deviations of the explanatory variables from their expected values.

So the natural estimators of the regression coefficients are based on replacing the cumulative distributions by the empirical distributions (which are calculated using ranks). The elements of $E(V'Y)$ and $E(V'X)$ are $cov(Y, F(X_k))$ and $cov(X_j, F(X_k))$, respectively. It is assumed that the rank of $V'X$ equals K, the number of explanatory variables.

Next the constant term can be estimated by minimizing a function of the residuals. The exact function used determines whether the regression passes through the mean, the median, or any other quantile. The multiple regression procedure, although it is not based on an optimization procedure, generates equivalents to the OLS's normal equations. By defining the error term and substituting for the multiple regression coefficients, it can be shown that $cov(\varepsilon, F_k(X)) = 0$ for $k = 1, \ldots, K$.

The Gini semi-parametric approach has the advantage of relying on a few assumptions, no linearity hypothesis is needed. The estimator β_N is less sensitive to extreme values since it is built on the matrices $V'X$. Among those concepts is R^2 of the regression, which can be considered as a measure to assess the share of the (square of the) GMD which is explained by the model:

$$_GR^2 = 1 - [cov(e, r(e))/cov(y, r(y))]^2 \tag{9}$$

$r(x)$ denotes ranks in the sample, where $e = y - x\beta_N$.

4 Application of Gini Regression in Portfolio Analysis

Companies from the WIG sector index were used for the analysis. Currently, the WIG includes all companies listed on the Warsaw Stock Exchange that meet the basic criteria for participation in indices. The WIG index follows the principle of diversification, aimed at limiting the share of a single company and the stock exchange sector. It is a total return index and its calculation takes into account both the prices of its shares and dividend and subscription rights income. The following 15 companies were selected for the analysis: PZU, PKN Orlen, KGHM S.A., Santander Polska, Cyfrowy Polsat, ING SK, mBank, Orange Polska, Kęty, Millenium, Kruk, Alior, Intercars, Kernel, Eurocash. This analysis was prepared based on data for the period from 18.02.2018 to 18.02.2020.

Step 1 Preliminary Analysis

The study presents rates of return and risk measures, and rankings of companies based on quantified risk measures. As indicated in Table 1, the company with the highest expected rate of return is ORANGE PL, while the lowest rate of return was achieved by KGHM SA. As many as 11 out of the 15 companies surveyed obtained negative expected rates of return, so investing in these companies would bring a loss. The standard deviation is the extent to which, on average, returns deviate from the expected rate of return.

The lowest fluctuation from the expected rate of return was reported by ING SK, while the largest one by KETY and KRUK. In the case of a semi-standard deviation, just as with a standard deviation, ING SK is the least risky company and

Table 1 Parameters of rate of return distribution[a]

Assets	PZU	PKNORLEN	KGHM SA	SANPL	CYFRPLSAT
R	−0.00013	−0.00025	−0.00008	−0.00031	0.00049
V	0.10782	0.18478	0.19475	0.17155	0.14341
S	0.32836	0.42986	0.44130	0.41419	0.37869
SV	0.00011	0.00018	0.00019	0.00017	0.00014
SS	0.01044	0.01342	0.01370	0.01288	0.01181
d	0.01103	0.01502	0.01507	0.01414	0.01256
sd	0.00552	0.00751	0.00754	0.00707	0.00629
Assets	INGSK	MBANK	ORANGEPL	KETY	MILLENNIUM
R	−0.00001	−0.00033	0.00064	0.00024	−0.00056
V	0.08477	0.17074	0.21538	0.15808	0.22731
S	0.29115	0.41321	0.46409	0.39760	0.47677
SV	0.00008	0.00017	0.00018	0.00016	0.00024
SS	0.00892	0.01312	0.01349	0.01249	0.01541
d	0.00943	0.01421	0.01476	0.01258	0.01564
sd	0.00471	0.00711	0.00739	0.00630	0.00782
Assets	KRUK	ALIOR	INTERCARS	KERNEL	EUROCASH
R	−0.00030	−0.00203	−0.00034	0.00019	−0.00038
V	0.30472	0.24146	0.17653	0.11367	0.21252
S	0.55201	0.49139	0.42016	0.33715	0.46100
SV	0.00027	0.00025	0.00018	0.00010	0.00021
SS	0.01655	0.01567	0.01334	0.00975	0.01456
d	0.01631	0.01593	0.01188	0.01104	0.01539
sd	0.00816	0.00790	0.00597	0.00554	0.00770

[a]R—expected return, V—variance of returns, S—standard deviation of returns, SV—semi-variance of returns, SS—standard semideviation of returns, d—mean absolute deviation of returns, sd—mean absolute semideviation of returns
Source own calculations

KRUK is the most riskiest. As far as average deviation and semi-average deviation are concerned, the results are almost the same as with standard deviation and semi-average deviation (Table 1) besides that there are used to describe only positive surpluses in relation to average return.

In order to verify the hypothesis of normal distribution, the Kolmogorov-Smirnov test and the Shapiro-Wilk test were performed. The significance level for the tests used was set at $\alpha = 0.05$. The hypotheses are as follows:

H_0: The distribution of the rates of return of the audited company is a normal distribution,
H_1: The distribution of the rates of return of the audited company is different from the normal distribution.

Table 2 Results test of normality tests

Assets	Kołmogorow-Smirnow	*p*-value	Shapiro-Wilk	*p*-value
PZU	0.045	0.017	0.987	0.000
PKNORLEN	0.036	0.173	0.996	0.198
KGHM SA	0.052	0.003	0.990	0.002
SANPL	0.041	0.048	0.989	0.001
CYFRPLSAT	0.051	0.004	0.981	0.000
INGSK	0.070	0.000	0.970	0.000
MBANK	0.041	0.048	0.994	0.034
ORANGEPL	0.064	0.000	0.924	0.000
KETY	0.067	0.000	0.958	0.000
MILLENNIUM	0.056	0.001	0.969	0.000
KRUK	0.088	0.000	0.897	0.000
ALIOR	0.062	0.000	0.956	0.000
INTERCARS	0.115	0.000	0.887	0.000
KERNEL	0.111	0.000	0.968	0.000
EUROCASH	0.054	0.002	0.984	0.000

Source own calculations

The Kolmogorov-Smirnov test and Shapiro-Wilk test were carried out and the asymmetry of the companies' rates of return distribution was checked in Table 2. On the basis of the Kolmogorov–Smirnov test, we reject the zero hypothesis in favor of the alternative hypothesis that almost all the above mentioned companies have a different distribution than the normal one, without PKNORLEN ($p < 0.05$).

The Shapiro-Wilk test showed, analogously to the Kolmogorov-Smirnov test, a different distribution than the normal one for all the examined companies. We can observe non-normality of the rate of returns by box-plot for chosen assets based on quantile (25–75%), additionally we have extreme value and outliers in the rate of return observations (Fig. 1).

Step 2 Portfolio Analysis

In the next step the portfolio was constructed based on the investor's requirements concerning the level of risk and return. Two portfolios were prepared for the analysis.

The first portfolio was built of all 15 companies, in line with the Markowitz approach. The share of each company in the portfolio was supposed to be positive. Criterion for determining the portfolio composition, for this one with risk minimization in mind (I portfolio). This portfolio had a loss: $R_p = -0.001555$ and had a risk as $S_p = 0.015025$.

To improve investment performance, the second portfolio was with an investor's expected rate of return imposed (II portfolio), which was set at least 0.0001. We see the highest rate of return ($R_p = 0.000164$) for a II portfolio, smaller after the optimization, the set of six companies. This is also the portfolio with the lowest risk in comparison with the first portfolio $Sp = 0.006898$ (results in Table 3).

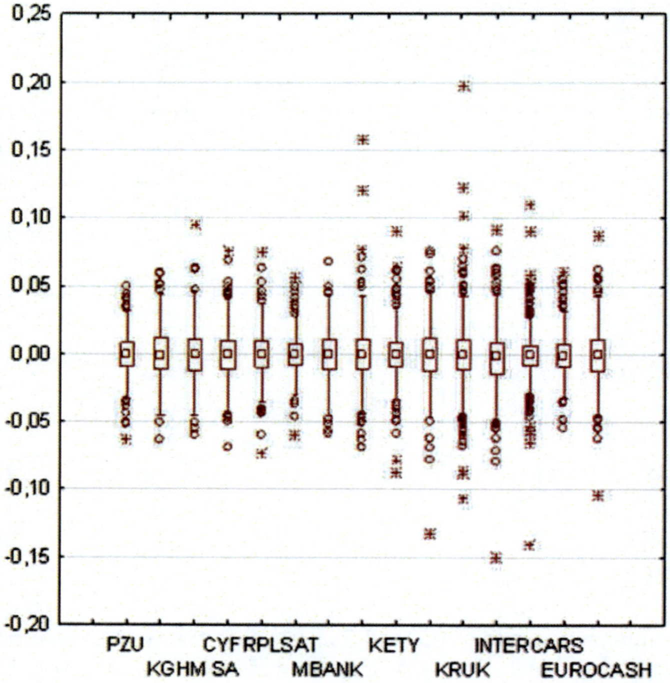

Fig. 1 Box-plot of rate of return chosen assets based on quantile (two extra sing means: *—extreme value, °—outliers). *Source* own calculations

Step 3 Portfolio Sensitivity Risk Analysis

Sensitivity risk analysis for two portfolios was made. We start the study assessment of the dependence on the rate of return of portfolio with the rate of return from the market by a scatter plot made for two-dimensional median (dark color). We can observe in contrast to normal ellipse many outliers oh the distribution of the rates of return of the audited portfolio.

The first portfolio one with risk minimization (I portfolio) was presented on Fig. 2. Significant majority of observations have non-positive coordinates, which finally translates into the result of the expected rate of return of the portfolio which is less than zero. The second portfolio with an investor's expected rate of return imposed (II portfolio), was presented on Fig. 3. Both scatter plot shows relation to the market, and both are given signals to the trader that the portfolio's return rate fluctuates with the market rate of return.

Two portfolios, described in Table 3, have been subject to the sensitivity analysis and we start to estimate beta, because the systematic risk is reflected by the beta coefficient. It measures average changes of an asset when the market index

Table 3 Parameters of portfolio

Assets	I portfolio	II portfolio
PZU	0.067	0.193
PKNORLEN	0.083	0
KGHM SA	0.000	0
SANPL	0.000	0
CYFRPLSAT	0.133	0.133
INGSK	0.245	0.269
MBANK	0.002	0.118
ORANGEPL	0.001	0
KETY	0.068	0.180
MILLENNIUM	0.016	0
KRUK	0.027	0
ALIOR	0.000	0
INTERCARS	0.107	0
KERNEL	0.190	0.180
EUROCASH	0.062	0
S_p	0.015025	0.006898
R_p	−0.001555	0.000164

Source own calculations

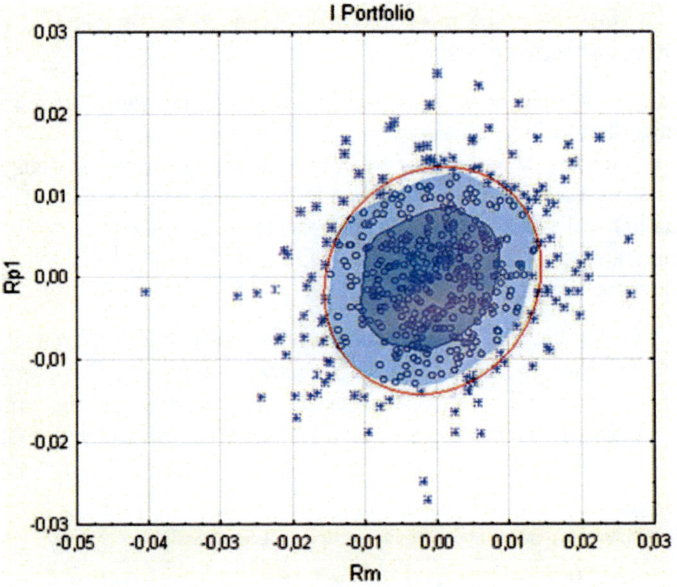

Fig. 2 Scatter plot of rate of return portfolio I by rate of return of market based on median and inter-quantile (sing means: *—outliers, dark color means 2D median) in relation to normal ellipse. *Source* own calculations

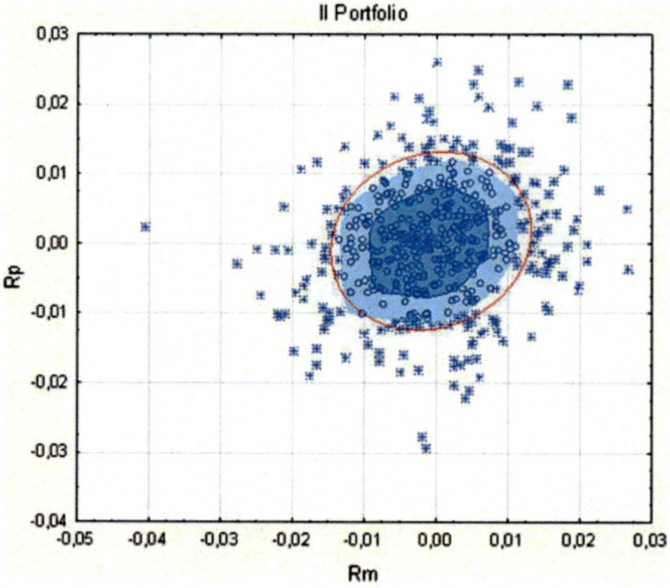

Fig. 3 Scatter plot of rate of return portfolio II by rate of return of market based on median and inter-quantile (sing means: *—outliers, dark color means 2D median) in relation to normal ellipse. *Source* own calculations

increases or decreases. The most important assumptions of this is a risk-return concept in a near-perfect market:

- market investors agree to increase the risk, but only the rate of return, and maximize their utility functions,
- investors have equal access to free information on the stock exchange,
- investors can take out loans and provide loans at a risk-free rate of return,
- the number and types of assets in the market are constant,
- all shares are perfectly liquid,
- βeta ratio for shares are stable over time.

We begin from classical basic market models, described in introduction is the Sharp model, which is used to position equity investments. We used the OLS (ordinary least squares) method was used for estimation of this model. Our empirical studies show that the estimator obtained by means of the OLS is not

Table 4 Parameters of market and beta for portfolio OLS methods

	E(M)	S(M)	R_f	Beta	R^2
I portfolio	−0.000105	0.00906	0.000074	0.2282	0.0384
II portfolio	−0.000105	0.00906	0.000074	0.1829	0.0295

Source own calculations

robust to the observed extreme values. The determination coefficient is low (Table 4) for both portfolio, that means that the models are unstable, insignificant. Beta indicators for both portfolios are positive, which signals to the trader that the portfolio's return rate fluctuates with the market rate.

To continue this analysis we used Gini regressions. Not proper for chosen set of assets OLS method we were replacing by a Gini regression model. For the second portfolio, which achieved better results, we compare the estimates of beta estimators using both methods. Gini beta values we obtained as a results on estimation based on Gini regression methods. Financial leverage is described more properly due to extreme and outliers observation in our rate of return distributions. We compare the results of two methods in Table 5.

Determination coefficients are low (Table 5) that means that the OLS model is unstable, insignificant. Beta indicators for some assets are positive, which signals to the trader that the portfolio's return rate fluctuates with the market rate. Financial leverage is not properly estimated into wrong method.

The final step was to apply the Gini regression methods for estimation beta for both portfolios. In Table 6, we present the final results: for I portfolio and for II portfolio, for which the evaluation of the model fit increased significantly after applying Gini regression (compare R^2 and GiniR^2, form Tables 4 and 6, respectively).

The Gini determination coefficient is significant, for both portfolios, so we can use this results for better financial leverage on our investments. The obtained results also indicate that it is worth investing in such created portfolios, because Sharpe's indicators in Gini regression model have values higher than the results obtained for the market portfolio.

Gini regression through a different way of assessing the changes in the distribution of values in the distribution of a random variable (in the distribution of the rate of return) allows to omit the assumptions of the classical model, in particular, we do not assume the linearity of the model associated with the assumption of a normal distribution. These assumptions are not met for the group of companies surveyed, as shown in the previous steps of the analysis. The results recorded in Table 6 are interpreted as follows: the classic beta coefficient (β OLS) determines the degree of sensitivity of a given stock to changes in the stock exchange index profit rate, when the reference point is the average change in the stock index return rate. The beta coefficient associated with the Gini regression (β GINI) determines

Table 5 Parameters beta for assets in II portfolio: OLS and Gini methods

	PZU	CYFRPLSAT	INGSK	MBANK	KETY	KERNEL
OLSbeta	0.0499	0.00063	0.0926	0.0899	0.0392	0.0513
R^2	0.0066	0.000001	0.0178	0.0338	0.0059	0.0073
GINIbeta	1.1343	−15.944	0.9917	1.1608	0.2894	15.614
GINI R^2	0.7689	0.0003	0.8942	0.5378	0.0673	0.0821

Source own calculations

Table 6 Parameters of market and beta for portfolio GINI methods

	E(M)	S(M)	R_f	GINI Beta	GINI R^2
I portfolio	−0.000105	0.009060	0.000074	1.37	0.8005
II portfolio	−0.000105	0.009060	0.000074	1.06	0.9977

Source own calculations

the degree of sensitivity of a given stock to changes in the stock index return rate when the reference point is the median change of the empirical cumulative distribution of the stock index return rate. The quality level of the systematic risk assessment is determined by the determination coefficient: R2 and GR2, respectively.

5 Discussion and Conclusion

The beta coefficient (the so-called stock aggressiveness coefficient) is a measure of the sensitivity of the income from a given stock to the statistical volatility of all paid markets, i.e., it is a measure of its sensitivity in terms of average risk. The beta factor tells you how many percent the output data return will approximately increase when the market index (market portfolio) returns by 1%. Having information about the correlation coefficient between the rate of return of a given share and the rate of return of the market portfolio (index) as well as information about the standard deviation for the rate of return of this share and the stock index.

In the last step of the analysis, the Gini regression was used to assess the systematic risk of the analyzed companies. The rationale for departing from the classical measurement using the central moments of the analyzed distributions in the definitions of measure structures is the lack of meeting the assumptions allowing for the correct application of these measures: no symmetry of distribution, no distribution consistent with the normal distribution, outliers and extreme observations in the distributions of rates of return of the studied companies.

The conclusion can be formulated as follows, for all models using the expected utility, the effect of increasing the number of selected shares on the increase in the expected utility value can be converted into the effect of the expected value of the return and systematic (beta) risk of the selected stock. Systematic risk is a covariance between the marginal utility of capital in the analyzed portfolio and the rate of return on shares. As shown, these concepts can be measured using the assumption of linearity—then we determine the moments of a random variable as well as departing from these assumptions. The choice of the regression method is also the choice of the marginal utility of the capital function as well as risk aversion.

References

Choi SW (2009) The effect of outliers on regression analysis: regime type and foreign direct investment. Quart J Political Sci 4:153–165

Koenker R, Bassett G (1978) Regression quantiles. Econometrica 46:33–50

Olkin I, Yitzhaki S (1992) Gini regression analysis. Int Stat Rev 60:185–196

Schechtman E, Yitzhaki S, Artzev Y (2008) Who does not respond in the household expenditure survey: an exercise in extended Gini regressions. J Bus Econ Stat 26(3):329–344

Schechtman E, Yitzhaki S, Pudalov T (2011) Gini's multiple regressions: two approaches and their interaction. METRON—Int J Stat LXIX(1):67–99

Sharpe WF (1981) Investments, second edition. Englewood Cliffs. Prentice Hall, New York

Trzpiot G (2007) Decomposition of risk and quantile risk measures. In: Dynamiczne Modele Ekonometryczne. Prace Naukowe Uniwersytetu Mikołaja Kopernika w Toruniu, pp 35–42

Trzpiot G (2008) O wybranej metodzie estymacji beta. In: Metody matematyczne, ekonometryczne i komputerowe w finansach i ubezpieczeniach. Prace Naukowe AE Katowice, pp 345–354

Trzpiot G (2019) Application quantile-based risk measures in sector portfolio analysis-warsaw stock exchange approach. In: Tarczynski W, Nermend K (eds) Effective investments on capital markets. Springer International Publishing

Yitzhaki S, Schechtman E (2013) The Gini methodology: a primer on a statistical methodology. Springer, Berlin

Application in Economics

Enterprise Dark Data

Katarzyna Raca

Abstract The increasing amount of digital data and the declining cost of data storage have led to the fact that companies began collecting any the data possible, regardless of its adequacy and usability. This results in increasingly diverse data, in terms of its structure, quality, availability and the source of origin. Dark data is one type of data that increases significantly as the volume of data expands. Scientific literature does not precisely define the term "dark data", while its interpretation among scientists is ambiguous. The aim of this article entails an attempt to define the dark data occurring in an enterprise, by identification of its essential features. The article presents an overview of the definitions of the term dark data, a proposal of its interpretation, and a classification of data in a company with regard to: usability, availability and quality. The analysis of the concept of dark data was carried out via a review of international journals and articles published on the Internet by Data Science practitioners. As part of the research, four universal features of dark datasets have been indicated (unavailability, unawareness, uselessness, and costliness). Based on data availability and its quality, four groups of enterprise data have also been distinguished. The data classification developed in this way allowed systematization of the term "dark data".

Keywords Dark data · Data classification · Missing data · Big data

1 Introduction

From year to year, the number of Internet users, social networks and mobile phone owners, along with the volume of the data generated by them, has been increasing. The "Digital 2020"[1] report shows that since 2019, 298 million new Internet users

[1]The Digital 2020 report is a global overview of Internet users, mobile devices, social networks and e-commerce, organized by Hootsuite and We are social. The statistics published on a quarterly

K. Raca (✉)
University of Gdańsk, Gdańsk, Poland
e-mail: katarzyna.raca@ug.edu.pl

K. Jajuga et al. (eds.), *Data Analysis and Classification*, Studies in Classification, Data Analysis, and Knowledge Organization,
https://doi.org/10.1007/978-3-030-75190-6_8

have joined the web, thus the increase was 7%, compared to the previous year. In 2020, more than half (59%) of the global population used the Internet, almost half of which (49%) were social network users. The top three countries in the ranking of Internet usage are India, China and Indonesia. These countries account for the largest share of the increase in the volume of digital data. The "Digital 2020" report shows that an average user daily spends 6 h and 43 min on the Internet, where 2 h and 24 min is the time devoted to social networks. Depending on the country and its culture, the time spent on the Internet varies considerably, beginning with Japan (4 h and 22 min) and ending with the Philippines (9 h and 45 min). Globally, the three most-used Internet platforms are Google, YouTube and Facebook.

The growing volume of data results from the technological development and the social changes that have been affecting the increasing popularity of the Internet and social networks. According to the Micro Focus data, in 2016 the global society generated 44 trillion GB of data daily by, inter alia, searching for information, watching movies, adding comments or photos on social networks. IDC forecasts that in 2025 this figure will increase to 463 trillion GB daily. In 2018, the global database (digital resources of the world) was 33 ZB. In its report, IDC forecasts that in 2025 all global data will reach the size of 175 zettabytes.[2] If we were to store all this data on DVDs, the pile of discs would be 23 times the distance from the Earth to the Moon, or it could encircle the Earth 222 times.

The growing volume of data also results from the decreasing cost of data storage. In 1980, 1 GB of disk space had cost $ 193 000, in 2009 it was only $ 0.07. This decrease in the cost of storage enables enterprises to collect virtually all the globally generated data they need.

2 Data Classification in Enterprises

Ever since data acquisition and storage have become easy and inexpensive, enterprises focus more on data quantity rather than its quality. Less and less attention is paid to the fact that the structure of the data should match traditional databases. The technological progress and the avalanche of data growth resulted in an increasing amount of data that is characterized by diversity in terms of structure, source of origin, quality and availability.

basis refer to global and national data (https://wearesocial.com/digital-2020. Accessed 20 Aug 2020).

[2]IDC report (https://www.seagate.com/files/www-content/our-story/trends/files/idc-seagate-dataage-whitepaper.pdf. Accessed 10 Sept 2020).

2.1 *Data Visibility*

Data on the core activities of enterprises have been collected for a long time. The data used by enterprises on an ongoing basis includes: financial data, data on customers or transactions. Usually it is structured data that is saved in a strictly defined format. An organized system of data recording and collection supports enterprise management. In 1970, Codd (1970) proposed a relational database that orders structured data into tables and allows it to be referenced using SQL (Structural Query Language) queries (Chamberlin and Boyce 1974). This is still used today, because it allows data to be organized by name, date or user. Administration of a company's data in this manner allows its easy identification and analysis.

Currently, most digital data is unstructured. According to IDC, in 2025 it will constitute 80% of all data in the world. Such data is created, inter alia, as a result of social network users' activity. Another important source of such data is the IOT (Internet of Things) devices, which generate such data by communicating with each other. Enterprises collect such unstructured data as: text, photos, videos, or sensor-generated data, because it constitutes an additional source of data. The manner in which such data is stored and analyzed is different and more complex. Enterprises often use NoSQL (not a relational database) to save and store such data (Eberendu 2016).

The literature also addresses the issue of semi-structured data, i.e., partially structured data. Its format does not allow it to be entered in traditional databases, nevertheless, such data contains metadata that allows easier analysis thereof (Abiteboul 1997).

Such division of enterprise data is presented, among others, by Grim (2019), who provided appropriate names for the above-mentioned datasets. He refers to structured data in relational databases as light data. Grey data is partially structured data with metadata. The last group consists of unstructured data, called by the author dark data.

2.2 *Data Quality*

Data quality is another feature that differentiates enterprise data. The first references to this issue appeared in the 1990s and concerned the definition and the methods of data quality measurement. The three characteristics of statistical data quality, mentioned by researchers, concerned: relevance of the data for users, timeliness and accuracy (Kordos 1988). R. Y. Wang and D. M. Strong paid particular attention to the former. Based on a two-stage questionnaire survey, they distinguished four categories, with fifteen data quality areas that are important from the perspective of the users (Wang 1996). Over the next years, scientists proposed new standards and data quality measures (Zhu and Cai 2015).

As a result of the rapidly growing big data of diverse structure, new challenges, in terms of data quality, have emerged. Accordingly, scientists began to extend the existing data quality standards and propose new ones. Li Cai and Yangyong Zhu propounded five dimensions of quality for big datasets: availability, usability, reliability, relevance, presentation quality. The Authors consider the first four categories to be imperative, while the last one is an additional dimension that strengthens the quality criterion (Zhu and Cai 2015). J. Maślankowski lists ten key dimensions of big data quality in more detail (Maślankowski 2015). The features listed by scientists with regard to this issue are mutually coherent and consistent with the "European Statistics Code of Practice". On November 16, 2017, the European Statistical System Committee adopted the "European Statistics Code of Practice", which includes five data quality characteristics[3]:

- usability—reflection of the need for data collection,
- accuracy and reliability—the data collected does not differ from reality,
- timeliness and punctuality—refers to the up-to-dateness of the data collected,
- accessibility and clarity—describes the lack of barriers in acquisition of data and its metadata,
- coherence and comparability—the ability to compare data in terms of time, geographical area, subject and source of origin.

Enterprise data, regardless of its dimensions, should meet the data quality requirements. It is not always possible, however, to meet all of them. In such case, the data obtained is the so-called dirty data. Unreliable data storage, the lack of metadata or the lack of an appropriate storage location, inter alia, lead to formation of datasets of insufficient quality. The degree of data "dirtiness" is affected by many factors, including human error (Taleb et al. 2016).

Increasingly often, data is perceived as a valuable asset in business, since strategic decisions are made on its basis. As such, important enterprise activity should entail continuous quality monitoring of the data held. Analysis of dirty data is possible, but it is a time-consuming and costly process that involves data profiling and cleaning, followed by prevention of contamination (Migdał-Najman and Najman 2018).

Enterprise data includes data that meets all the quality standards, i.e., clear data. Typically, it is structured data that does not require any complex data preparation prior to analysis, such as inter alia, missing data imputation, transformation, or data aggregation over time.

[3]These principles refer to the national statistical authorities as well as the EU statistical authority (Eurostat) and constitute a set of features characterizing the data quality for official statistics (https://ec.europa.eu/eurostat/web/products-catalogues/-/KS-02-18-142. Accessed 14 Sept 2020).

2.3 *Data Availability*

Enterprises generate and store internal data as part of their activity. This can be customer data, transaction histories or sales results (structured data), but also meeting reports, e-mails or photos of products (unstructured data). Additionally, companies may collect external data, generated by other entities. In most cases, the source of such data is the Internet, e.g., the posts generated by social network users, which are broadly available and can be used by companies to develop marketing strategies that are tailored to potential customers.

Both internal and external data can be accessed by the enterprises. This means that data can be stored or collected, while companies can use it at any time. Such data is called light data. The data unavailable to enterprises is called dark data. Unavailability may result from various factors, which will be described in more detail in the next chapter.

The division of enterprise data with regard to availability was presented, among others, by Lugmayr et al. (2017). According to the authors, enterprise data can be divided into four types: light data—available data that can be used at any time; data spots—light data that is taken into account in analyzes; grey data—constitutes part of all company data and reliable conclusions could be drawn based on that data, but the data is not available; dark data—unknown data, which cannot be classified or specified in any way.

3 Dark Data Definitions—Literature Overview

Increasingly often, enterprises collect data somewhat in reserve. They do not use that data in their day-to-day operations, hoping that in the future it will provide some potentially useful information. This additional data, usability of which is unknown at the time of its collection, accounts for a significant part of all the data collected. This data is called dark data.

Most likely, the term dark data was first used by the Gartner IT Glossary consulting company, which has defined dark data as a set of information that a company collects, processes and stores as part of its regular business activities but does not use it for other purposes (e.g., for business relationship analysis). The data collected by enterprises on a daily basis can be used to achieve additional direct benefits, but its cursory analysis prevents this. For instance, when historical data that was used during regular business operations ends up in the archives and is never used again. In its definition, Gartner's definition also points out that storage of such data generates significant costs. According to Veritas, averagely, 52% of enterprise data is dark data, while the average cost of storing 1 petabyte of data (2.3 trillion files) for one year is $ 5 million. If a company has 10 petabytes of data, it spends about $ 26 million annually on collecting data of unknown value. 80% of it is data that no one has been interested in for at least 3 years. Collection of data without a well thought out purpose puts a strain on the company's systems and processes as well as poses a risk of losing important information.

According to D. Tranajov, dark data is unused, hidden and can take various structures (structured, partially structured or unstructured) (Trajanov et al. 2018). A. Banafa (https://www.bbvaopenmind.com/en/technology/digital-world/understanding-dark-data/. Accessed 18 Oct 2020), on the other hand, describes it as unstructured, untagged, untapped data that is neglected by business and IT administrators. Ce Zhang also draws attention to the unstructured nature of data (text, tables, images, video and audio files), which prevents its analysis via the use of traditional database tools (Zhang et al. 2016). The differences in these scientists' definitions may result from the existence of many types of "dark data".

D. J. Hand distinguishes 15 types of dark data, which differ depending on the source of origin. The first type of data (DD-Type 1) refers to known but lost data ('known unknowns'). Such data is hidden in the data gaps that occur when data is saved. Such data emerges, for example, when a customer refuses to provide personal information. Lack of consent is the information, while the data lost is dark data. The second type of data (DD-Type 2) is data that we know nothing about ('unknown unknowns'). We are not aware of its lack or loss. Such data can be obtained, for example, when an anonymous online survey is carried out and the actual list of respondents is unknown. As such, there is no information about the persons taking part in the study (Hand 2020). The above-mentioned examples have been known in statistics for a long time. The increasing pace of data growth did not contribute to the creation of dark data but increased its quantity. Dark data can appear in both small and big datasets. Depending on the source, data can take various forms. In some cases, it will result from unconscious actions or an unsuitable measurement tool and sometimes from deliberate fraud.

Backups and server logs constitute a significant proportion of this type of data. With today's reliability of computer systems, such data is almost unused, which in turn leads to the lack of due diligence in its creation. Other examples of dark datasets are: information on clients or previously employed employees, digital marketing information, e-mail correspondence, Internet calls, financial statements, notes, presentations, or old versions of important paper documents. Companies collect this data because it is relatively cheap to keep such data. They hope that some valuable information can be extracted from it in the future.

D. J. Grim points to the significant confidence that companies place in the potential future usability of this data. He also points out that often the term information assets accompanies the term dark data. The fact of equating the two concepts suggests that, at present, unanalyzed data, dormant on servers and storage clouds, is valuable for enterprises, even if the value is only potential (Grim 2019).

Bansari Trivedi draws attention to the risk that enterprises bear when disregarding dark data processing. He argues that it negatively affects the data sources and the rest of the data in an enterprise (Trivedi 2017). Unawareness of the hidden value of dark data can lead to ineffective or wrong decisions. One example of such negative consequences caused, inter alia, by the lack of dark data awareness is the crash of the space shuttle Challenger in 1986. During a teleconference, the day before the take-off, the NASA and representatives of the company manufacturing the rocket engines discussed the impact of temperature on the rubber seals connecting the rocket engine blocks. Both the opinion of the chief engineer responsible for the rocket engines as well as the empirical data showing the impact of low temperatures on the seals were ignored. A tragic decision was made that led to the explosion of the space shuttle and the death of the crew (Hand 2020).

Another risk mentioned by scientists is emergence of legal problems in association with data security. D. J. Grim points to a wide range of invisible legal risk (Grim 2019). The lack of awareness of sensitive data makes it more difficult for entrepreneurs to comply with the provisions regulating personal data processing. In Poland, absence of adequate data protection may result in a fine of up to 4% of the total annual global turnover in the previous year.

Heidron (2008) points out that dark data results from improper data saving and storage. Enterprises often do not pay attention to the lack of metadata, which causes valuable information to be hidden. In this way, data also becomes inaccessible to the data analysts who could assign new meaning to the forgotten data. As a consequence, data is sometimes lost, and nothing is known about it.

TRUE Global Intelligence and Splunk conducted a survey at the turn of 2018 and 2019 in seven countries (USA, UK, Germany, France, China, Japan, Australia), to obtain information on enterprise data collection, management and use. The survey respondents were global business and IT managers. It turns out that 60% of the surveyed declared that half or more of the data in their organization is dark data. 77% of the respondents also agreed that the processing of such data should be prioritized in companies. This means that more and more companies become aware of the significance the growing dark data bears.

4 Propositions and Results

Each analysis begins with data. If the data is not trusted, reliability of statistical analysis results is not granted. According to the colloquial term "garbage in—garbage out", poor quality of the data entered results in unreliable conclusions. The definition of dark data propounded in the article provides a broader perspective on the quality and availability of enterprise data.

Some data is easy to classify as dark data, but there is no simple rule to do so. Classification should be made after a careful analysis of the dataset. Conversely, the unawareness of the concept's multidimensionality may result in incorrect classification of such data. For easier identification of dark data, the following set of characteristics has been formulated:

- **Unavailability**. Dark data is always unreachable and hidden. The reason for this may be, for instance, the data format that does not allow for its traditional analysis. Another source of this feature may be the unknown data storage location or the lack of knowledge about the value and the possible use of the data.

- **Costliness**. Collection and storage of dark data involves significant costs, since such data usually constitutes the largest share of all company data. Analysis of such data is another costly process that involves steps to increase its availability, whereas its processing is time-consuming.
- **Unawareness**. Dark data results from unawareness of its existence or a lack of knowledge about the value of that data for the enterprise. It is the incomprehension of the adequacy and the purpose the data is collected for that makes the data a mystery. This feature is closely related to inaccessibility. The more incomprehensible the data is, the more inaccessible it will be.
- **Uselessness**. The data collected and stored by enterprises without a specific purpose constitute untapped potential. In such form, dark data can be considered useless. Dark data can become adequate via appropriate actions. When data becomes useful, it must, at the same time, be available, known, while its costliness slowly turns into profit.

4.1 Location of Dark Data in Enterprise

Sets of enterprise data can be illustrated using Fig. 1. The image area to the left of the dashed line describes the data used and analyzed by the company, called light data. The area on the other side describes a dark dataset. Each of the above-mentioned sets may contain data of both insufficient quality (dirty data) and good quality (clear data). The share of clear and dirty data in the light and dark data usually has the opposite proportions. This is due to the fact that the data used on a daily basis must be of better quality than the data collected without a specific purpose.

The above diagram is meant to show the difference between the usability and the quality of enterprise data. The proportions of the data types are selected subjectively and may differ, depending on the individual case of a given enterprise. The share of enterprise dark data may depend, inter alia, on the specificity of the business activity, the existing knowledge of the data, the manner of information management, and even the employees' diligence in assigning metadata.

Discovery of enterprise dark data is an important but a difficult task. This endeavor can be facilitated by division of data according to its availability and usability. In this article, two diagrams, in the form of a coordinate system, have been developed, which represent such enterprise data classification. On the first of those diagrams (Fig. 2), the vertical axis contains the awareness feature, which defines the knowledge of data and the possibilities of using it, while the horizontal axis contains the "availability" feature, which determines the accessibility of enterprise data. By classifying data in this way, dark data can be assigned to the second, third and fourth quadrants of the coordinate system. Good quality

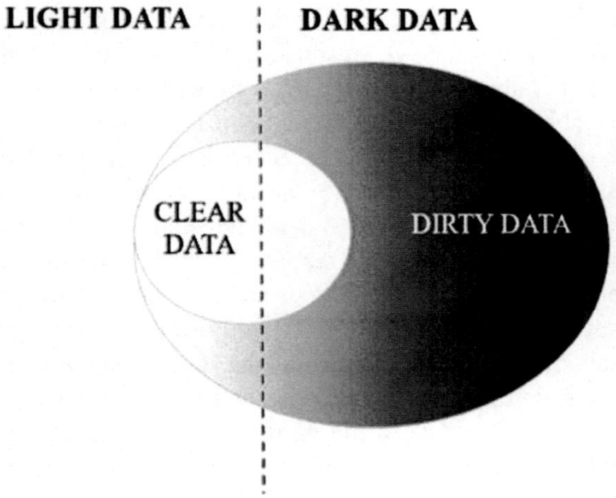

Fig. 1 Graphical representation of dark data in an enterprise. *Source* own elaboration

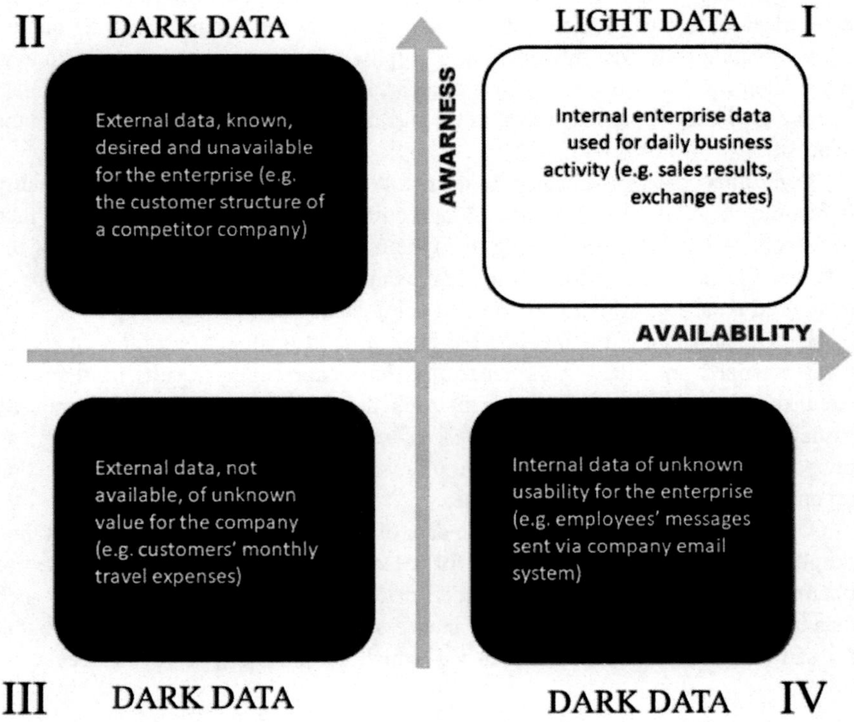

Fig. 2 Classification of enterprise data in terms of availability and quality. *Source* own elaboration

structured data collected and processed by a company as part of its core business[4] (light data) is presented in the first quadrant of the following coordinate system.

Enterprises should aim to obtain the largest possible volume of the good quality data that is available. Depending on the data held, in accordance with the classification presented, this aim can be approached in appropriate ways. Data availability should be increased with regard to the data entered in the second quadrant of the coordinate system. One of the ways is to cooperate with the public institutions and other enterprises operating on the market or to incur the costs associated with, e.g., conduction of a survey. Conversion of the data from the fourth quarter of the coordinate system can be a difficult task. The only way to increase prospects in this area is to educate both the data analysts and the managers, with regard to the data held, or to hire a company that would recognize the value of the company's information resources. The data in the third quadrant of the coordinate system is currently unavailable for enterprises. One possible solution would be to create an artificial intelligence algorithm that would detect information relevant for the enterprise.

The next diagram (Fig. 3) shows a division of internal enterprise data based on quality and availability. The vertical axis contains the "quality" of availability and quality.

"Light-clear" data is the data contained in structured databases and analyzed by enterprises on a daily basis. It does not require any processing, since its quality allows reliable statistical inference. The purpose of collecting such data is known, most often related to the company's core business. Examples of this type of data include the history of sales, orders, or the tracking of customer Internet traffic on the Web site.

'Light-dirty' data is the enterprise data related to the core business, but its quality is insufficient, i.e., a cleanup process is required for its analysis. Enterprises take into account the cost of processing this type of data, because they are aware of the real benefits associated with its use. One example of such data can be the gaps in structured databases which can be filled using statistical methods.

'Dark-clear' data is the inaccessible data of good quality, constituting a significant untapped potential for the enterprise. Its inaccessibility results from insufficient knowledge regarding the use of such data or its complicated accessibility associated with faulty file saving (lack of metadata, inadequate file naming, an unopenable file format). One example of such data may be the company meeting reports written for no specific purpose.

'Dark-dirty' data is the inaccessible data of insufficient quality. This means that, despite its possible significant value for the enterprise, costly activities related to obtaining access and then cleaning such data are required. One example of such data is employee e-mails. This data is inaccessible, due to the privacy policy, but at the same time its format is unstructured, which automatically lowers its quality.

---

[4]A more extensive explanation of good quality data can be found in (European Statistics Code of Practice 2017).

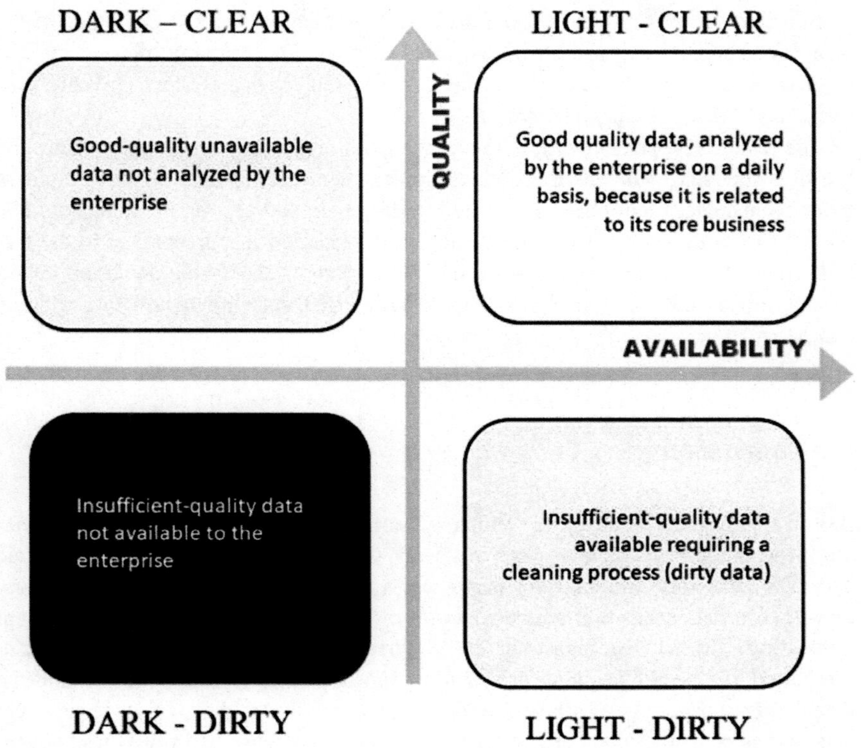

Fig. 3 Classification of internal enterprise data in terms of availability and quality. *Source* own elaboration

Table 1 Examples of InPost internal enterprise data, with regard to availability and quality

Light-clear	Light-dirty	Dark-clear	Dark-dirty
Collection of orders from customers, owing to which the company generates a specific profit	Customers' online comments regarding the services provided by InPost. This data is available to the enterprise but needs to be processed. The data is incomplete, duplicative and could be generated by bots	Continuous real-time location of your customers—it could allow reminders about the parcel when a given customer is in the vicinity of the parcel locker, which would help clear the parcel lockers	Customer feedback regarding the company posted on unknown Internet forums

Source own elaboration

In order to better reflect the data division presented, in Table 1 examples of the data developed for a specific company are presented. A company offering logistic services, i.e., parcel sending and collection via self-service lockers (InPost), was selected for detailed characterization.

The examples presented are theoretical and apply to a company from the logistics industry, whereas the data examples mentioned could also be found in other companies. Identification of possible dark data sources requires considerable knowledge of the company's operations and information resources. Due to the fact that each industry has different needs in relation to the implementation of its mission, data classification should always be carried out with regard to a specific company.

5 Conclusions

Before making any strategic decisions, an enterprise should become familiar with the information resources it possesses. Dark data is data that grows at a dynamic pace. Its discovery and analysis is not yet a popular topic among entrepreneurs. Just-in-case data collection generates costs of both the storage and the subsequent processing. Knowledge regarding the occurrence of dark data needs to be disseminated, due to the growing information gap between the data collected and the data that is of actual use. Failure to undertake activity in this regard may obscure the data that is potentially important for the enterprise, in favor of information noise.

The characteristics of dark data presented in this article (unavailability, costliness, unawareness, uselessness) are closely interrelated, while elimination of one eliminates the rest. Awareness of the possession of enterprise dark data and of the possible ways of finding and eliminating it is thus important. Emergence of dark data may be caused, inter alia, by the lack of due diligence in data collection and storage, which is a business area companies are able to supervise. Decreased data quality increases its unavailability, therefore, it is also important to employ data analysts who care for its quality. This allows reduction of the costs of data storage and the subsequent data analysis.

Enterprises holding dark data can use the tools available, such as deepdive (http://deepdive.stanford.edu/. Accessed 10 Aug 2020), to find value in the data. Another solution is to develop own software, which is time-consuming and requires appropriate expert knowledge. The main dark data characteristics distinguished should facilitate attempts to implement statistical methods for detection and analysis of dark data. Discovery of value in dark data can affect both marketing and finances as well as the business itself. Scientists should disseminate the knowledge regarding dark data as well as propose solutions for detecting and analyzing dark datasets. Dissemination of such solutions would most likely increase the amount of the good quality data available. This can be beneficial not only for businesses but also for society.

References

Abiteboul S (1997) Querying semi-structured data. Lecture notes in computer science (including subseries lecture notes in artificial intelligence and lecture notes in bioinformatics), vol 1186, pp 1–18. https://doi.org/10.1007/3-540-62222-5_33

Banafa A (2015) Understanding dark data. https://www.bbvaopenmind.com/en/technology/digital-world/understanding-dark-data/. Accessed 18 Oct 2020

Chamberlin D, Boyce R (1974) Sequel: a structured english query language. Indo-US nuclear deal: seeking synergy in bilateralism, pp 209–224. https://doi.org/10.4324/9781315816166-20

Codd EF (1970) A relational model of data for large shared data banks. Commun ACM 26(1):64–69. https://doi.org/10.1145/357980.358007

DeepDive. http://deepdive.stanford.edu/. Accessed 20 Feb 2020

Eberendu AC (2016) Unstructured data: an overview of the data of big data. Int J Comput Trends Technol 38(1):46–50. https://doi.org/10.14445/22312803/ijctt-v38p109

European Statistics Code of Practice (2017) https://ec.europa.eu/eurostat/web/products-catalogues/-/KS-02-18-142. Accessed 14 Sep 2020

Grim DJ (2019) The dark data quandar. Am Univ Law Rev 68(76):761–822

Hand DJ (2020) Dark data: why what you don't know matters. Princeton University Press, Princeton

Heidorn BP (2008) Shedding light on the dark data in the long tail of science. Libr Trends 57 (5):280–299

IDC report (2018) https://www.seagate.com/files/www-content/our-story/trends/files/idc-seagate-dataage-whitepaper.pdf. Accessed 10 Sept 2020

Kordos J (1988) Jakość danych statystycznych. PWE, Warszawa

Lugmayr A et al (2017) Cognitive big data: survey and review on big data research and its implications. What is really "new" in big data? J Knowl Manage 21(1):197–212. https://doi.org/10.1108/jkm-07-2016-0307

Maślankowski J (2015) Analiza jakości danych pozyskiwanych ze stron internetowych z wykorzystaniem rozwiązań Big Data. Roczniki Kolegium Analiz Ekonomicznych 38:167–177

Migdał-Najman K, Najman K (2018) Dirty data—profiling, cleansing and prevention. Prace Naukowe Uniwersytetu Ekonomicznego We Wrocławiu. https://doi.org/10.15611/pn.2018.508.15

Taleb I et al (2016) Big data quality: a quality dimensions evaluation. Intl IEEE. https://doi.org/10.1109/UIC-ATC-ScalCom-CBDCom-IoP-SmartWorld.2016.145

Trajanov D et al (2018) Dark data in internet of things (IoT): challenges and opportunities. In: Proceedings of the 7th small systems simulation symposium, February, pp 1–8

Trivedi B (2017) Research on dark data analysis to reduce data complexity in big data. Int Educ Res J 3(5):361–362

Wang RY (1996) Beyond accuracy: what data quality means to data consumers. J Manage Inf Syst 12(4):5–34. https://doi.org/10.1080/07421222.1996.11518099

We Are Social, Hootsuite (2020) Global digital report 2020. https://wearesocial.com/digital-2020. Accessed 3 Jan 2020

Zhang C, Shin J, Ré C, Michael Cafarella, FN (2016) Extracting databases from dark data with DeepDive. In: SIGMOD '16 Proceedings of the 2016 International Conference on Management of Data, pp 847–859. https://doi.org/10.1145/2882903.2904442

Zhu Y, Cai L (2015) The challenges of data quality and data quality assessment in the big data era. Data Sci J 14(2):1–10. https://doi.org/10.5334/dsj-2015-002

The Significance of Medical Science Issues in Research Papers Published in the Field of Economics

Urszula Cieraszewska, Monika Hamerska, Paweł Lula, and Marcela Zembura

Abstract The analysis of issues related to medical science in research papers in the field of economics is the main goal of the publication. The publication presents the results of research including the analysis of abstracts of scientific articles in the field of economics, prepared in English by authors from 36 European countries and registered in the Scopus database in the years 2011–2020. The ontology-based approach will be used for identification of concepts related to medical science and economics. The Journal of Economic Literature (JEL) system will be used to describe the domain knowledge in the field of economics. The MeSH ontology developed by the national library of medicine will be used to describe knowledge in the field of medical science. The description of relationships between medical science and economics issues will be based on bipartite graph model. The paper will also present the results of research on the relationship between the interdisciplinary nature of research in the field of economics and the number and degree of internationalization of authors' teams. All analysis will be performed with the use of programs prepared by authors in R language.

Keywords Interdisciplinarity · Medicine science · Economics · Scientific papers · JEL · MeSH · Bipartite graph

---

U. Cieraszewska · M. Hamerska · P. Lula (✉)
Cracow University of Economics, Cracow, Poland
e-mail: pawel.lula@uek.krakow.pl

U. Cieraszewska
e-mail: cieraszu@uek.krakow.pl

M. Zembura
Medical University of Silesia, Katowice, Poland

K. Jajuga et al. (eds.), *Data Analysis and Classification*, Studies in Classification, Data Analysis, and Knowledge Organization,
https://doi.org/10.1007/978-3-030-75190-6_9

1 Introduction

The importance of interdisciplinary research has grown in the last decade. This is due, among other things, to the increasingly complex nature of the research, which requires the combination of knowledge, skills and resources of scientists from various disciplines. The growing importance of interdisciplinary research also has a direct impact on the development of scientific cooperation and internationalization of research teams.

In the literature on the subject, we can find following definitions of interdisciplinary research, which are presented in Table 1.

Taking into account the presented definitions and on the basis of the literature review, it is possible to point out some feature's characteristic of interdisciplinary research (Aboelela et al. 2007):

1. There are two or more distinct academic fields.
2. It describes or defines in language of at least two fields, using multiple models or intersecting models.
3. It is drawn from more than one, with multiple data sources and varying analysis of same data.
4. The results of the research are share publications, in a language understandable to all involved areas.

The transition towards interdisciplinary collaboration is manifold. It emanates from within the scientific community as research progresses and new scientific problems emerge that are not confined to a single disciplinary perspective (Pedersen 2016). Interdisciplinary knowledge strengthens connections between disciplines, and in that process, it weakens the division of labour in disciplines, exposes gaps

Table 1 Definitions of interdisciplinary research

Definitions	Source
Direct or indirect use of knowledge, methods, techniques, devices (or other products) as a result of scientific and technological activities in other fields	Tijssen (1992)
A mode of research by teams or individuals that integrates from two or more bodies of specialized knowledge or research practice: • Perspectives, concepts, theories and/or • Tools, techniques and/or • Information, data	Porter et al. (2006)
Any study or group of studies undertaken by scholars from two or more distinct scientific disciplines. The research is based upon a conceptual model that links or integrates theoretical frameworks from those disciplines, uses study design and methodology that is not limited to any one field and requires the use of perspectives and skills of the involved disciplines throughout multiple phases of the research process	Aboelela et al. (2007)
Analyses, synthesizes and harmonizes links between disciplines into a coordinated and coherent whole	Choi and Pak (2006)

and creates new field of focus for knowledge inquiry (Klein 2002) Rpt. in (Chettiparamb 2007).

Interdisciplinary research leads to the integration of knowledge from various scientific disciplines and, consequently, to the creation of new knowledge. In the knowledge-based economy, the integration of research from various disciplines is a potential source of competitive advantage and innovation. The growing interest in the integration of knowledge and, consequently, the succession of interdisciplinary research has led to the creation of programs aimed at financial support related to the development of science and technology on a global scale.

Summing up the essence of interdisciplinary research, its essential features can be indicated:

1. Higher interdisciplinarity is then often assumed to be associated with higher research impact (Okamura 2019).
2. Interdisciplinary research has a positive influence on knowledge production and innovation (Rijnsoever and Hessels 2011).
3. Interdisciplinary research is essential for the development of scientific cooperation and the internationalization of research teams.
4. Interdisciplinary research allows for solving complex research problems.
5. Interdisciplinary research brings together the knowledge, skills and resources of researchers from different disciplines.

2 Interaction of Economic and Medical Sciences

In 2017, spending on health care in the European Union stood at 9.6% of gross domestic product. Among the EU member states, seven had spending on health at 10% or more of GDP, with France (11.5%) and Germany (11.3%) having the highest shares of GDP spent on health. At the other end of the scale, the share of health spending in GDP was lowest in Romania (5.2%), Luxembourg (6.1%), Latvia and Lithuania (both at 6.3%) (OECD 2017).

Economics offers unique insight into management of healthcare system, hospital funding, drug treatment programs and medical research. Research conducted in the field of medical economics can lead to development of improved treatment protocols, reduce costs and improve effectiveness of the treatment. Medical economics research is used in various fields of medicine, e.g. oncology, hematology, bariatrics, rheumatology and infectious disease.

Health economics as well as pharmacoeconomic are playing an increasingly important role in clinical development and market access decisions of new innovative medicines (Kumar and Baldi 2013). There are journals devoted entirely to this topic. *Journal of Medical Economics* specializes in the publication of studies that determine the effectiveness of medical treatment, involving measurements of therapeutic and/or preventative outcomes. *Journal of Health Economics* seeks articles related to the economics of health and medical care.

The aim of health economics is to identify the interventions that produce the best health output with the available resources. There are five commonly used forms of economic evaluation of medical procedures: cost-benefit analysis (CBA), cost-effectiveness analysis (CEA), cost-utility analysis (CUA), cost-minimisation analysis (CMA) and decision analysis (DA) (Brockhuis et al. 2002).

The term CBA is used to refer to analysis used in decision-making that compares the expected costs and benefits (both in monetary terms) of an investment. The cost-benefit analysis (CBA) was used in the study of HPV vaccination conducted in UK. Vaccine cost was defined as the maximum vaccine cost per person (including the administration cost) at which HPV vaccination has a benefit-to-cost ratio above one (i.e. the vaccination programme is cost-beneficial). The direct benefits of vaccination included all medical cost avoided due to reduced screening for and treatment of cervical cancer and pre-cancerous lesions (Park et al. 2018).

The study conducted in Egypt evaluated the cost-effectiveness of 6-month versus 1-year trastuzumab treatments in HER positive breast cancer from payer perspective over a 10-year time horizon. Direct medical costs including cost of treatments, day-care, surgery, health states and follow-up visits were collected. Findings of this study provided costs reduction with the improvement of the patient's outcomes, results confirmed the dominance of 6-month trastuzumab treatment (Elsisi 2020).

Cost-utility analysis compares results of medical interventions mentioned in utility units. Usually, it expresses the number of additional years of life produced by medical intervention, with special regard to the quality of these years (Brockhuis et al. 2002). According to the analysis of cataract surgery performed in the US for the year 2018, cataract surgery in both the first eye and second eye when analysed by cost-utility analysis is highly cost-effective. First-eye cataract surgery resulted in a 2.523 quality-adjusted life year (QALY) gain, a 33.3% patient value gain and 25.5% quality-of-life gain (Brown 2019).

If the health effects of two alternative interventions are known to be equal and only the costs need to be analysed, we use cost-minimisation analysis (CMA) (Brockhuis et al. 2002). Robot-assisted hysterectomy was evaluated using cost-minimisation analysis. The aim of this study was to compere robot-assisted hysterectomy with a combination of traditional open and conventional laparoscopic surgery in the publicly funded healthcare system in Ireland. The outcomes of this study showed that robot-assisted hysterectomy is more costly than the current mix of open and traditional laparoscopic surgery (Teljeur 2014).

Decision analysis is applied when only the health effects of medical intervention are important. Alternative intervention has the same cost, or costs are not important in a particular decision situation (Kernick 2003). Decision analysis was used in this study to assess the potential utility gains/losses and costs of adding bilateral inferior turbinoplasty to tonsillectomy/adenoidectomy (T/A) for the treatment of obstructive sleep-disordered breathing (oSDB) in children (Baik and Brietzke 2019).

Pharmacoeconomics is a branch of health economics. Pharmacoeconomics identifies measures and compares the cost and consequences of pharmaceutical products and services and describe the economic relationship involving drug research, drug production, distribution, storage, pricing and used by the people

(Rawlins and Culyer 2004). It has been suggested that innovation and advances in new and high-cost pharmaceuticals, appearance of new diseases, and enhancing patients' expectations are of the main causes of a sharp increase in pharmaceutical expenditures worldwide. Usage of pharmacoeconomic analysis leads to recognition which treatment may be the most efficient strategy for treating patients. Using the efficient treatment strategies could, in turn, increase the efficiency of the pharmaceutical services (Davari 2012). In the USA, the Food and Drug Administration is considering requiring studies in pharmacoeconomic in addition to the standard studies of the safety and efficacy of drugs (Arbuckle 2002).

Lack of health can be a cause of social exclusion, loss of work ability, whether short term or long term (Tomeš 2011). Therefore, the health policy can be considered as a specific area of social policy (Pekarová 2017). Health care of citizens is not only medical care, since it consists also from other, predominantly societal factors. In this manner, societal aspects of health care can be described as the socio-economic aspect, socio-educational aspect and socio-environmental aspect (Jusko 2002).

Health system performance assessment is a widely recognised tool, which supports the decision-making process in the health system and monitors the progress in achieving its goals. Various international and national organizations and institutions offer more or less comprehensive frameworks for health system performance assessment (HSPA) (Rohova et al. 2017). At EU-level, the European Commission has created the European Core Health Indicators Initiative, which assembles 88 indicators relevant to health system performance assessment (European Core Health Indicators, https://ec.europa.eu/health/indicators_data/echi_en).

The ongoing COVID-19 pandemic is a health and economic crisis. Virus has heavily hit the global economic activity, and the world's economy started falling even before nationwide lockdowns were implemented (Romei and Burn-Murdoch 2020). The COVID-19 pandemic has led to high rates of unemployment across advanced economies (Holzer 2020). To stop the ongoing epidemic, it is urgent for governments to increase funding for research concerning the development of vaccine.

Apart from the COVID-19 crisis, the late Twentieth and early Twenty First centuries have seen the rapid transmission and difficulty of containing various diseases such as HIV/AIDS, malaria, tuberculosis, dengue fever, Ebola, H1N1 influenza and Zika. In general, the economic effects of epidemics depend on heterogeneities in three dimensions: (i) disease-specific heterogeneities in terms of mortality, morbidity, infectiousness, and prospects for recovery, (ii) Heterogeneities at the population level in disease susceptibility, such as the population share of older adults and (iii) cross-country heterogeneities are decisive in the sense that richer countries might be able to sustain lockdown measures and social distancing for a longer time period than poorer countries (Bloom et al. 2020).

3 Description of Classifications

The pursuit of ever greater specialization and at the same time internal coherence is the basis for the development of science. This development gained particular momentum in the XIX century, leading to the identification of a large number of detailed disciplines on the one hand, and on the other hand, showed the bonds occurring in various planes between disciplines previously separated as independent. This was due to the discovery of common aspects for various phenomena, the dissemination of mathematical methods used in various fields of science (natural, social). The manifestation of this process is the increase in interdisciplinary links.

Currently, an interdisciplinary approach to the study of business processes can be seen. Economics is most often treated as social science, although there is no shortage of researchers who make extensive use of advanced mathematics—science.

With such connections, the question arises of how to build classification schemes of research disciplines and fields, to reflect the nature of modern science, and how to reconcile the hierarchical nature of classification with links in science, which are often bilateral. This article will discuss two classifications Journal of Economic Literature—JEL (Journal of Economic Literature, https://www.aeaweb.org/econlit/jelCodes.php?view=jel) and Medical Subject Headings (MeSH) Medical Subject Headings (MeSH) (https://www.nlm.nih.gov/mesh/meshhome.html).

First classification the JEL is system originated with the Journal of Economic Literature and is a standard method of classifying scholarly literature in the field of economics. It is used in many of the AEA's published research materials. Second classification the MeSH is a hierarchically organized terminology for indexing and cataloguing of biomedical information. It serves as a thesaurus that facilitates searching. Created and updated by the United States National Library of Medicine (NLM).

JEL classification has 20 primary JEL categories, and each JEL primary category has secondary and tertiary subcategories. Below is an excerpt from the JEL classification (© American Economic Association; reproduced with permission of the Journal of Economic Literature, https://www.aeaweb.org/econlit/jel-Codes.php?view=jel):

A General Economics and Teaching
B History of Economic Thought, Methodology, and Heterodox Approaches
C Mathematical and Quantitative Methods
D Microeconomics
E Macroeconomics and Monetary Economics
F International Economics
G Financial Economics
H Public Economics
I Health, Education, and Welfare
J Labor and Demographic Economics
K Law and Economics
L Industrial Organization
M Business Administration and Business Economics • Marketing • Accounting • Personnel Economics
- M00 General
- M1 Business Administration
  - M10 General
  - M11 Production Management
  - M12 Personnel Management • Executives; Executive Compensation
  - M13 New Firms • Startups
  - M14 Corporate Culture • Diversity • Social Responsibility
  - M15 IT Management
  - M16 International Business Administration
  - M19 Other
- M2 Business Economics
  - M20 General
  - M21 Business Economics
  - M29 Other
- M3 Marketing and Advertising
  - M30 General
  - M31 Marketing
  - M37 Advertising
  - M38 Government Policy and Regulation
  - M39 Other

N Economic History
O Economic Development, Innovation, Technological Change, and Growth
P Economic Systems
Q Agricultural and Natural Resource Economics • Environmental and Ecological Economics
R Urban, Rural, Regional, Real Estate, and Transportation Economics
Y Miscellaneous Categories
Z Other Special Topics

Structure of MeSH is strictly synonymous with each other and is grouped in a category called a “Concept”. Each MeSH record consists of one or more concepts, and each concept consists in one or more synonymous terms. For example:

- Anatomy [A]
- Organisms [B]
- Diseases [C]
- Chemicals and Drugs [D]
- Analytical, Diagnostic and Therapeutic Techniques, and Equipment [E]
- Psychiatry and Psychology [F]
    - Behavior and Behavior Mechanisms [F01]
        - Adaptation, Psychological [F01.058]
        - Attitude [F01.100]
        - Behavior [F01.145]
        - Child Rearing [F01.318]
        - Defense Mechanisms [F01.393]
        - Emotions [F01.470]
        - Human Characteristics [F01.510]
        - Human Development [F01.525]
        - Mental Competency [F01.590]
        - Motivation [F01.658]
        - Neurobehavioral Manifestations [F01.700]
        - Personality [F01.752]
        - Psychology, Social [F01.829]
        - Psychosocial Functioning [F01.872]
        - Temperance [F01.914]
    - Psychological Phenomena [F02]
    - Mental Disorders [F03]
    - Behavioral Disciplines and Activities [F04]
- Phenomena and Processes [G]
- Disciplines and Occupations [H]
- Anthropology, Education, Sociology, and Social Phenomena [I]
- Technology, Industry, and Agriculture [J]
- Humanities [K]
- Information Science [L]
- Named Groups [M]
- Health Care [N]
- Publication Characteristics [V]
- Geographicals [Z]

4 Research Methodology

4.1 Research Scope and Goals

The research presented here is focused on the analysis of interdisciplinarity in the field of economics, in particular to the analysis of the participation of medical science-related issues in scientific papers on economics.

The authors defined the following detailed objectives:

1. Development of research methodology and tools allowing for the analysis of the interdisciplinarity of research papers in the field of economics based on the automatic analysis of abstracts of scientific publications.
2. The analysis of relationships between subareas defined within economics and medical science areas in the light of content analysis of abstracts.

The research was planned according to the following stages:

1. Preparation of a set of abstracts of scientific articles.
2. Identification of concepts in the field of economics.
3. Identification of concepts in the field of medicine.
4. Modelling and analysis of relationships between concepts.

The total number of papers taken into account in the analysis was 124,460. The distribution of papers over selected countries in presented in Table 2.

Table 2 Distribution of research papers in the area of economics over European countries in the period 2011–2020

Country	N	Country	N	Country	N	Country	N
Albania	602	Finland	2841	Lithuania	1351	Romania	2344
Austria	3405	France	14,448	Luxembourg	776	Serbia	925
Belgium	5141	Germany	18,837	Malta	146	Slovakia	605
Bulgaria	358	Greece	3420	Montenegro	164	Slovenia	858
Croatia	1150	Hungary	400	Netherlands	9591	Spain	13,435
Cyprus	849	Iceland	225	North Macedonia	161	Sweden	5454
Czech republic	3948	Ireland	2369	Norway	3609	Switzerland	6393
Denmark	3660	Italy	11,724	Poland	4069	Turkey	4643
Estonia	343	Latvia	267	Portugal	3247	United Kingdom	23,977

4.2 Identification of Topics Occurring in Abstracts and Related to Main Subareas of Economics and Medical Science

Identification of topics occurring in abstracts in field of economics was performed using the following algorithm:

1. For every abstract from the corpus.
2. Its content is spit into phrases with the use of the sliding window technique.
3. For every phrase, a contribution of every concept from the JEL ontology is calculated.
4. Coefficients described above form for every abstract a phrase-concept matrix:

$$\mathbf{K} = \begin{array}{c} \\ ph_1 \\ \cdots \\ ph_P \end{array} \begin{array}{c} \begin{array}{ccc} c_1 & \cdots & c_M \end{array} \\ \begin{bmatrix} k_{11} & \cdots & k_{1M} \\ \cdots & \cdots & \cdots \\ k_{P1} & \cdots & k_{PM} \end{bmatrix} \end{array}$$

   maximum value calculated for every column of the **K** matrix determines the contribution of the c_j concept in a given abstract
1. Finally, for the whole set of abstracts, a matrix E is defined:

$$\mathbf{E} = \begin{array}{c} \\ a_1 \\ \cdots \\ a_N \end{array} \begin{array}{c} \begin{array}{ccc} c_1 & \cdots & c_M \end{array} \\ \begin{bmatrix} e_{11} & \cdots & e_{1M} \\ \cdots & \cdots & \cdots \\ e_{N1} & \cdots & e_{NM} \end{bmatrix} \end{array}$$

   where e_{ij} element informs about the contribution of the c_j concept in the a_i abstract.

Main steps realized by the proposed algorithm are presented in Fig. 1. The system was implemented by the authors in the R language.

Next, the same set of abstracts was analysed in terms of the occurrence of concepts related to medical area. This step was based on the Medical Subject Headings (MeSH) ontology with pyMeSHSim package which can work as a wrapper for the MetaMap system which is delivered by the national library of medicine (https://www.nlm.nih.gov/) as a main tool for recognition medical concepts in texts, whereas the pyMeSHSim (http://pymeshsim.systemsgenetics.cn/index.html) is a Python package which facilitates the process of text file processing with the use MetaMap system.

As a final result of the analysis of abstracts in terms of medical concepts, the matrix **M** was formed:

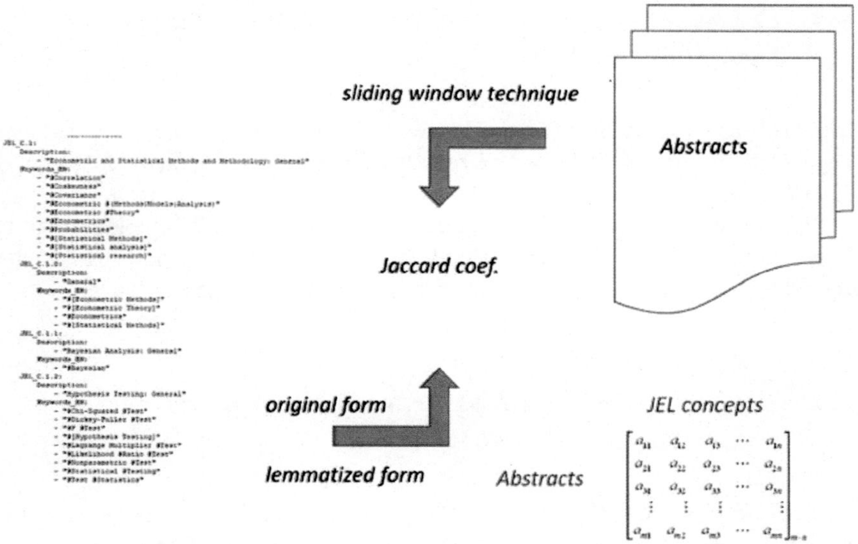

Fig. 1 Structure of the ontology-based analyser

$$
\mathbf{M} = \begin{array}{c} \\ a_1 \\ \cdots \\ a_N \end{array} \begin{array}{c} \begin{array}{ccc} h_1 & \cdots & h_R \end{array} \\ \begin{bmatrix} m_{11} & \cdots & m_{1R} \\ \cdots & \cdots & \cdots \\ m_{N1} & \cdots & m_{NR} \end{bmatrix} \end{array}
$$

where m_{ij} element informs about the contribution of the h_j concept from the MeSH ontology in the a_i abstract.

Identification of topics occurring in abstracts and related to medical science was performed by the analyser presented in Fig. 2.

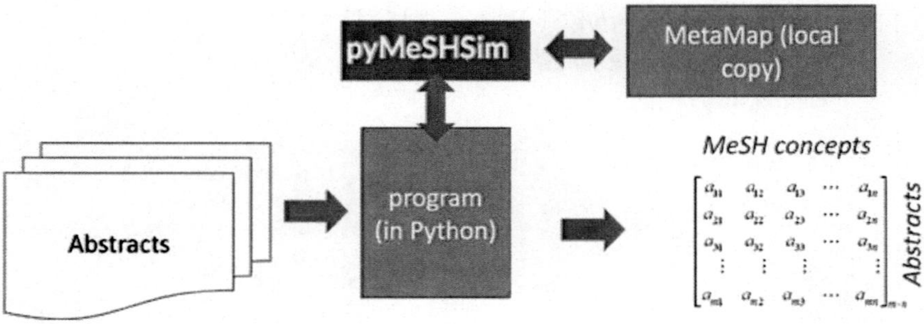

Fig. 2 System for MeSH concepts identification in abstracts

4.3 Projection of Identified Topics into Main Concepts from JEL and MeSH Ontologies

The analysis was conducted on the highest level of generality. It means that concepts identified during analysis were replaced by topics existing on the first level of classification (Fig. 3).

It is worth mentioning that generalization process in MeSH ontology may be based not on hierarchical relations among concepts. In empirical part of the research, a generalization was based on values of *semantic type* attribute.

4.4 Analysis of Relationships Between Concepts Related to Economics and Medical Science

Having two matrices **E** and **M** describing the occurrence of concepts related to economics and medicine, the matrix of co-occurrence **G** was defined:

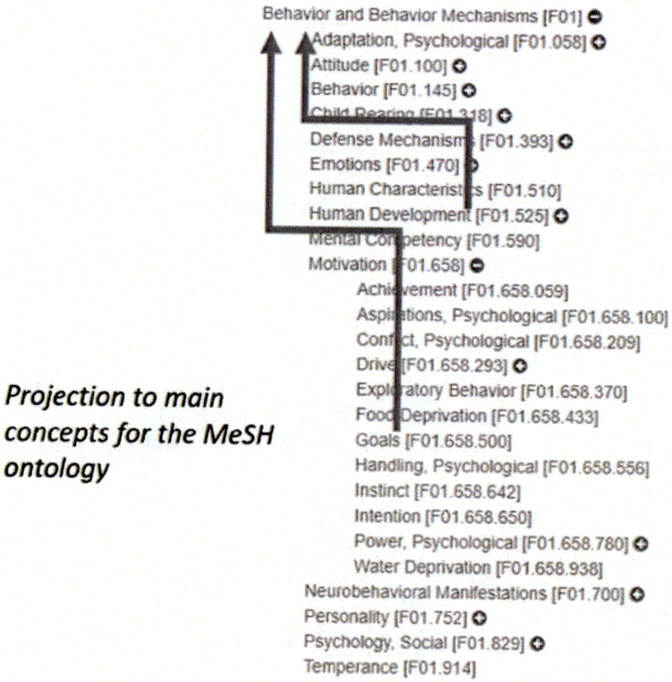

Fig. 3 Projection of concepts identified during the analysis on the first level concepts

$$\mathbf{G} = \begin{array}{c} \\ c_1 \\ \cdots \\ c_M \end{array} \begin{array}{c} \begin{array}{ccc} h_1 & \cdots & h_R \end{array} \\ \begin{bmatrix} g_{11} & \cdots & g_{1R} \\ \cdots & \cdots & \cdots \\ g_{M1} & \cdots & g_{MR} \end{bmatrix} \end{array}$$

where the element g_{ij} indicates how many times the concept c_i related to economics and concept h_j related to medicine appeared in the same abstract. The **G** matrix can be treated as a definition of the bipartite graph model of relationships between concepts related to these disciplines.

The **G** model together with matrices **E** and **M** allows to express:

(a) The significance of concepts belonging to scientific areas:

- the significance of concepts belonging to economics (defined as a number of occurrences of the concept c_j):

$$I(c_j) = \sum\nolimits_{i=1}^{N} e_{ij} \tag{1}$$

- the significance of concepts belonging to medicine (expressed as a number of occurrences of the concept h_j):

$$I(h_j) = \sum\nolimits_{i=1}^{N} m_{ij} \tag{2}$$

- the strength of concepts belonging to economics:

$$(c_i) = \sum\nolimits_{j=1}^{R} \frac{g_{ij}}{\sum\nolimits_{m=1}^{M} g_{mj}} \tag{3}$$

- the strength of concepts belonging to medicine:

$$S(h_j) = \sum\nolimits_{i=1}^{M} \frac{g_{ij}}{\sum\nolimits_{r=1}^{R} g_{ir}} \tag{4}$$

(b) the character of relationships between concepts belonging to two different scientific areas:

- the degree value calculated for every node representing scientific area. This measure informs about the number of partners from the second scientific area. The degree value for concepts taken into account in the research is formulated as:

$$D(c_i) = \sum\nolimits_{j=1}^{R} \operatorname{sgn}(g_{ij}) \tag{5}$$

and:

$$D(h_j) = \sum_{i=1}^{M} \operatorname{sgn}(g_{ij}) \tag{6}$$

where sgn(.) is a signum function;

- the normalized degree defined as a degree value divided by the number of possible partners:

$$ND(c_i) = \frac{D(c_i)}{R} \tag{7}$$

and:

$$ND(h_j) = \frac{D(h_j)}{M} \tag{8}$$

- specificity index which informs about the diversity of interactions between a given concept and concepts from another discipline. Low diversity can be interpreted as low specificity, whereas high diversity indicates high specificity. According to Poisot et al. (2012), the specificity index is defined as:

$$SP(c_i) = \frac{\sqrt{\sum_{j=1}^{R} (g_{ij} - \mu_{c_i})^2}}{\mu_{c_i}\sqrt{R}\sqrt{R-1}} \tag{9}$$

whereas the specificity index for concepts from medical science is formulated as:

$$SP(h_j) = \frac{\sqrt{\sum_{i=1}^{M} \left(g_{ij} - \mu_{h_j}\right)^2}}{\mu_{h_j}\sqrt{M}\sqrt{M-1}} \tag{10}$$

- the specialization index H_2' can be treated as an aggregated measure of specificity. For calculating its value, first all elements of **G** matrix should be transformed into probabilities:

$$p_{ij} = \frac{g_{ij}}{\sum_{k=1}^{M} \sum_{l=1}^{R} g_{kl}} \tag{11}$$

and then, the two dimensional Shannon entropy is calculated:

$$H_2 = -\sum_{i=1}^{M} \sum_{j=1}^{R} (p_{ij} \ln p_{ij}) \tag{12}$$

Finally, the specialization index H_2' is calculated as a normalized version of the H_2 index:

$$H_2' = \frac{H_2^{max} - H_2}{H_2^{max} - H_2^{min}} \tag{13}$$

where H_2^{max} and H_2^{min} are, respectively, the maximum and minimum value of H_2 index calculated for the matrix having M rows and R columns.

The H_2' index always belongs to [0; 1] range and is close to 0 for low specialization and close to 1 for high specialization.

5 The Analysis of the Contribution of Medical Science Issues in Research Papers Published in the Field of Economics

The analysis was performed with the use of abstracts of research papers published in the field of economics, prepared by authors from 36 European countries and registered in the Scopus database from 2011 to 2020 year. For further steps of the analysis from the whole corpus of abstracts, the 10,000 abstracts were chosen randomly.

For the description of the area of economics, the JEL ontology was used. The medical science area was defined by the MeSH ontology.

The number of occurrences of concepts representing all main subareas of the JEL ontology is presented in Fig. 4.

For the same set of abstracts, the identification of concepts belonging to the semantic type *mental process* from the MeSH ontology was performed. The choice of this semantic type was motivated by its importance for the area of economics and management. Within *mental process* semantic type, the following list of subtypes can be identified:

- *menp_1*—Behaviour and behaviour mechanisms.
- *menp_2*—Behavioural disciplines and activities.
- *menp_3*—Characteristics, population.
- *menp_4*—Diagnoses.
- *menp_5*—Humanities.
- *menp_6*—Information science.
- *menp_7*—Investigate technique,
- *menp_8*—Musculoskeletal and neural physiological phenomena.
- *menp_9*—Psychological phenomena.
- *menp_10*—Therapeutics.

The number of occurrences of concepts related to every semantic subtype is presented in Fig. 5.

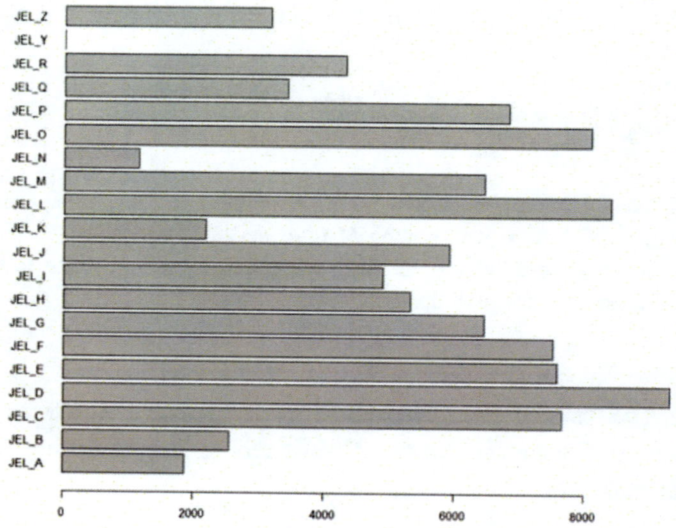

Fig. 4 Number of occurrences of the concepts in the JEL classes

The bipartite graph showing relationships between main subareas of *economics* discipline and subtypes defined within *mental process* semantic type is presented in Fig. 6.

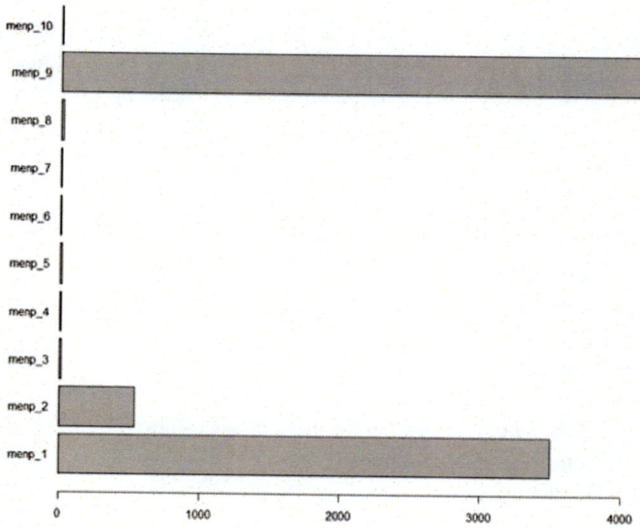

Fig. 5 Number of occurrences of concepts related to subtypes of semantic-type mental process defined within the MeSH ontology

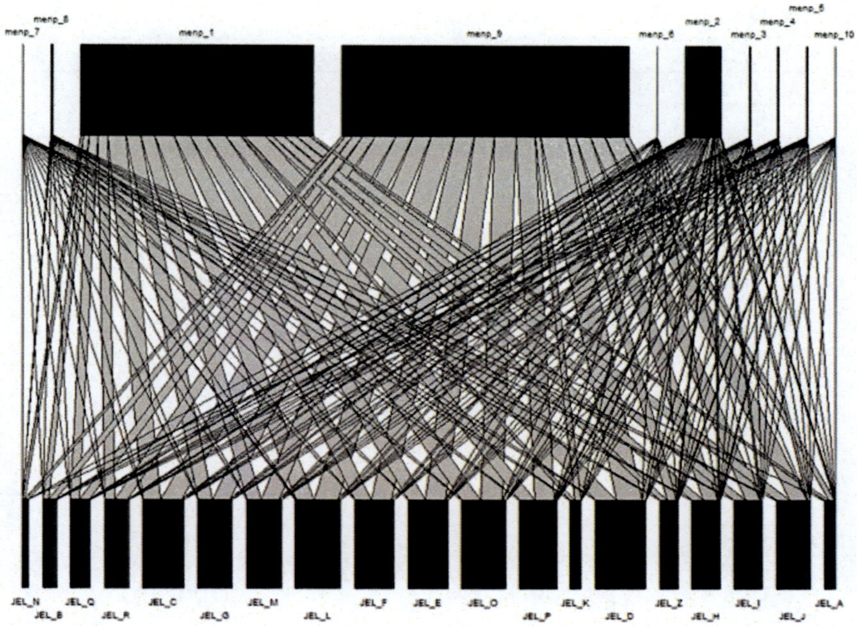

Fig. 6 Bipartite graph showing relationships between main subareas from the JEL and concepts belonging to subtypes of *mental process* semantic type defined within the MeSH ontology

Subarea's strength is represented by a width of a block representing a given group of concepts. And the width of an edge indicates the strength of a connection.

Specificity indexes for subareas are presented in Table 3 for MeSH ontology and Table 4 for JEL ontology.

Specificity indexes calculated for concepts from economics and medical science areas indicate that the diversity of connection strength from medical area concepts is lower than the diversity of connection strength of concepts within economic area.

Table 3 Specificity indexes for MeSH subareas

	Degree	Normalised degree	Species strength	Resource range	Species specificity index
menp_1	19	1.00000	7.86136	0.00000	0.10232
menp_2	19	1.00000	1.21369	0.00000	0.10484
menp_3	18	0.94737	0.03464	0.05556	0.12624
menp_4	17	0.89474	0.01515	0.11111	0.14051
menp_5	19	1.00000	0.02600	0.00000	0.10739
menp_6	14	0.73684	0.00626	0.27778	0.16378
menp_7	18	0.94737	0.01369	0.05556	0.10531
menp_8	19	1.00000	0.05057	0.00000	0.08770
menp_9	19	1.00000	9.77399	0.00000	0.09903
menp_10	13	0.68421	0.00465	0.33333	0.17799

Table 4 Specificity indexes for JEL subareas

	Degree	Normalised degree	Species strength	Species specificity index
JEL_A	8	0.8	0.311990	0.604787
JEL_B	7	0.7	0.191234	0.613387
JEL_C	9	0.9	0.650320	0.613497
JEL_D	10	1.0	0.907169	0.613196
JEL_E	10	1.0	0.663211	0.611468
JEL_F	9	0.9	0.602732	0.610509
JEL_G	9	0.9	0.501230	0.612166
JEL_H	10	1.0	0.572645	0.611853
JEL_I	10	1.0	0.540931	0.613130
JEL_J	10	1.0	0.673033	0.612355
JEL_K	9	0.9	0.243854	0.620286
JEL_L	10	1.0	0.788808	0.617848
JEL_M	10	1.0	0.650620	0.617337
JEL_N	6	0.6	0.110365	0.620876
JEL_O	10	1.0	0.795027	0.613182
JEL_P	10	1.0	0.669273	0.611042
JEL_Q	10	1.0	0.327745	0.628256
JEL_R	8	0.8	0.358607	0.618076
JEL_Z	10	1.0	0.441207	0.618017

But within these two groups of concepts, the specificity for every concept is almost the same. Calculated for the whole network the H_2' index is equal to 0.001 and indicates low specificity in the graph.

6 Conclusions

In the paper, the analysis of the contribution of topics related to medical science area to economics was presented. It seems that the fusion of ontology-based topic identification and bipartite graph models form a useful tool for this task.

In most cases, the identification of economic concepts was correct. The MetaMap system, in some cases, incorrectly analyzes texts in the field of economics (it treats non-medical terms as MeSH terms).

The research presented in the paper concerned the identification of relations between the concepts defined in the MeSH ontology, belonging to the group of "mental processes" and the concepts of economic sciences. The specificity of individual concepts was also assessed.

It seems that interesting results could be obtained by reducing the level of generality of the concepts.

It should be emphasized that the importance of interdisciplinary research has been increasing recently. This is due to the growing complexity of research problems that extend beyond a single scientific discipline. The concept of interdisciplinary is particular importance in the case of economics because economic aspects should be taken into account in analysis of many contemporary problems. Additionally, research methodology developed in many various disciplines is very often used in the area of economics. Interdisciplinarity is also helpful for increasing the possibility of publishing obtained results as publications in economics often concern problems of local communities or economies and are not distributed internationally, and the change on the research perspective allows transform their character to more global.

In authors' opinion, the approach used in this work can be used for monitoring and analysis of interdisciplinarity present in research publications. This type of analysis may be useful for identification and observation and prediction the most innovative approaches and ideas and proper allocation of financial support for research institutions and teams.

Acknowledgements The research has been carried out as part of a research initiative financed by the Ministry of Science and Higher Education within "Regional Initiative of Excellence" Programme for 2019–2022. Project no.: 021/RID/2018/19. Total financing: 11 897 131,40 PL.

References

Aboelela SW, Larson E, Bakken S, Carrasquillo O, Formicola A, Glied SA, Janet H, Gebbie MK (2007) Defining interdisciplinary research: conclusions from a critical review of the literature. HSR: Health Serv Res 42:329–346

Arbuckle RB, Adamus AT, King KM (2002) Pharmacoeconomics in oncology. Expert Rev Pharmacoecon Outcomes Res 2:251–260. https://doi.org/10.1586/14737167.2.3.251

Baik G, Brietzke SE (2019) Cost benefit and utility decision analysis of turbinoplasty with adenotonsillectomy for pediatric sleep-disordered breathing. Otolaryngol Head Neck Surg 161:343–347. https://doi.org/10.1177/0194599819841882

Bloom DE, Kuhn M, Prettner K (2020) Modern infectious diseases: macroeconomic impacts and policy responses. NBER Working Paper

Brockhuis B, Lass P, Popowski P, Scheffler J (2002) An introduction to economic analysis in medicine—the basics of methodology and chosen terms. Examples of results of evaluation in nuclear medicine. Nucl Med Rev 5:55–59

Brown GC, Brown MM, Busbee BG (2019) Cost-utility analysis of cataract surgery in the United States for the year 2018. J Cataract Refract Surg 45:927–938. https://doi.org/10.1016/j.jcrs.2019.02.006

Chettiparamb A (2007) Interdisciplinarity: a literature review. The interdisciplinary teaching and learning group, subject centre for languages, linguistics and area studies. School of Humanities, University of Southampton

Choi BC, Pak AW (2006) Multidisciplinarity, interdisciplinarity and transdisciplinarity in health research, services, education and policy: definitions, objectives, and evidence of effectiveness. Clin Investig Med 20:351–364

Davari M (2012) Pharmacoeconomics; an appropriate tool for policy makers or just a new field of research in iran? Iran J Pharm Res 11:1–2

Elsisi GH, Nada Y, Rashad N, Carapinha J, Noor AO, Almasri DM, Zaidy MA, Foad A, Khaled H (2020) Cost-effectiveness of six months versus 1-year adjuvant trastuzumab in HER2 positive early breast cancer in Egypt. Null 23:575–580. https://doi.org/10.1080/13696998.2020.1724682

European Core Health Indicators. https://ec.europa.eu/health/indicators_data/echi_en

Holzer HJ (2020) The COVID-19 crisis: how do U.S. employment and health outcomes compare to other OECD countries? https://www.brookings.edu/research/the-covid-19-crisis-how-do-u-s-economic-and-health-outcomes-compare-to-other-oecd-countries/

Journal of Economic Literature. https://www.aeaweb.org/econlit/jelCodes.php?view=jel

Jusko P (2002) Základy sociálnej politiky. Banská Bystrica

Kernick DP (2003) Introduction to health economics for the medical practitioner. Postgrad Med J 79:147. https://doi.org/10.1136/pmj.79.929.147

Klein G (2002) It takes more than a passport: interdisciplinarity in study abroad. In: Haynes C (ed) Innovations in interdisciplinary teaching. American Council on Education/Oryx Press, Washington, pp 201–220

Kumar S, Baldi A (2013) Pharmacoeconomics: principles, methods and economic evaluation of drug therapies. Pharm Tech Med 2:362–369

Medical Subject Headings (MeSH). https://www.nlm.nih.gov/mesh/meshhome.html

OECD/Eurostat/WHO (2017) A system of health accounts 2011: revised edition. OECD Publishing

Okamura K (2019) Interdisciplinarity revisited: evidence for research impact and dynamism. Palgrave Commun 5:141. https://doi.org/10.1057/s41599-019-0352-4

Park M, Jit M, Wu JT (2018) Cost-benefit analysis of vaccination: a comparative analysis of eight approaches for valuing changes to mortality and morbidity risks. BMC Med 16:139. https://doi.org/10.1186/s12916-018-1130-7

Pedersen DB (2016) Integrating social sciences and humanities in interdisciplinary research. Palgrave Commun 2:16036. https://doi.org/10.1057/palcomms.2016.36

Pekarová D (2017) Health Policy as a specific area of social policy. Challenges Future 2:102–120

Poisot T, Canard E, Mouillot D, Mouquet N, Hochberg ME (2012) A comparative study of ecological specialization estimators. Methods Ecol Evol 3:537–544. https://doi.org/10.1111/j.2041-210X.2011.00174.x

Porter AL, Roessner JD, Cohen AS, Perreault M (2006) Interdisciplinary research: meaning, metrics and nurture. Interdisc Res Meaning Metrics Nurture 15:187–195

Rawlins MD, Culyer AJ (2004) National institute for clinical excellence and its value judgments. BMJ 329:224–227. https://doi.org/10.1136/bmj.329.7459.224

Rijnsoever FJ, Hessels LK (2011) Factors associated with disciplinary and interdisciplinary research collaboration. Res Policy 40(3):463–472

Rohova M, Atanasova E, Dimova A, Koeva L, Koeva S (2017) Health system performance assessment—an essential tool for health system improvement. J IMAB 23:1778–1783. https://doi.org/10.5272/jimab.2017234.1778

Romei V, Burn-Murdoch J (2020) Real-time data show virus hit to global economic activity. https://www.ft.com/content/d184fa0a-6904-11ea-800d-da70cff6e4d3

Teljeur C, O'Neill M, Moran PS, Harrington P, Flattery M, Murphy L, Ryan M (2014) Economic evaluation of robot-assisted hysterectomy: a cost-minimisation analysis. BJOG 121:1546–1553. https://doi.org/10.1111/1471-0528.12836

Tijssen RJW (1992) A quantitative assessment of interdisciplinary structures in science and technology: co-classification analysis of energy research. Res Policy 21:27–44

Tomeš I (2011) Obory sociální politiky. Portál, Praha

Application of Duration Analysis Methods in the Study of the Exit of a Real Estate Sale Offer from the Offer Database System

Ewa Putek-Szeląg and Anna Gdakowicz

Abstract The aim of the article is to present selected methods of duration analysis to assess the probability of exitc from the real estate sale offer system, taking into account various types of competing risk (the year of submitting the property for sale). The study used the cumulative frequency function and the complement to unity of the Kaplan-Meier estimator. Using estimators, the authors compared the probability of withdrawing the real estate sale offer from the offer database due to: the sale of real estate and suspension or withdrawal of the offer. The analysis was carried out on the basis of data obtained from the West Pomeranian Association of Real Estate Brokers in Szczecin regarding the sale of residential real estate on the underdeveloped market—Szczecin. The survey is innovative because the calculation of the offer duration takes into account the properties that have been sold and are still current (on the day of the end of the survey). The probability of selling a residential property decreased significantly after 180 days in the MLS system. Apartments registered in 2018 and then in 2019 were the fastest to be sold. Due to a large number of censored observations, it was not possible to determine the survival function quartiles for 2017 and 2020.

Keywords Residential real estate market · Duration analysis · Time on market

1 Introduction

When deciding to sell a property, the owner usually has two requirements: to do it quickly and get the highest possible price. However, these two elements are often beyond the seller's control (Lin and Liu 2008). In markets with a well-developed sales network (the so-called warm markets), real estate is sold in a short period of

---

E. Putek-Szeląg · A. Gdakowicz (✉)
University of Szczecin, Szczecin, Poland
e-mail: anna.gdakowicz@usz.edu.pl

E. Putek-Szeląg
e-mail: ewa.putek-szelag@usz.edu.pl

K. Jajuga et al. (eds.), *Data Analysis and Classification*, Studies in Classification, Data Analysis, and Knowledge Organization,
https://doi.org/10.1007/978-3-030-75190-6_10

time and reaches high prices, while in less developed markets (cold ones), prices tend to fall and the sales time is extended (Krainer 2001).

After deciding to sell the property, choosing the method of selling it (selling it yourself or using the services of an estate agent), the following questions arise: how long does it take to sell the property? How long will the owner wait for the transaction to be finalised? Real estate market specialists make the sale dates dependant on many factors, such as the type of property, price, location, functional layout of the property, neighbourhood, promotion method, as well as the economic situation of the region and the country and many other economic and social factors of the real estate market environment (Miller 1978; Lippman and McCall 1986; Yavas and Yang 1995; Ong and Koh 2000; Knight 2002; Leung et al. 2002; Anglin et al. 2003; Haurin et al. 2013).

Different types of property are sold at different times. Polish real estate agents indicate that the average time of sale of a flat is about 3–6 months, a house—6 to 9 months, a building plot—6 to 12 months (Nowak 2019). An analysis carried out in 2018 for selected Polish cities showed that the fastest way to sell a flat was in Lublin—about 1.5 months, while the longest time to wait for a buyer was in Poznań—almost six months (Fig. 1).

The National Bank of Poland, when publishing its cyclical report on flat prices and the situation on the residential real estate market, also mentions the time of real estate sale (NBP 2020). On the primary market, in the first quarter of 2020, the fastest buyers were in Warsaw and Krakow (over 2 quarters), the longest—in Poznań—over 3 quarters. On the secondary market, the average time of selling a residential property for all cities was similar and amounted to approximately

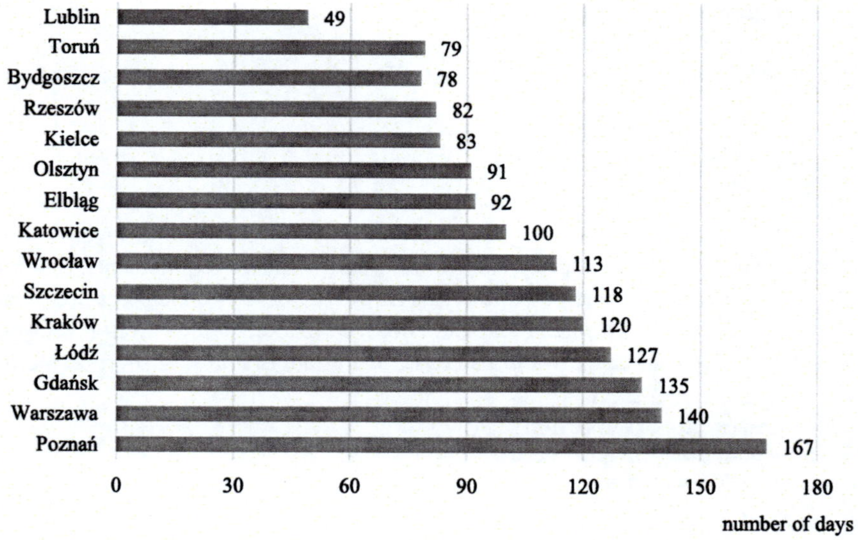

Fig. 1 Average time of flat sales in selected Polish cities in 2018. *Source* Walczak (2018)

4 months. However, it should be noted that the authors of the study are aware that the given time of sale of real estate is underestimated, as it is calculated on the basis of real estate sold and does not include real estate that has been submitted for sale and still remains on offer.

In the article, the authors attempted to estimate the time of sale of residential property based on the properties sold, but also taking into account the properties that were still in the offer database. The aim of the article is to present a method of duration analysis to assess the probability of the real estate sale offer coming out of the system, taking into account the year of introducing the offer into the system.

One of the characteristics of the real estate market is its heterogeneity, caused by a diversified spatial range (Kucharska-Stasiak 1997). The local, regional, national and even international market stands out. The residential real estate market is a local market and all analyses concerning the mechanisms of its functioning should be conducted in relation to such a market. Different behaviour will be observed in big cities, different in small ones. The article fits into the trend of analysing small and imperfect and constantly developing markets (Cirman et al. 2015; Gdakowicz and Putek-Szeląg 2020). The study was conducted on the residential real estate market in Szczecin.

2 Data Used in the Study

The study was conducted on individual data obtained from the West Pomeranian Association of Real Estate Agents (ZSPON). The information concerned offers to sell residential properties located in Szczecin. ZSPON is an organisation which brings together almost all real estate agents operating in Szczecin. The association is the administrator of the West Pomeranian Offer Exchange System (ZSWO)—the local format of multiple listing system or multiple listing service, i.e. MLS.

The analysis included offers of residential properties (dwellings), which were submitted for sale (and to the ZSWO system) by real estate agents in the period from 1.01.2017 to 30.06.2020. The database consisted of 11,816 observations. The properties were described by a set of variables: offer number, date of offer issuance, transaction date, district, street name, current offer status, offer price, last offer current price, transaction price, right to premises, area, number of rooms, number of floors on which the flat in the building is located, kitchen type, finishing standard, year of construction of the building, number of floors in the building, whether there is monitoring, whether there is a lift, and in which Internet portal/press the agent places data.

In the analysed period, 1988 properties were sold (nearly 17%), 3289 were still up to date (as of June 30, 2020), 350—blocked, which means that for various reasons the real estate agent blocked the offer, e.g. by reserving it for a potential buyer (Table 1). The last category (the most numerous, over 50% of all offers) was properties that were withdrawn (removed) from the system. The reasons for their removal varied: from a change of mind about the owner's willingness to sell the

Table 1 Structure of residential property offers submitted for sale in the period from 1 January 2017 to 30 June 2020

Current offer status	Number of offers
Current	3288
Blocked	350
Withdrawn	6189
Closed	1988
Total	11,816

Source Own elaboration

property to the owner selling the property himself. The latter cases are common because many real estate agents conclude open-ended contracts with the sellers, so the property owner can sell the apartment himself or through another agent. At the time of the transaction, the owner may (but does not have to) inform the other offices about the withdrawal of the offer, without giving any reason.

Ultimately, further analysis was carried out on the basis of current and closed offers, i.e. 5626 offers. Most flats were submitted for sale in 2017—nearly 2800, of which nearly 23% were sold (Fig. 2). In the following years, the number of offers introduced to the ZSWO system ranged from 824 in 2018 to 1079 in 2019. Despite the specific economic situation in 2020 (the COVID-19 pandemic), the number of property offers submitted for sale has not decreased, on the contrary, it should be expected to increase, as the data refers only to the first half of 2020. The freezing of

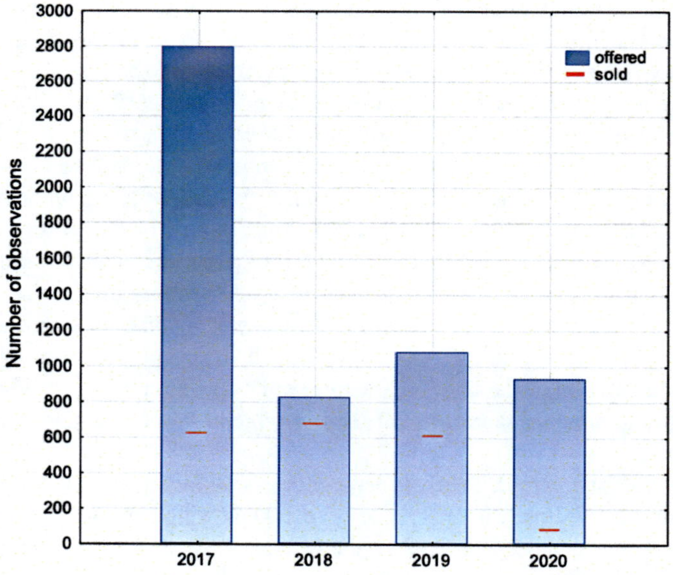

Fig. 2 Residential real estate sales offers submitted to the ZSWO system and sold in 2017–2020. *Source* Own elaboration

the economy in the spring period was reflected in the number of properties sold—only 77 flats (8.3% of the reported properties) found buyers in the first half of 2020.

For properties submitted for sale in subsequent years, basic statistical characteristics were calculated and the results are presented in the form of distributions in the analysed years. The flats were described in terms of their area and price per m^2. The analysis was conducted for properties offered for sale and sold (Figs. 3, 4, 5 and 6).

The median area of flats offered for sale in the analysed years remained at a similar level of 52–55 m^2 (Fig. 3). The median area of flats sold was slightly lower: 49–52 m^2 (Fig. 4). While the average size of flats offered in the subsequent years increased, the median of flats sold decreased.

The largest number of flats offered for sale and sold was between 40 m^2 and 60 m^2, which indicate 2 or 3 room flats. On the other hand, very large flats—over 140 m^2—were sold poorly—they are reported to real estate agents, but they do not find buyers. One of the reasons for the low interest in such large apartments is that you can buy a house surrounded by a green area for a similar price (and a similar size). Having a private space in the open air became especially important in the spring of 2020, during the period of the country's lockdown.

Both the median of the offer price and the transaction price of 1 m^2 increased in the analysed years. Sellers' expectations increased from PLN 4714/1 m^2 in 2017 to

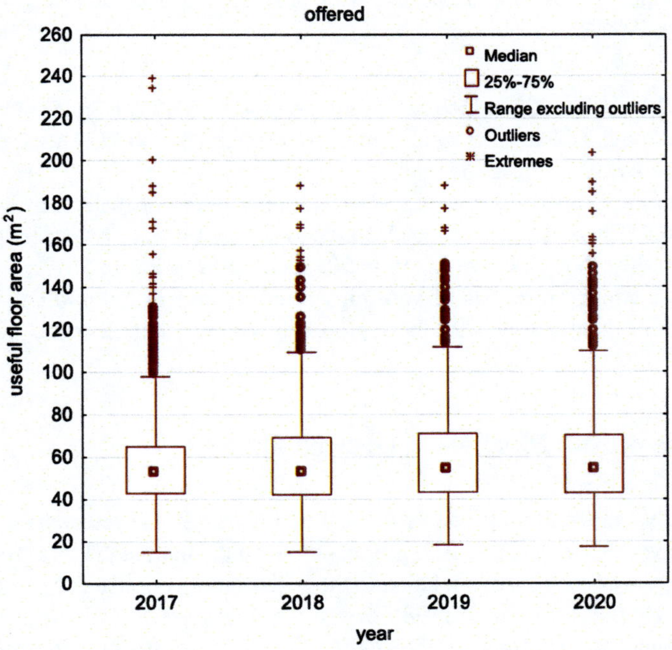

Fig. 3 Descriptive statistics of usable floor space of flats offered in Szczecin in 2017–2020 (m^2). *Source* Own elaboration

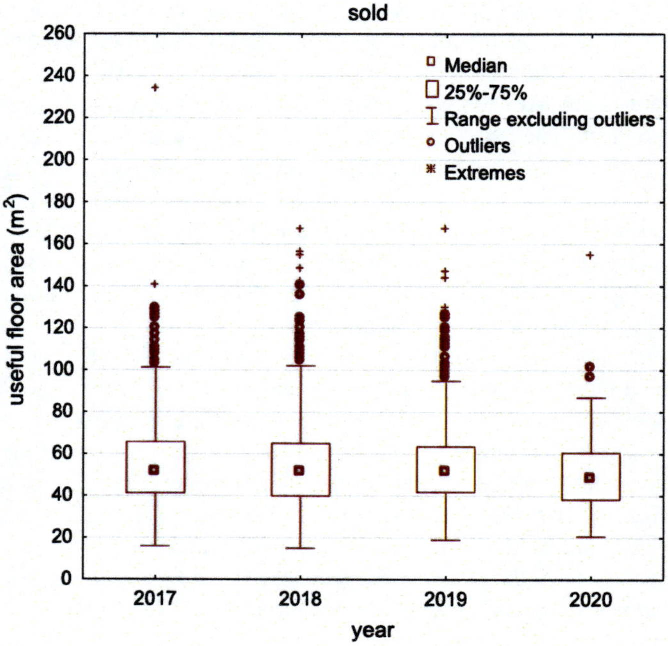

Fig. 4 Descriptive statistics of usable floor space of flats sold in Szczecin in 2017–2020 (m^2). *Source* Own elaboration

PLN 6500/1 m^2 in 2020, which means an increase by 37.9% (Fig. 5). The median transaction price of 1 m^2 increased from PLN 4339 in 2017 to PLN 6062 in 2020—an increase by 39.7% (Fig. 6).

To sum up: property owners expected a higher price for their apartments from year to year. The transaction price also increased year by year, but each year it was lower than the offer price—from 4% in 2018 and 2019 to 8% in 2017. In all years, there was a strong correction of very high prices (above PLN 8000/1 m^2)—either there were no transactions with these properties or the sellers lowered the price.

3 Time on the Market—Censored Data

When analysing the supply side of the real estate market, the researchers took into account, inter alia, real estate selling time—time on the market (TOM). Time to sell the property was examined in relation to the seller's motivation and the selling price (Springer 1996; Glower et al. 1998), or just the selling price (Cubbin 1974; Miller 1978; Asabere and Huffman 1993; Anglin et al. 2003), or the offered price and the impact of its possible overstatement (Yavas and Yang 1995; Jud et al. 1996; Ong and Koh 2000; Knight 2002; Anglin et al. 2003). Factors influencing TOM were

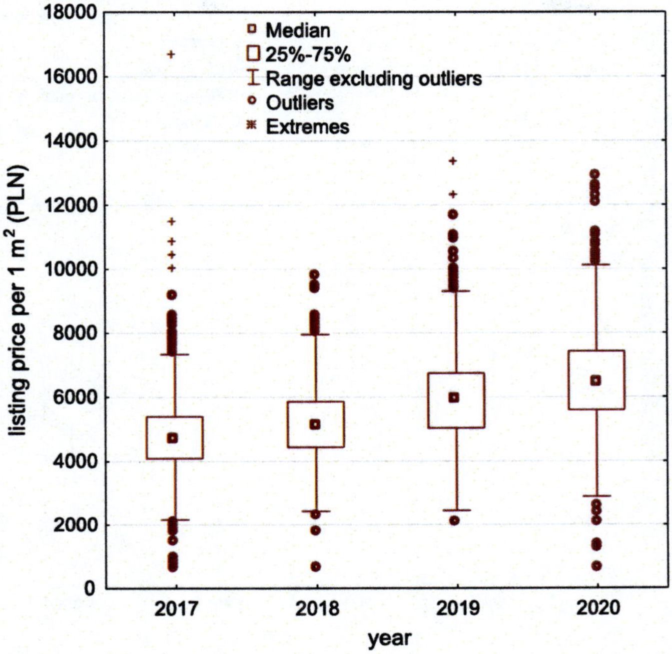

Fig. 5 Descriptive statistics of the price of 1 m^2 flats offered in Szczecin in 2017–2020 (PLN). *Source* Own elaboration

searched for (Cirman et al. 2015); TOM was combined with low liquidity of the real estate market (Lippman and McCall 1986). It was confirmed that the time of sale of real estate is related to the condition of the housing and financial markets (Haurin 1988; Yavas and Yang 1995; Anglin et al. 2003, Filippova and Rehm 2014) and the supply and demand factors indicated by (Leung et al. 2002), as well as with the sales strategy used by sellers (Haurin et al. 2013). The time of sale was analysed in relation to the risk and profit related to investing in real estate (Lin and Liu 2008) and the impact of new residential stamp duty (Liang et al. 2018).

Figure 7 presents the basic characteristics of the time of sale of residential properties in Szczecin in 2017–2020. Only properties that were sold during the period considered are taken into account.

In 2017, it took about 3 months (84 days) from the submission of the property to the database for sale, but there were also flats that waited more than 2.5 years for sale. In 2018, the average waiting time for a buyer was even longer—to 91 days. In 2019, a subtle shortening of the time to sell the property to 85 days was observed, while the year 2020 brought a significant reduction of this time to 47 days.

In all of the above studies, the time of sale of the property was analysed taking into account only the properties that were sold. However, properties which were still offered by real estate agencies and in relation to which no final event, i.e. sale,

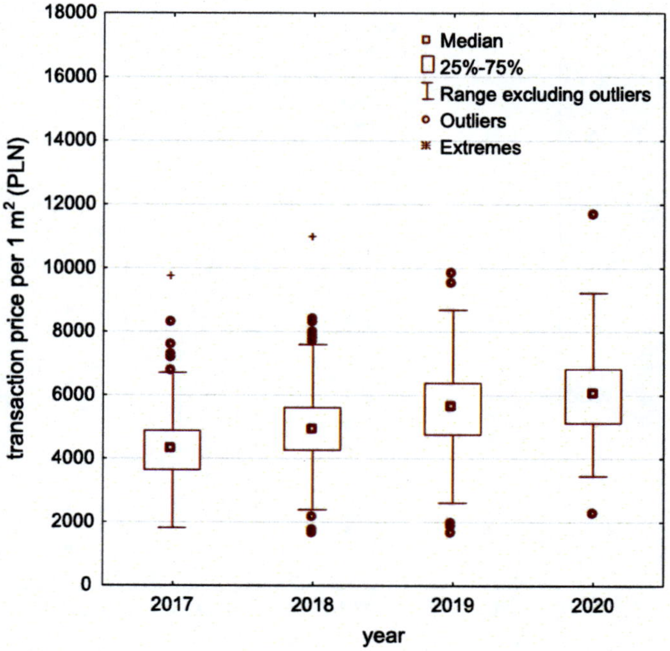

Fig. 6 Descriptive statistics of the price of 1 m^2 flats sold in Szczecin in 2017–2020 (PLN). *Source* Own elaboration

occurred were not taken into account. Such cases are referred to as incomplete, censored observations (Balicki 2006). In the case of censored observations, the duration of the phenomenon cannot be determined because there is no information about the beginning, end or beginning and end of the event for a given unit. Such data are called left-, right- or two-sided censored respectively. Figure 8 shows complete data for which the beginning (d_0) and the end of the observation (d_n) can be identified—units 3 and 4 (marked with a black dot at the end), and incomplete data for which the end of the observation cannot be identified—units 1, 2 and 5 (marked with an empty circle). Observation 1 and 5 remain in the database (the property is still offered for sale), but the observation time has ended, while in the case of observation 2—the unit has been removed from the database for reasons other than sale.

For the purpose of the study, the duration of the sale process of each property at the time of its appearance is assigned the value of the duration $t = 0$. The same study units are shown in Fig. 9 as in Fig. 8, but they differ in the date of the initial event—the observation time is converted into duration. Random censorship takes place when individual units enter the observation field at different calendar time and the observation ends with a specific date (30.06.2020 in our case).

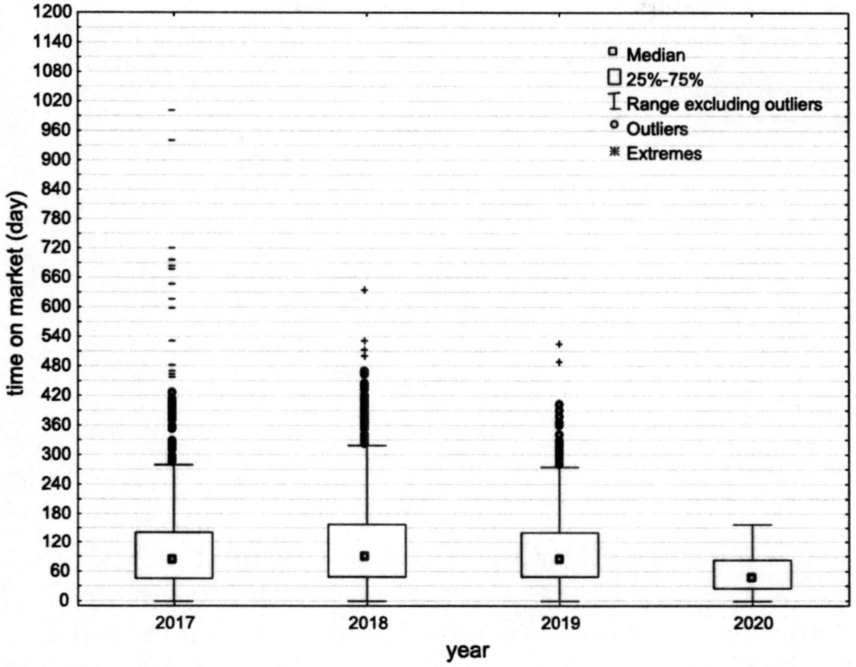

Fig. 7 Descriptive statistics of the time on the market flats sold in Szczecin in 2017–2020. *Source* Own elaboration

Fig. 8 Right-hand data censored (observation time)—random censorship. *Source* Own preparation based on Markowicz (2012)

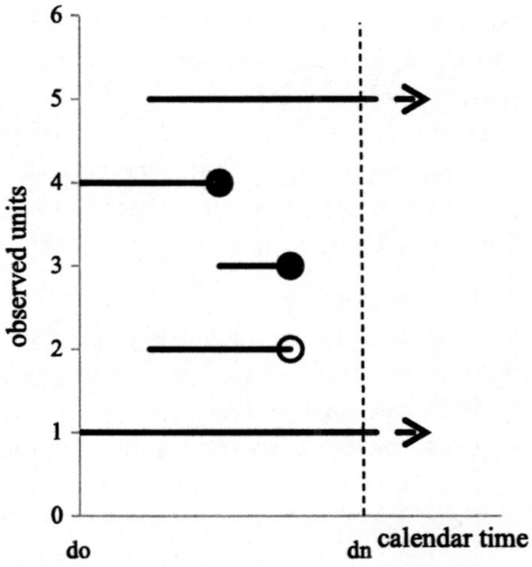

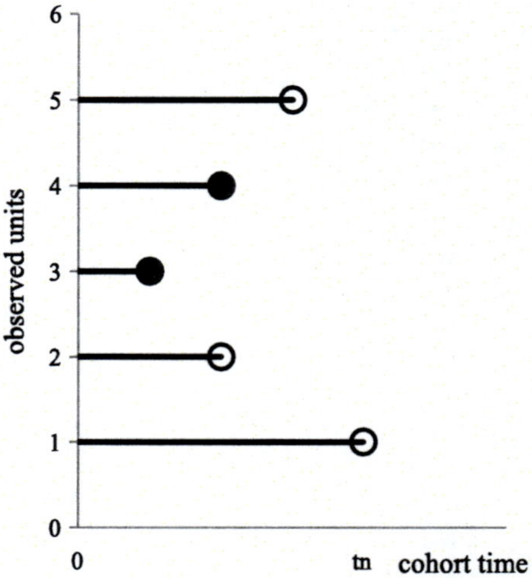

Fig. 9 Right-hand data censored (duration)—random censorship. *Source* Own preparation based on Markowicz (2012)

4 Duration Analysis of the Real Estate Offer

The analysis of event history, also known as survival analysis or duration analysis, is a set of statistical methods that are used in various scientific disciplines. Originally these methods were used in actuarial statistics and demography. Then they were developed and introduced by different researchers to other scientific disciplines, hence the name of the same methods in different disciplines may differ: in medicine, demography and biology they are called survival analysis, in economics and social sciences they are called duration analysis or transition analysis, and in engineering, technology and industry they are called reliability analysis or failure time analysis.

In the duration analysis, time is the subject of the study. The time that elapses from the beginning of the observation to the occurrence of a specific event terminating the observation on a given unit—that is, the duration time in a given state. Time may be observed:

- from birth to death,
- from birth to the diagnosis of cancer,
- from the beginning of disease to death,
- from marriage to divorce,
- from the establishment of the company to its bankruptcy (Markowicz 2012; Bieszk-Stolorz and Markowicz 2019),
- from losing a job to resuming it (Bieszk-Stolorz 2013; Landmesser 2013; Bieszk-Stolorz and Dmytrów 2019),
- from the introduction of a real estate offer to the MLS system to its sale.

Time is a random variable *T*. The survival function can be defined as:

$$S(t) = P(t < T) = 1 - F(t) \tag{1}$$

where

T duration of the phenomenon,
F(*t*) distribution function of the random variable *T*

The function (1) determines the probability that the duration (duration of the offer validity) for a given unit (real estate) will be longer than *t*. The second function that is very important in the survival analysis is the hazard function describing the intensity of an event at time *t* under the condition of survival until *t*. The hazard function is defined by the formula (Kleinbaum and Klein 2012):

$$h(t) = \lim_{\Delta t \to 0} \frac{P(t \le T < t + \Delta t | T \ge t)}{\Delta t} \tag{2}$$

One of the most frequently used function estimators (2) is the Kaplan-Meier estimator (Kaplan and Meier 1958):

$$\hat{S}(t) = \prod_{j:t\,j \le t} \left(1 - \frac{d_j}{n_j}\right) \tag{3}$$

where

d_j number of events at the moment, t_j,
n_j number of units at risk until, t_j.

The duration function (1) indicates the probability of the event not happening at least until time *t*. Distributor *F*(*t*) expresses the probability that an event will occur at the latest by time *t*. In the conducted research, the event is the sale of the real estate (and its removal from the MLS database), the estimator of the duration function (3) informs about the probability of staying in the offer database - not selling the real estate, and the estimator of the distributor allows to determine the probability of selling the real estate (removal from the database). In the d_j study, the number of removals of the real estate from the database for a given reason at the moment, t_j (sale of the real estate or change of the owner's decision to sell the real estate).

5 Empirical Research

Table 2 presents the data on the properties submitted and sold to the offer exchange system, as well as properties that as at June 30, 2020 have not been sold and the owners still expressed their will to sell the property. The latter are referred to as censored observations because no end event has been recorded for them.

Most properties were entered into the MLS system in 2017. Of these, 22.4% were sold in the analysed period, and 77.6% of the offers until June 30, 2020 were active in the system, i.e. the owners of these properties still wanted to sell them. The lowest number of censored properties was in 2018. There were only 142 of them. In 2019, more than half of the offers changed their owner. It can be noted that in 2017–2019, approximately 600–680 purchase and sale transactions of residential real estate were concluded. Data for 2020 differ from the observations in previous years. The number of offers submitted for sale was very high (more offers were submitted in the first half of 2020 than in the entire 2018), however, very few apartments sold were recorded—which was due to the prohibition to move around in the spring period, limitation of direct work of real estate agents, notaries and banks, and the prevailing mood of uncertainty. Due to the epidemic situation, more than 90% of the offers submitted for sale in 2020 were defined as censored observations.

Using formula 1, the overall survival curve was estimated for all properties reported throughout the analysis period. The results are presented in Fig. 10.

The survival function presented in Fig. 10 indicates that in the first 180 days (6 months) from the moment the property was issued by an MLS agent, there was a rapid decrease in the probability of the offer remaining in the system, i.e. the property changed the owner. If the property was not sold within six months from the moment of its issue, the rate of probability of sale of the property decreased significantly in the following months. If the property has been listed in the system for more than 15 months, its probability of sale was the same as that of properties listed in the system for even 3 years.

In the next stage of research, a similar analysis of the offer's survival was carried out, but with a breakdown for particular years of the introduction of the offer into the MLS system. The results of the research are presented in Table 3.

Table 2 Structure of residential real estate offers submitted for sale in the MLS system in 2017–2020

Year	Dwellings submitted	Dwellings sold	Censed observations	
			Number	Share (%)
2017	2795	626	2169	77.6
2018	824	682	142	17.2
2019	1079	603	476	44.1
2020	928	77	851	91.7
Total	5626	1988	3638	64.7

Source Own elaboration

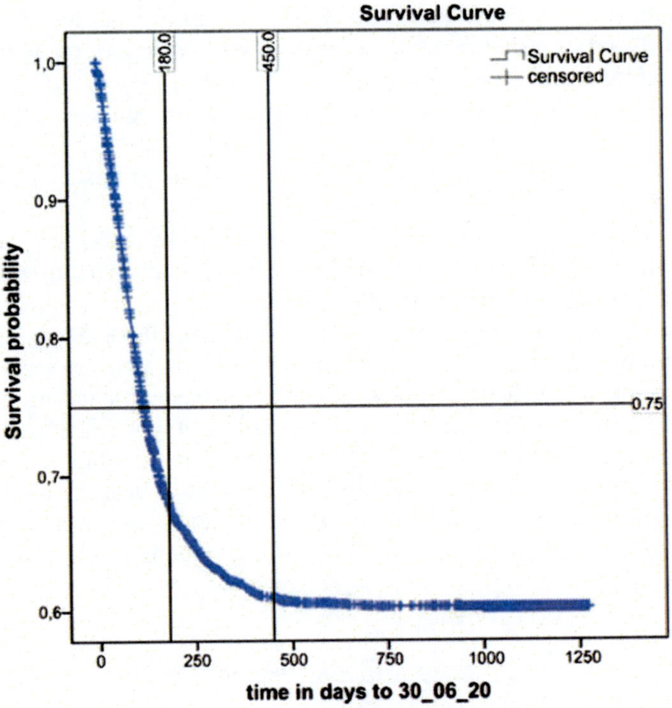

Fig. 10 Survival curve of apartments offered for sale in the MLS in Szczecin system in 2017–2020. *Source* Own elaboration

Table 3 Structural parameters for the survival function in 2018 and 2019

Year	Quartile 1	Median	Quartile 3
2018	58.0	111.0	263.0
2019	80.0	194.0	
Total	117.0		

Source Own elaboration

In 2017, open offers (censored observations) accounted for as much as 77.6% of all offers. This has its consequences as it is not possible to determine the value of quartiles as a function of survival for this year. 25% of the properties put up for sale in 2018 changed owners within the 58th day, i.e. after about two months. 50% of them found buyers within 111 (approximately 4 months) days, and 75% within 263 days (approximately 9 months).

In 2019, however, 44.1% of real estate offers were unsuccessful, i.e. sales. In the analysed year, 25% of the properties changed owners within 80 days, and 50% within 194 days (6.5 months). The values of quartiles in 2019 were higher than in the previous year, so when submitting a property for sale in 2019, one should expect a buyer longer than in 2018.

As mentioned above, 2020 was a very unusual period of analysis. As a result, 91.7% of the offers were censored observations and, as for 2017, a quartile of the survival function could not be determined.

In the analysed period, 25% of properties submitted for sale found buyers within 117 days.

A graphical presentation of the calculated survival curves broken down by year is shown in Fig. 11.

Depending on the year the offer was introduced to the market (application for sale), the time of sale of the property differed. Properties submitted for sale in 2018 were subject to sale the fastest, and then in 2019.

The sale of real estate reported to the MLS system in 2018 took up to 250 days (over 8 months). Up to 15 months, the likelihood of sales was also quite high, but after this period the likelihood of sales stabilised, remaining at a not very high level.

In turn, the probability of selling properties proposed for sale in 2019 was high for properties that were in the system for up to six months. After this period, the probability of sale was actually constant and much lower than for the offers submitted in 2018. However, the properties reported in 2017 remained the longest in the MLS system. The probability of selling such a property is very low.

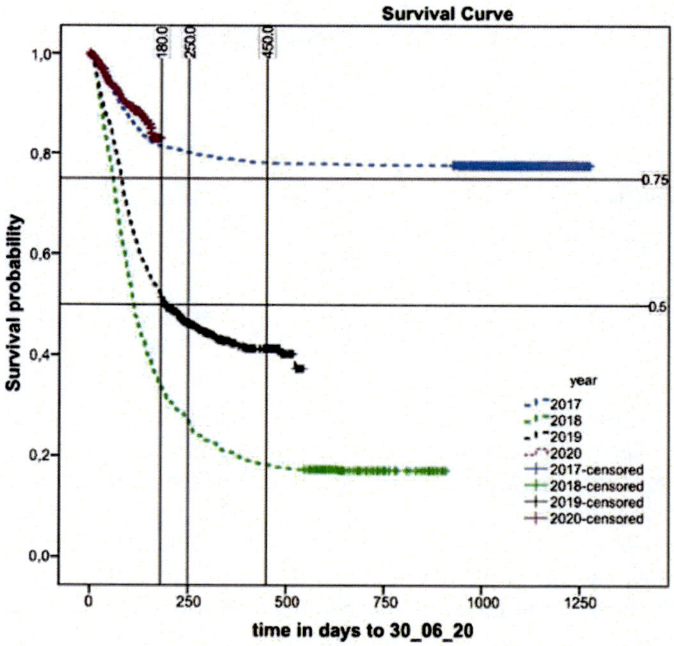

Fig. 11 Survival curves of residential property sales offers submitted in 2017–2020 in Szczecin. *Source* Own elaboration

To confirm the differences in the probability of the course of the survival curves for the offers submitted in the following years, the Log-rank Test was conducted. It confirmed that the survival curves calculated for the apartment sales offers that appeared on the market in the following years differed significantly ($\chi^2 = 237.112$; $p = 0.000$).

6 Conclusions

When analysing the time of sale of the property or the duration of the offer (time on the market), researchers have so far only taken into account properties that have been sold. In the presented study, the authors, when determining the time necessary to sell a residential property, also took into account the offers that remained in the MLS database, and the owners of the properties still declared their willingness to sell them. An additional advantage of the study is the analysis of the imperfect market, i.e. the local residential market in Szczecin.

When calculating TOM, the survival function and Kaplan-Meier estimator were used. The analysis was carried out for all properties introduced to the MLS system in the period from 01.01.2017 to 30.06.2020 in total and by year of introduction of the offer to the system. The analysis of all the offers together showed that the chances of selling the real estate were rapidly decreasing after 180 days of remaining in the MLS system. The survival functions calculated depending on the year of offer introduction indicated that the time of sale was shorter for offers introduced in 2018: 25% of offers were sold within 58 days of the offer remaining in the system, while 50% of offers were sold within 111 days. For offers submitted for sale in 2019, quartiles of the survival function were, respectively: quartile 1—80 days and median—194 days. Due to a large number of observations censored in 2017 and 2020, it was not possible to determine quartiles of the survival function for properties submitted for sale in those years. It was confirmed that TOM of residential properties submitted for sale in particular years was different.

The analyses carried out have a practical aspect—they can help real estate agents—indicating that for properties remaining in the system for more than 15 months, the probability of their sale is significantly decreasing. Such an "overdue" offer requires additional efforts, e.g. reduction of price or change of the prepared marketing offer.

The presentation of the study is a preliminary study from the planned cycle, which should allow to indicate the factors affecting TOM of residential properties in Szczecin.

References

Anglin PM, Rutherford R, Springer TM (2003) The trade-off between the selling price of residential properties and time-on-the-market: the impact of price setting. J Real Estate Finance Econ 26(1):95–111. https://doi.org/10.1023/A:1021526332732

Asabere PK, Huffman FE (1993) Price concessions, time on the market, and the actual sale price of homes. J Real Estate Finance Econ 6:167–174. https://doi.org/10.1007/BF01097024

Balicki A (2006) Analiza przeżycia i tablice wymieralności. PWE, Warszawa

Bieszk-Stolorz B (2013) Analiza historii zdarzeń w badaniu bezrobocia. Volumina.pl Daniel Krzanowski, Szczecin

Bieszk-Stolorz B, Dmytrów K (2019) Prawdopodobieństwo wyjścia z bezrobocia rejestrowanego na przykładzie Szczecina. The Polish Statistician 64(11):7–24. https://doi.org/10.5604/01.3001.0013.7585

Bieszk-Stolorz B, Markowicz I (2019) Analiza trwania w badaniach ekonomicznych. CeDeWu, Warszawa

Cirman A, Pahor M, Verbic M (2015) Determinants of time on the market in a thin real estate market. Inzinerine Ekonomika-Engineering Economics 26(1):4–11. https://doi.org/10.5755/j01.ee.26.1.3905

Cubbin J (1974) Price, quality and selling time in the housing market. Appl Econ 6:171–187. https://doi.org/10.1080/00036847400000017

Filippova O, Rehm M (2014) Market conditions, marketing time, and house prices. J Hous Res 23(1):45–56. http://www.jstor.org/stable/24862555. Accessed 14 Sept 2020

Gdakowicz A, Putek-Szeląg E (2020) The demand and supply analysis and comparison of dwellings in Szczecin. In: Bilgin MH, Danis H, Demir E, Tony-Okeke U (eds) Eurasian economic perspectives: proceedings of the 28th Eurasia business and economics society conference 15(1):169–184. http://doi.org/10.1007/978-3-030-48531-3_12

Glower M, Haurin DR, Hendershott PH (1998) Selling time and selling price: the influence of seller motivation. Real Estate Econ 26(4):719–740. https://doi.org/10.1111/1540-6229.00763

Haurin D, McGreal S, Alastair A, Brown L, Webb JR (2013) List price and sales prices of residential properties during booms. J Hous Econ 22:1–10. https://doi.org/10.1016/j.jhe.2013.01.003

Jud DG, Seaks TG, Winkler DT (1996) Time on the market: the impact of residential brokerage. J Real Estate Res 12(3):447–458

Kaplan EL, Meier P (1958) Non-parametric estimation from incomplete observations. J Am Stat Assoc 53

Kleinbaum D, Klein M (2012) Survival analysis: a self-learning text, 3rd edn. Statistics for biology and health. Springer Science + Business Media. http://doi.org/10.1007/978-1-4419-6646-9_1

Knight JR (2002) Listing price, time on market, and ultimate selling price: causes and effects of listing price changes. Real Estate Econ 30(2):213–237. https://doi.org/10.1111/1540-6229.00038

Krainer J (2001) A theory of liquidity in residential real estate markets. J Urban Econ 49:32–53. https://doi.org/10.1006/juec.2000.2180

Kucharska-Stasiak E (1997) Nieruchomość a rynek. Wydawnictwo Naukowe PWN, Warszawa

Landmesser J (2013) Wykorzystanie metod analizy czasu trwania do badania aktywności ekonomicznej ludności w Polsce. Wydawnictwo SGGW, Warszawa

Leung CKY, Leong YCF, Chan IYS et al (2002) TOM: why isn't price enough? Int Real Estate Rev 5(1):91–115

Liang C, Hui ECM, Yip TL (2018) Time on market (TOM): the impact of new residential stamp duty. Phys A 503:1117–1130. https://doi.org/10.1016/j.physa.2018.08.126

Lin Z, Liu Y (2008) Real estate returns and risk with heterogeneous investors. Real Estate Econ 36(4):753–776. https://doi.org/10.1111/j.1540-6229.2008.00229.x

Lippman S, McCall J (1986) An operational measure of liquidity. Am Econ Rev 76(1):43–55

Markowicz I (2012) Statystyczna analiza żywotności firm. Wydawnictwo Naukowe Uniwersytetu Szczecińskiego, Szczecin

Miller NG (1978) Time on the market and the selling price. J Am Real Estate Urban Econ Assoc 6:164–174. https://doi.org/10.1111/1540-6229.00174

Narodowy Bank Polski (2020) Informacja o cenach mieszkań i sytuacji na rynku nieruchomości mieszkaniowych i komercyjnych w Polsce w I kwartale 2020 r. https://www.nbp.pl/home.aspx?f=/publikacje/rynek_nieruchomosci/index2.html. Accessed 14 Sept 2020

Nowak B (2019) Ile czasu trwa sprzedaż nieruchomości? https://www.morizon.pl/blog/czas-sprzedazy-nieruchomosci/. Accessed 14 Sept 2020

Ong SE, Koh YC (2000) Time on-market and price trade-offs in high-rise housing sub-markets. Urban Stud 37(11):2057–2071. https://doi.org/10.1080/713707223

Springer TM (1996) Single-family housing transactions: seller motivations, price, and marketing time. J Real Estate Finance Econ 13:237–254. https://doi.org/10.1007/BF00217393

Walczak A (2018) Ile czasu trwa sprzedaż mieszkania—analiza. https://gratka.pl/blog/nieruchomosci/ile-czasu-trwa-sprzedaz-mieszkania-analiza/14802/. Accessed 14 Sept 2020

Yavas A, Yang S (1995) The strategic role of listing price in marketing real estate: theory and evidence. Real Estate Econ 23(3):347–368. https://doi.org/10.1111/1540-6229.00668

Is Society Ready for Long-Term Investments?—Profiles of Electricity Users in Silesia

Sylwia Słupik and Joanna Trzęsiok

Abstract The objective of this article is to identify and characterize electricity users in terms of their attitudes towards energy saving. The analyses applied data from the proprietary survey conducted among the inhabitants of the Silesian Province. The respondents were asked, inter alia, about saving electricity through short- and long-term investment actions. Model types of users were defined, so that it was possible to assign the surveyed people to appropriate groups using distance measures dedicated to non-metric variables. Moreover, the user classes formed in this way were characterized; and for that purpose, the measures to study the dependence of qualitative variables—the chi-square test and the Cramer's *V* coefficient—were used. As expected, it turned out that the respondents' actions were significantly impacted by financial considerations, i.e. income, which determined the nature of the undertaken investments, and thus significantly influenced the result of the classification. However, as the research results have shown, also aspects related to environmental protection significantly differentiate the investors' attitudes. People focused on long-term actions, aimed at reducing energy consumption, more often declare their interest in subsidies for environmental protection, spending higher amounts on energy-saving devices, or are even guided by ecological signs when doing shopping. The authors' original contribution is the proposed segmentation of respondents and their characterization based on the obtained empirical data.

Keywords Energy consumption · RES · Social ecological awareness · Linear ordering · Chi-square test

S. Słupik · J. Trzęsiok (✉)
University of Economics in Katowice, Katowice, Poland
e-mail: joanna.trzesiok@ue.katowice.pl

S. Słupik
e-mail: sylwia.slupik@ue.katowice.pl

K. Jajuga et al. (eds.), *Data Analysis and Classification*, Studies in Classification, Data Analysis, and Knowledge Organization,
https://doi.org/10.1007/978-3-030-75190-6_11

1 Introduction

Global warming of the climate is a fact and an issue that should define the way we function in the coming years; if we do not want our grandchildren, or maybe even our children, to wage wars over food and water. Another pro-ecological activity is saving electricity, which translates into lower emissions of carbon dioxide into the atmosphere. Such activities do not always have to originate from our beliefs, they are often the result of financial calculations, but what counts is the end result—reducing the negative impact on the natural environment.

Therefore, in recent years, we have been observing major changes in the global and European energy market. The energy sector is undergoing a low-emission transition and the role of the consumer in this process is significantly increasing. It is widely believed that the way to reduce environmental pressures and mitigate climate change is to increase energy efficiency while reducing energy demand. Owing to technological progress and established legal regulations, protecting the environment and limiting the impact of consumption on its degradation, it is possible to increase energy savings (Steg et al. 2005). Nevertheless, already now, more and more often in the studies of economists and behaviourists (Zhou and Yang 2016; Lutzenhiser 1993), it is recognized that behavioural factors are of great importance for this process. It is the encouraging of ecological mindsets of individual consumers that seems to be crucial in the fight for the future shape and volume of total energy consumption in the world. It is also becoming the goal of energy and environmental policies, both regionally and locally, of many countries, especially European ones (Gardner and Stern 1996). Hence, it is vital to know the needs of consumers, the level of their demand for electricity, and to be aware of the changes taking place in the attitudes of energy end users.

According to researchers of sustainable consumption in households (Frederiks et al. 2015; Clancy and O'Loughlin 2002; Hille 2016), the most important determinants of behaviour change and the emergence of the so-called *greening of consumption* (Matel 2016, p. 56) can be summarized as follows (Nagaj 2018, pp. 4–5; Matel 2016, pp. 56–57):

- consumers are increasingly aware of the value and need for sustainable energy practices and climate change issues,
- the sense of responsibility of consumers for the choices they make increases, they prefer to buy raw materials and energy-saving products as well as products that are safe for human health (Kiełczewski 2005, 2015, p. 55),
- energy-saving people, apart from their propensity to save and ecological awareness, are also characterized by such features as high aversion to consumption and interest in investing in energy-saving technologies (Clancy and O'Loughlin 2002),
- consumers often give up purchasing products manufactured in a way that pollutes the environment, striving to minimize the use of non-renewable resources,
- very often, they limit or give up gadget consumption (Dąbrowska et al. 2015, p. 43),

- it is observed that consumer actions, including those related to the will to save energy, are the result of behavioural rather than rational or physical factors (Nakamura 2016),
- at the same time, there is often a significant discrepancy between consumers' declared knowledge, values, attitudes and intentions, and their observable behaviour.

Taking into account the above reflections, it seems imperative to anticipate and understand the behaviour of energy consumers and to evaluate them. The achievements of behavioural economics and psychology can be of help, where in combination with the experience of consumption theory, environmental economics and natural resources, the development of incentives for the use of renewable energy and sustainable energy consumption will be developed.

This work aims to identify and characterize electricity users in terms of their attitudes towards energy saving. The authors of the article have based their analysis on the results of the proprietary research conducted among households in the Silesian Province in 2018 and on a review of the literature on profiling individual energy consumers. In the article, the authors also characterize the obtained segments and identify fundamental factors influencing the respondents' behaviour towards save energy.

2 Study of Energy Consumers' Behaviours and Their Impact on Shaping Pro-ecological Attitudes

Currently, household consumption patterns, including energy consumption, are shaped by various factors. Patterns are individualized under the influence of socio-economic changes, such as, on the one hand, an increase in income; an increase in the number of one-person households and, on the other hand, demographic ageing, which forces a change in lifestyles and consumer choices (cf. OECD 2002). Greater autonomy in the actions of consumers allows to form a full identity with the use of consumer goods and services available on the market, and it even enables the co-creation of these goods on the basis of presumption (Popczyk 2014; Bylok 2014). At the same time, in the subject literature much attention has been drawn to develop behavioural and psychological models of consumers. Also, the behaviour of households in terms of energy consumption and factors influencing these behaviours have been scrutinized (Zhou and Yang 2016). Efforts have also been made to design effective intervention strategies aimed at stimulating the behaviour of households and improving energy efficiency, as well as leading to a significant reduction in energy.

Many years of observation and research have shown that in order to encourage the use of measures and instruments to improve energy efficiency, it is necessary to understand how consumers behave and how they use energy in their everyday personal and professional lives. It is necessary to answer the questions why people stop taking action, even when the economic calculation shows the possibility of obtaining potential benefits, and why are they not interested in optimizing their consumption and reducing the negative impact of excessive energy consumption on the environment. A full understanding of the motives of behaviour and the elimination of barriers will contribute to better communication and prepare decision-makers to create interventions that will successfully bridge the gap between pro-ecological knowledge, values, attitudes and intentions and the daily energy-related behaviour of consumers (Frederiks et al. 2015). It will also allow consumers to receive better recommendations on how to become more energy efficient. Yet, in order to devise effective behavioural change interventions, it is imperative to tailor interventions to different target groups and to consider the differences in their willingness to change behaviour (Seidl et al. 2017).

Table 1 presents an overview of selected profiles of individual energy consumers along with their brief characteristics. These segmentations have been developed on the basis of empirical research, often conducted at regular intervals. The literature on the subject is dominated by studies carried out in Great Britain, the USA and other Western countries. The table also includes Polish examples, but it should be noted that so far in Poland little research has been performed in which attempts have been made to segment energy consumers (cf. Słupik 2015; Ropuszyńska-Surma and Węglarz 2018a). The absence of such analyses and comprehensive studies is perceived by the authors of the article as a research gap and a field for further studies.

One of the classifications quoted in the table is segmentation developed as part of the currently implemented project "Personalised ICT-tools for the Active Engagement of Consumers Towards Sustainable Energy" abbreviated as "eco-bot"[1]. It is a 43-month project co-financed by the European Commission under the Horizon 2020 program: "Reducing energy consumption and carbon footprint by smart and sustainable use", implemented in 2017–2021 as part of an international consortium which includes the University of Economics in Katowice, and the authors of this chapter compose the core of the research team.

The aim of the project is to create a personalized virtual assistant. It will provide a consumer with information about his/her current energy consumption, disaggregated to the level of individual electrical devices. In addition, it will play an educational and advisory role by providing recommendations tailored to the user's needs regarding energy efficiency measures in order to motivate and encourage the users of the program to behave more energy-efficiently. Due to the fact that the project is in the pilot-verification phase, the behavioural model of energy consumers developed as part of the study along with their segmentation (presented in Table 1) is being tested in partner countries; hence, the authors of the article do not currently have the full results of the study and have not made comparisons with the studies obtained in 2018.

[1]Detailed information about the project is available at www.eco-bot.eu.

Table 1 Review of selected profiles of energy consumers

Source	Identified segments	Brief characteristics
Albert and Maasoumy (2016)	Behavioural greens	Think and act in an environmentally friendly way
	Think greens	Have a positive attitude towards environmental issues, but are less likely to act if it requires effort on their part
	Potential greens	Willing to take environmentally friendly actions if it is in their interest
	True browns	They are not interested in environmental issues and energy efficiency
SECC (2019)	Green innovators	Care for the environment is their driving force behind behavioural changes in order to save energy
	Tech-savvy proteges	They are very interested in saving energy and using technology for it, but keeping the same comfort and lifestyle
	Movable middle	They do not reject the idea of energy saving, but the actions they take are usually easy to perform, such as installing energy-saving lighting
	Energy indifferent	Environmental issues have a low priority here, and they are less interested in how to save energy
Pluskwa-Dąbrowski (2016)	Prosumers	Actively participate in the electricity market and become energy producers
	Aware consumers	They are not active and committed, but are aware (at least generally) of their rights in the energy market
	Passive consumers	They are not interested in their entitlements or possibilities of action
Accenture (2010)	Proactives	Focused on taking action to reduce the use of home appliances, low interest in environmental issues
	Eco-rationals	Interested in environmental issues and increasing energy efficiency
	Cost conscious	Focused on saving on electricity bills
	Pragmatics	Ready to change products and brands, but not to implement new technologies
	Skepticals	Less sensitive to savings on bills, little influence of social groups, they seek and use professional advice
	Indifferents	Not interested in pro-efficiency activities, reducing the use of home appliances and reducing energy consumption
Słupik et al. (2019)	Ecological idealist	They care about the environment, and it motivates them to save energy, they have high ecological awareness
	Aspiring ecologist	Follow current trends, (currently "eco") fashion and choices in line with their own lifestyle
	Dedicated saver	Motivated to save energy by financial matters
	Opportunist	Only save energy when it is easy to do
	Indifferent	Uninterested in any change of behaviours

The research presented in the table was mainly aimed at finding out about consumers' perception of their own energy consumption and their readiness and willingness to take pro-ecological as well as pro-efficiency measures. Nonetheless, the resulting segmentations differ due to the contracting subject material scopes, geographic scope and methodology applied. Among other things, consumer attitudes towards electricity management programs; awareness and level of knowledge; involvement and activity in the energy market; motivations; habits, opinions and preferences regarding the application of energy efficiency measures have been studied. The obtained characteristics of individual segments have been largely influenced by the ongoing socio-cultural changes, as well as the emergence of new opportunities related to, for example, the increase in the use of IT tools to manage energy consumption at homes and companies. From the analysis of the segments obtained, a general conclusion emerges that consumer behaviour related to energy efficiency is very complex and is characterized by a large variety of perceptions, attitudes and preferences. It can be noticed, however, that in each study there has been a segment of consumers that are uninvolved and not interested in any change in behaviour, as well as the segment of "green leaders" consumers for whom concern for the environment has become a priority and motivation to savings and change of behaviours and to increase energy efficiency. Each group of respondents also included consumer segments motivated by financial issues and interested in saving money. Research has also shown that consumers have more control over energy use and savings over the years.

3 Characteristics of the Surveyed Electricity Users and Description of the Methods Applied in the Study

The analysis performed in this article uses data from the original survey conducted in 2018 among the inhabitants of the Silesian Province. The Silesian Province is located in southern Poland and covers an area of 12,333 km^2, which accounts for almost 4% of the country's area. The region is the second most populous in Poland, while the Mazowieckie Province ranks first. It has 4.5 million inhabitants, with 368 inhabitants per 1 km^2. Accordingly, the area of the province has the highest population density index in Poland. In addition, as many as 77% of the residents of the province live in towns and cities (the average for Poland is 60.1%) (GUS 2020; Polityka gospodarki niskoemisyjnej dla Województwa Śląskiego. Regionalna polityka energetyczna do roku 2030-projekt 2020). The Silesian Province has a long-standing tradition of using coal as the main energy fuel, and today, it is the last large hard coal mining area in the European Union. Its high degree of urbanization and industrialization has caused serious environmental damage, which has resulted in deteriorating living conditions. The region's energy industry is built on conventional energy sources. Although the region ranks high in electricity generation, only a small share is produced from renewable energy sources, which amounted to mere 3.2% in 2018. In 2018, 24,905.9 GWh of electricity were produced in the Silesian Province (14.64% of domestic production), which ranked the region third in the country after the Łódzkie Province (22.7%) and the Mazowieckie Province (17.9%). In the same year, 27,273 GWh of electricity was consumed in the region, which accounted for 16.3% of the energy consumed in the country, and, apart from the Mazowieckie Province, it was the highest consumption in Poland. Due to the industrial character of the Silesian Province, the largest amount of electricity generated in 2018 was used by the industrial sector (9,107 GWh) and the energy sector (6,973 GWh). Substantial electricity consumption was also reported for households, which—in 2018—consumed 3,520 GWh of energy, which placed the region in the second place behind the Mazowieckie Province (4,828 GWh). (GUS 2020; Polityka gospodarki niskoemisyjnej dla Województwa Śląskiego. Regionalna polityka energetyczna do roku 2030-projekt 2020). The main air pollutants in the region are the emissions generated by the industry, households and transportation. Point emissions are related to the operations of the main industries in the Silesian area, such as mining, iron, zinc and lead metallurgy, and electricity generation. Surface emissions have a decisive impact on air pollution in the Silesian Province and they are mainly associated with local boiler plants, small- and medium-sized enterprises using coal for heating and technological purposes, and coal heaters used in households (Polityka gospodarki niskoemisyjnej dla Województwa Śląskiego. Regionalna polityka energetyczna do roku 2030-projekt 2020).

During the authors' research in 2018, information was collected from 1,237 people representing households, although due to the missing values of some key variables for this analysis, the answers of 1,147 respondents were ultimately used. A non-random sample selection was applied. The study was carried out by the method of a diagnostic survey with the use of a questionnaire distributed by the snowball sampling method. The survey questionnaire contained questions with a semi-open cafeteria. The questions in the survey concerned the attitudes and level of environmental awareness of energy consumers. However, taking into account the purpose of this paper, the study covered only those questions that allowed to classify respondents in terms of their attitude to energy saving. Therefore, the respondents when asked *How do you save electricity*? could choose any number of responses from the following options:

- I turn off unnecessary lighting,
- I replace light bulbs with energy-saving ones,
- I buy energy-saving kitchen equipment,
- I do not leave the equipment in standby mode,
- I cook in an energy-saving manner,
- I have two energy tariffs: day and night,
- I buy energy-saving household appliances (TV, computer, etc.),
- I invest in thermal modernization of the building,
- I invest in systems for obtaining energy from renewable sources.

Marking a statement from the above list indicated that an activity related to energy saving was performed, and therefore, each surveyed person indicated their attitude to energy use.

Technically, each of the listed statements represented a separate variable (X_1–X_9). Indication of the statement number j (for $j = 1,\ldots,9$) by i—this respondent was assigned the implementation x_{ij} with a value "1". Otherwise, when the respondent did not perform and did not mark a given answer, "0" was entered as the value of the appropriate variable. This means that in the study aimed at determining the respondent's profile, nine nominal, dichotomous variables were used.

In further analyses, to profile the formed classes, variables were used that characterized:

- the respondent in terms of his/her attitude to green energy as well as problems related to environmental protection,
- the respondent's attitude to the issue of energy prices, but also to environmental protection programs and subsidies,
- the surveyed person's household.

The list of questions representing the variables, on the basis of which the mentioned class profiling was performed, is presented in Table 2.

As already mentioned, the aim of the article was to classify electricity users in terms of attitudes towards energy saving. But, based on the analysis of the literature on the subject (Accenture 2010; Słupik et al. 2019; Pluskwa-Dąbrowski 2016; Albert and Maasoumy 2016) as well as the research on attitudes and awareness of energy consumers in the Silesia Province (Słupik 2015) carried out in previous years, it was assumed that energy consumers are divided into two basic groups, i.e.:

- people who only perform easy and ad hoc activities (so-called short-term ones) aimed at saving energy (SI),
- people interested in long-term investments (more difficult ones, requiring greater financial and time outlays), which will bring significant energy savings, but at the same time the financial return on investment will take longer (LI).

Moreover, the authors of the study put forward a hypothesis, which will be verified later, that the above division was mainly conditioned by financial possibilities or motives, indicated as a significant factor influencing energy saving by the surveyed respondents.

The research procedure was planned and carried out in four successive stages:

Stage 1. Defining two reference energy consumers: long-term and short-term investor.
Stage 2. Calculating the distance between each respondent and reference objects.
Stage 3. Assigning the respondent to the appropriate class (SI or LI).
Stage 4. Characteristics of the classes obtained.

Table 2 Variables used for class profiling

Type	Question in the survey representing the relevant variable
Financial aspects	What is your view on the cost of electricity?
	Are you ready to pay more for green energy?
	Have you used any co-financing to change the heating or thermal modernization of the building?
	Are you interested in a subsidy for environmental protection?
	Have you used the prosument program?
Aspects related to environmental protection	Do you heat your house with coal?
	Do you have renewable energy sources?
	How much do you spend on devices related to environmental protection?
	Do you pay attention to the eco-label when shopping?
Household characteristics	What is the average monthly income per one member of your household?
	How many people make up the household?
	How many people in your household work?
	What type of apartment/house do you have?
	What municipality do you live in?

The assumed classification of energy users determined the definition of "ideal consumers" in terms of their attitude to energy saving. It has been assumed that the actions determining the attitude of a long-term investor who is willing to take effort and bear financial costs, despite a much longer repayment period, are:

- purchase of energy-saving kitchen appliances (X_3) as well as household appliances (X_7),
- switching to two energy tariffs: day and night (X_6),
- carrying out thermal modernization of the building (X_8),
- installation of a system for obtaining energy from renewable sources (X_9).

The performance of the remaining activities is not characteristic of any of the adopted types. A short-term investor will probably decide to replace light bulbs with energy-saving ones much sooner than, for example, to buy a refrigerator with a higher energy class, however, a person who thinks about saving energy in the long term, usually also changes light bulbs or pays attention to turning off unnecessary lighting. Therefore, the statements represented by the variables: X_1, X_2, X_4 and X_5 are not essential for defining reference objects.

This means that the reference long-term investor will be represented by observation:

$$P_{LI} = (*, *, 1, *, *, 1, 1, 1, 1), \tag{1}$$

where, as previously, "1" means indicating a given option, and "*"—providing any answer.

However, the reference energy consumer with a short-term approach, i.e. “anti-ideal consumer”, will have the form:

$$P_{SI} = (*, *, 0, *, *, 0, 0, 0, 0). \tag{2}$$

For reference objects defined in this way, the distances have been calculated between:

- every observation O_i (for $i = 1, \ldots, 1147$), showing the answers of the respondent with the number i, and as a reference for a long-term investor

$$d_{iLI} = d(O_i, P_{LI}), \tag{3}$$

- every observation O_i (for $i = 1, \ldots, 1147$) and the anti-ideal of the short-term investor

$$d_{iSI} = d(O_i, P_{SI}). \tag{4}$$

Due to the scale of the variables' measurement X_1–X_9, the Sokal-Michener metric was used to calculate the distance (Rogers and Tanimoto 1960; Walesiak 2011)—a weighted variant of this measure was used.

Moreover, it was decided to use weights to calculate the distance, due to the varying degree of difficulty and the amount of potential investment costs. The purchase of new, energy-efficient home appliances or the switch to two energy tariffs requires much less investment on the part of the investor than the thermal insulation of the house or the installation of a renewable energy system. It was found that the indication of the latter two activities determines the respondent's behaviour and clearly proves the nature of the investment. Therefore, when calculating the aforementioned distances between the observations and the reference points, a vector of weights was adopted:

$$w = (1, 1, 1, 1, 1, 1, 1, 2, 2). \tag{5}$$

Consumer number i has been classified to the group of long-term investors if the observation that represents him/her O_i lies closer to the reference/ideal consumer P_{LI} than the anti-ideal consumer P_{SI}

$$d(O_i, P_{LI}) < d(O_i, P_{SI}). \tag{6}$$

Similarly, the respondent i was assigned to the class of short-term investors, if

$$d(O_i, P_{LI}) \geq d(O_i, P_{SI}), \tag{7}$$

where the distance d is calculated using the weighted Sokal-Michener metric, implemented by the presented proprietary, original procedure.

After all respondents were assigned to one of two groups (LI or SI), the last stage of the research procedure was conducted, i.e. these groups were characterized, and then, class profiling was performed. For this purpose, it was examined what influence on the classification result had the following variables:

- used to segment energy consumers (X_1–X_9),
- characterizing the respondents (presented in Table 2).

Taking into account that all analysed variables are measured on poor measurement scales, chi-square statistics was used to investigate whether the considered relationships are significant. In addition, a Cramer's *V* coefficient was also used in order to determine the strength of this dependency.[2]

4 Results and Discussion

The procedure described in the previous chapter allowed for the classification of the respondents into two groups. On the basis of the obtained results, it was found that among the respondents from the Silesian Province, 326 electricity users (28.4% of the sample) were assigned to the group of long-term investors (LI), and 821 (71.6%) were designated as short-term investors (SI) (Fig. 1).

When studying the behaviour of individual electricity consumers (cf. Ropuszyńska-Surma and Węglarz 2018b), it can be stated that in households, also in the Silesian Province, activities aimed at achieving energy savings are divided into two basic groups: (1) using energy-saving devices and changing the habits of energy use to more efficient ones, and (2) investing in thermal modernization of buildings and installing renewable energy sources, which is associated with the transition from the function of a passive electricity consumer to an active participant in the energy market, and acting as a prosumer. Already existing research (Ropuszyńska-Surma and Węglarz 2018b; Zhou and Yang 2016), on the one hand, confirms that behavioural factors have a significant impact on energy consumption in households, and on the other, indicate that financial factors and the possibility of long-term savings are the greatest incentives to invest in RES systems. The cited studies also point to the potential barriers to becoming a prosumer, the most important of which seem to be: high costs and a complicated installation process as well as the lack of technical possibilities of their application. It should also be remembered that the social acceptance of installing this type of energy sources is also important and translates into an increase in investments.

The analysis of the subject literature and the conducted empirical research confirm the correctness of the division adopted by the authors into two groups of

---

[2]As it was already mentioned, in conducted study, a non-random sample selection was applied. However, the authors used the tools of mathematical statistics, hoping to select a sample close to a random one. Considering the use of only descriptive statistics measures in the characteristics of the obtained groups would significantly reduce the value of this part of the study.

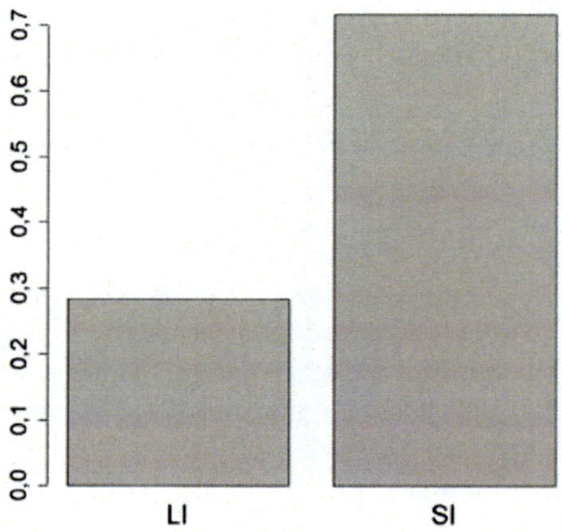

Fig. 1 Division of energy users into short-term and long-term investors

energy consumers in terms of short-term or long-term investments and potential energy savings, while the obtained results and disproportion in the number of segments are not surprising. It can be assumed that the reason for this is the tendency of consumers to *discount future energy savings* (Cabinet Office Behavioural Insights Team 2011, pp. 6–10), i.e. greater attention to achieving immediate effects without incurring high costs, even when the long-term effects of the investment may turn out to be much higher. Moreover, one could risk the statement that consumers focus more on short-term ad hoc measures, as especially in the case of energy efficiency they find it difficult to understand the long-term benefits that are likely to be achieved. Investments in energy efficiency can be perceived as unattractive or requiring a lot of knowledge and patience.

4.1 Characteristics of the Short-Term and Long-Term Investor Classes

Taking advantage of the division of respondents into two classes: short- and long-term investors, the distribution of answers to individual questions from people from both groups was analysed.

First of all, it is worth paying attention to the fact that increasing the weights for the last two variables, representing questions regarding investments in thermal modernization of the building (X_8), and in systems for obtaining energy from renewable sources (X_9), caused that all respondents who declared such actions were assigned to the long-term investors (LI) class. Among the short-term investors (SI), however, there are not those who, in response to the question about the forms of energy saving, indicate the eighth or ninth statement (Fig. 2). It was these investments that required the greatest financial outlays and long-term energy-saving way of thinking. Thus, they clearly show the nature of the investment, and, in some

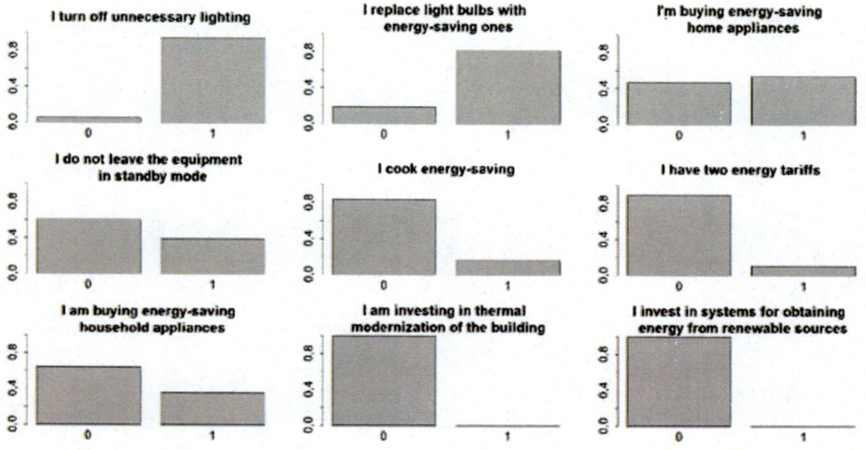

Fig. 2 Distribution of answers into individual segmentation questions of respondents classified as short-term investors

measure, determine the behaviour of the respondents. The increase in weights was aimed at classifying people with a long-term strategy of activities to the group of long-term investors, which was successful.

Comparing the distribution of short-term and long-term responses of investors (Figs. 2 and 3), it can be noticed that people from the second group more often buy energy-efficient kitchen appliances (71.5% of LI people and only 53.4% SI) and household appliances (55.2% LI and 35.4% SI). Moreover, a larger percentage of people with a long-term attitude to limiting energy consumption declare an energy-saving manner of cooking (24.9% LI vs. 16.3% SI). Such people are also more likely to decide on two energy tariffs (24.7% LI vs. 10.8% SI).

Many surveyed respondents indicated performing activities such as turning off unnecessary lighting (93.3% LI and 94.5% SI), or replacing light bulbs with energy-saving ones (83.4% LI and 81.6% SI). However, in this case, the fractions of people who marked these responses were very similar, suggesting that these activities will not significantly affect the obtained classification. Such a high percentage of declarations of this type of behaviour (in both classes) may be the result of information and education campaigns in the media related to energy saving as well as the methods of this saving. The most promoted and popular methods, also available to every consumer, are only illuminating the rooms where somebody stays in, as well as the use of energy-saving light sources (cf. Murawska and Mrozińska 2016).

In the next step, it was verified which of the considered energy-saving statements represented by the X_1–X_9 variables significantly affect the result of the obtained classification. For this purpose, the chi-square statistics (χ^2) and the Cramer's V coefficient were applied (Table 3). It is quite obvious, due to the previously adopted research assumptions, that the greatest impact on which class the respondent will be

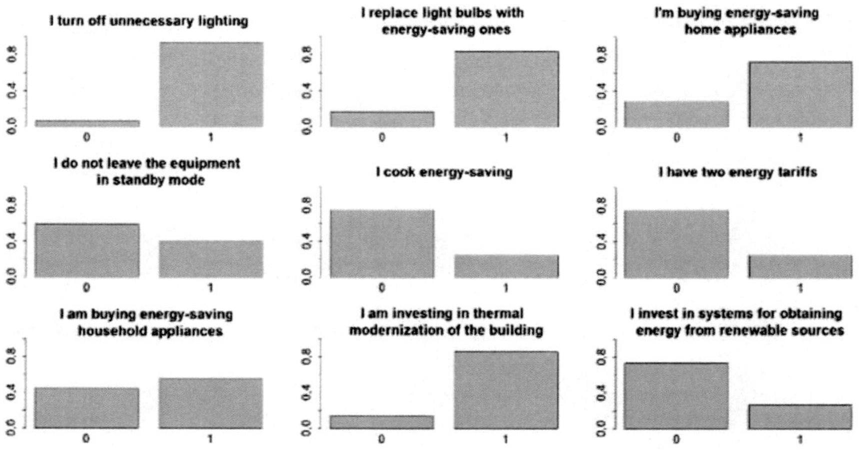

Fig. 3 Distribution of answers into individual segmentation questions of respondents classified as long-term investors

Table 3 Research results of the influence of individual questions on the obtained classification

Question	χ^2	p–value	V–Cramer
I invest in thermal modernization of the building	932.88	0	0.902
I invest in systems for obtaining energy from renewable sources	240.04	0	0.457
I buy energy-saving household appliances (TV, computer, etc.)	37.69	<0.001	0.181
I have two energy tariffs: day and night	36.26	<0.001	0.178
I buy energy-saving kitchen equipment	31.57	<0.001	0.166
I cook in an energy-saving manner	11.13	0.001	0.099
I turn off unnecessary lighting	0.68	0.409	–
I replace light bulbs with energy-saving ones	0.53	0.467	–
I do not leave the equipment in standby mode	0.21	0.650	–

assigned to has the answer that the thermal insulation of a building was performed (0.902). We observe a moderate dependency between the grouping result and the declared investments in systems for obtaining renewable energy (0.457). The purchase of energy-saving household appliances (0.181) and kitchen equipment (0.166), using two energy tariffs (0.178) or proper cooking (0.099) has an even weaker, but still significant influence on the obtained classification. As expected, behaviours such as turning off unnecessary lighting, replacing light bulbs with energy-saving ones, or not leaving the equipment in standby mode did not significantly differentiate the respondents.

4.2 *Profiling of the Short-Term and Long-Term Investor Classes*

At a further stage of the analysis, the influence of factors that could potentially affect the respondents' behaviour, and thus determine their belonging to the short-term and long-term investors class, was studied.

It has already been hypothesized before that the division of the respondents into the two groups in question is largely due to financial reasons, as promising, long-term investments require adequate cash. This hypothesis was verified again using the chi-square test. It was verified whether the average monthly income per one member of the respondent's household affects the result of the classification. The obtained *p*–values are the evidence of a significant dependence, though are weak in strength ($V = 0.149$) (Table 4). Therefore, it can be assumed that income is a factor that plays an important role on the actions of the respondents, and thus also on the result of clustering.

It was also interesting to check whether the fact that the respondents belonged to a certain class is related to the willingness to pay more for green energy (WTP). Research on the propensity of households to bear additional charges when using green energy seems to be extremely important in order to identify the existing patterns of individual energy consumption. However, it should be remembered that the obtained results indicate only the respondents' statements and not their actual behaviour. In Poland, little research of this type has been carried out so far. One of them is Profiling End User of Renewable Energy Sources among Residential Consumers in Poland (Ropuszyńska-Surma and Węglarz 2018a) conducted in 2018 and indirectly related to WTP research on the level of environmental awareness of energy consumers in the Silesian Province (Słupik 2015) carried out in 2015. The authors of the first document also made a short review of the existing subject literature on WTP for green energy, which shows a clear conclusion that there is a positive correlation between WPT for renewable energy and such variables as household or monthly income, electricity consumption and place of residence (rented or own flat/house). Moreover, the revealed other factors influencing specific attitudes of respondents in this respect turned out to be: individual beliefs; age of the respondents and their previous experiences with energy consumption, and the level of education. Research conducted in Poland also confirms these observations.

The question about willingness to pay in the current study combines financial and pro-environmental aspects. The environmental awareness of some respondents

Table 4 Study results of the impact on the obtained classification of the most important questions characterizing the respondent

Question	χ^2	*p*–value	*V*–Cramer
What is the average monthly income per one member of your household?	25.39	<0.001	0.149
Are you ready to pay more for green energy?	1.25	0.534	–

is not necessarily correlated with a sufficiently high income, which may significantly prevent such preferences. However, the people classified in the group of long-term investors declared that they had funds to carry out the relevant investments. It was also shown that their activities were significantly influenced by income, which leads to the conclusion that these people can probably afford to bear higher costs, if some of the energy supplied to them is the so-called green energy. Nevertheless, the result of the chi-square test shows no significant relationship (Table 4). The willingness to pay higher for renewable energy is therefore not related to the type of investor.

Due to the fact that 84% of all respondents indicated that they are not willing or not sure if they are willing to pay more for renewable energy, the reasons for such decisions were asked. In both the LI and SI segments, financial issues turned out to be the main motive. The respondents believe that the price of electricity is already too high (64% of indications in the LI segment and 58.53% in SI), or they say that they cannot afford to pay larger amounts (38.56% and 36.63%, respectively). It can be assumed that in 2018, there were still relatively few regional initiatives or programs enabling consumers to apply for co-financing of pro-ecological activities related, for example, to changing the heating sources of residential buildings to pro-ecological ones, which also influenced the price of electricity. It is also likely that the ecological awareness of the inhabitants of the Silesian Province was lower than it is currently observed. On the other hand, a positive phenomenon is the fact that in 2018, only a small number of respondents did not see the benefits of using green energy (9.71% LI and 12.42% SI), and only 14.29% of long-term investors and 9.89% of short-term investors believe that the environment, or rather caring for it, is not a sufficient reason to pay more for energy. Analysing other factors influencing WTP to green energy, it can be concluded that respondents willing to pay more for green energy are mainly:

- people living in a detached or terraced house (55.09% of all respondents declaring their willingness to pay more for green energy),
- people living in municipalities (84.52%),
- people living in households consisting of 3 (24.55%), 4 (23.35%) or 5 (28.14%) persons in the household,
- people declaring achievement of an average net monthly income per 1 person in a household in the amount higher than or equal to PLN 1,000 (63.7%), of which as many as 21.4% declare achieving an average net income in the amount exceeding PLN 2,000 net per 1 person in the household.

The financial aspect of the research was extended with four additional questions from the questionnaire (Table 5). Namely the respondent was asked whether s/he intends to use co-financing to change the heating system or to modernize the building. In addition, what is his/her relation to electricity costs and whether s/he used the prosument program. After the analysis, it turned out that the answers to each of these questions significantly differentiate the respondents, although once again it is only a weak correlation (Table 5). It can therefore be confirmed that the financial

Table 5 Study results of the impact on the resulting classification of questions related to financial aspects

Question	χ^2	*p*–value	*V*–Cramer
Have you used any co-financing to change the heating or thermal modernization of the building?	56.43	<0.001	0.223
Are you interested in a subsidy for environmental protection?	23.04	<0.001	0.143
What is your view on the cost of electricity?	10.51	0.015	0.097
Have you used the prosument program?	6.42	0.040	0.075

aspects influence the affiliation of the respondents to the groups of short- and long-term investors.

The respondents classified in the LI segment mostly (62.7% of responses) believe that the price of electricity is too high. The response was similar in case of 56.52% of energy consumers characterized by short-term investments (SI segment). This confirms the hypothesis put forward earlier that it is the financial issues that are an important motive for the division into individual segments. However, in the second group of respondents, at the same time, more than 40% of respondents accept the current price of electricity.

It should be noted that proper informing consumers on energy prices and its daily consumption plays an extremely important role here. When residents rarely receive energy bills, reports or information from their suppliers, most people are not able to realize which of their daily behaviour contributes most to their energy bills or which simple changes need to be made to lower their bills. A simple solution in this case seems to be the widespread introduction of smart meters or the dissemination of IT tools for managing energy consumption at homes, such as the aforementioned *eco-bot*. By using this type of opportunity, the consumer will receive real-time feedback on energy consumption and the impact of the introduced behaviour changes on its consumption, and will even be able to obtain other support in the form of opinions or individually tailored recommendations.

As the experimental research of behavioural economics indicates, consumers are not very often guided by rational motives in their decisions. On the other hand, there was a large discrepancy between people's values and material interests and their actual behaviour (Frederiks et al. 2015). It also influences investing in RES. The sense of uncertainty regarding electricity supplies, market prices, government and local government policy additionally occurs; but there is also uncertainty as to the rules for obtaining possible support and long-term financial repayments. This state of affairs makes investing in energy-efficient products and services seem like a risky decision for many consumers (Frederiks et al. 2015).

In Poland, prosumer investments in renewable energy sources can develop mainly thanks to the support of EU funds. Primarily, the funds distributed on the basis of regional operational programs (ROPs) implemented by provincial self-governments should be mentioned here. Although natural persons cannot apply for aid directly from the ROP, they use funds distributed, e.g. by the municipality from specific programs such as: "Prosumer"; "Prosumer 2"; "Clean Air" or "My Electricity".

As in the case of the survey conducted in 2018, the respondents in both segments generally indicated that they had never used a funding or subsidy to change the heating system of the house/flat, or to modernize their buildings (72% LI; 79.46 SI), and had not used the prosumer program operating in 2015–2017 (51.39% LI; 58.48% SI). On the other hand, the majority of respondents in each of the pointed out segments was interested in receiving subsidies for environmental protection purposes (75.78% LI; 61.49% SI). The interest was in changing the method of heating; building insulation, etc. But at the same time, a very large percentage of respondents (in each segment over 40% of respondents) did not hear about the prosument program, which may indicate insufficient information campaigns by decision-makers at that time and the existence of very large information gaps influencing decisions made by consumers.

There is also the question of how the obtained classification is influenced by the respondents' attitude towards environmental protection. Four subsequent questions from the survey questionnaire were used in this study (Table 6). The obtained results mean that both the expenditure on devices related to environmental protection, the use of renewable energy sources, and paying attention to the ecological signs significantly influence the obtained classes of investors. Even heating a house with coal has a significant (albeit again slight) influence on group membership (Table 6).

The surveyed respondents are significantly differentiated by almost each of the factors presented above. For the purchase of products and devices that directly protect the environment, such as energy-saving light bulbs; building insulation and changing the heating method, relatively larger amounts are spent annually by people classified as long-term investors (LI). Consumers in this segment usually spend amounts in the range of PLN 150–299 (26.65%); PLN 300–900 (25.7%), and even 11.91% of people declare eco-expenditures in amounts exceeding PLN 1000. The expenditures of short-term investors (SI), on the other hand, oscillate mainly in the range of PLN 100–299 (55.57%). A significantly lower percentage of these people declare expenses exceeding PLN 1000 (3.13%) or in the range of PLN 300–900 (21.53%). For comparison, based on the respondents answers concerning their received income, it can be estimated that half of the surveyed households had per

Table 6 Study results of the impact on the resulting classification of questions related to environmental protection

Question	χ^2	*p*–value	*V*–Cramer
How much do you spend on devices related to environmental protection?	45.77	<0.001	0.202
Do you have renewable energy sources?	28.10	<0.001	0.159
Do you pay attention to the eco-label when shopping?	12.71	<0.001	0.107
Do you heat your house with coal?	6.53	0.011	0.077

capita income that did not exceed PLN 1412. Moreover, 25% of the households with the strongest financial situation in the group reported per capita income not lower than PLN 1950.[3]

Also, a higher percentage of consumers from the LI segment (48.10%) pays attention to the eco-label when purchasing goods, such as Energy Star, Ekoland, Zielony Punkt and others, compared to the SI segment, where such attention is declared by 36.49% of respondents. Being familiar and paying attention to ecological signs, especially those relating to energy efficiency, in everyday purchasing choices is a manifestation of the so-called *greening of consumption* (Matel et al. 2018, p. 405). This phenomenon is most often perceived as changing consumer behaviour by making more responsible and environmentally friendly choices. On the one hand, it may result from both care for the state of the environment, as well as socio-cultural conditions (such as demonstrating a specific lifestyle (Bylok 2014, p. 30), caring for health, slowing down the pace of life, deconsumption) and economic conditions (such as the increase in energy and raw material prices). It can also be noticed that people's behaviour is generally not contrary to their environmental concerns and obligations only because they also want to satisfy their material needs or look for non-ecological benefits.

At this stage of the study, the respondents also declared having renewable energy sources systems. Here, too, we can see significant differences between the segments. 5.97% LI and only 0.76% of SI have such systems, while the main source of energy in households is still coal (44.97% LI and 36.7% SI), so it can be assumed that the investment potential is quite large. On the other hand, already existing research results (Frederiks et al. 2015) indicate that many consumers—even when faced with the clear profitability of investing in energy-saving systems or measures contributing to energy saving—remain reluctant to introduce them into their lives and homes. Perhaps this is due to the belief of consumers that such activities require considerable effort, technical knowledge on the systems or concerns about the inability to operate them later. Therefore, it is very important to inform consumers reliably about the existing possibilities and methods of energy reduction, dispel all uncertainties and fears, as well as introduce comprehensive support systems; also financial ones or combined with other possible benefits (e.g. of non-financial nature).

The final stage is an analysis on which factors characterizing the respondent's household affect the classification result. The conducted statistical tests showed that membership in clusters significantly depends on the number of people composing the household, the type of premises inhabited and the type of commune in which the respondent lives. Only the number of people working in the household has no significant impact (Table 7).

[3]Some respondents were reluctant to provide information concerning their financial situation (24.5% of respondents refused to share information on income). Moreover, the last position in the rating scale question on declared income was open, so the remaining group of the respondents had only quantiles determined in the distribution of income.

Table 7 Study results of the impact on the obtained classification of the questions characterizing the household

Question	χ^2	p–value	V–Cramer
How many people make up the household?	69.64	<0.001	0.247
What type of apartment/house do you have?	24.28	0.002	0.146
What municipality do you live in?	9.89	0.007	0.093
How many people in your household work?	9.59	0.213	–

In the surveyed sample, consumers classified in the LI segment usually indicated an urban municipality as their place of residence (75.93%), then a rural municipality (14.51%) and the smallest number indicated an urban and rural municipality (9.57%). People in this segment live mainly in a detached or terraced house (74.3%), and their household usually consists of 3–5 people (75.39%). The smallest percentage of one-person households (1.25%) of all surveyed households is also in this segment. In the SI segment, the respondents live mainly in urban municipalities (83.9%) in tenement houses or in flats in a block of flats (53.01%), and their households consists mostly of 2–4 people (72.82%).

Individual energy consumption patterns vary widely, as different factors influence consumer decisions. As J. Popczyk rightly points out, "social (lifestyle) changes are slow, because they are more profound, compared to the technological ones", but thanks to technological changes "*homo economicus* is transformed into a behaviourist" (Popczyk 2014, p. 30), and the importance of individualism increases, which influences the growth of autonomy in the actions of consumers on the market (Bylok 2014). The experience of other researchers (Zhou and Yang 2016; de Almeida et al. 2011) also reveals that behavioural factors have a significant impact on household energy consumption, and that these households have a high potential for savings in this regard. According to the estimates of the European Commission, this potential may even be 27% (European Commission 2006 after Zhou and Yang 2016, p. 811). Therefore, it is extremely important to understand the behaviour of energy consumers and the possibility of shaping these behaviours in the direction of increasing pro-ecological activities and improving energy efficiency. Promoting more sustainable consumption requires an approach engaging all interested parties, including government policy, market innovation, the mobilization of consumer groups through NGOs, and individual consumer initiatives alone (OECD 2002). This can be facilitated by the development of comprehensive intervention and information strategies with appropriate financial support programs that could stimulate house and flat owners to change their behaviour, promote sustainable development and effectively communicate the influence of energy consumption on the climate change.

5 Conclusions

One of the most important reasons why energy consumers do not invest in energy efficiency (whether through changing their behaviours, habits or lifestyles, or through green investments) is the lack of awareness of the high energy waste issue (Bator and Kukuła 2016). Moreover, although many energy consumers declare their concern for the natural environment and support limiting the negative impact of consumption or limiting the emission of harmful compounds into the atmosphere, this does not always translate into taking practical steps to reduce consumption (Frederiks et al. 2015)

As the research conducted in 2018 showed, for households in the Silesian Province, financial issues were a significant motivation for taking actions related to the reduction of energy consumption. As indicated in the analysis, consumer behaviour may result from both ecological (high level of ecological awareness or the will to care for the natural environment) and non-ecological reasons (where financial issues dominate, but also following existing trends or influences of social groups). Thinking and caring for the environment significantly differentiate the obtained segments. As revealed by the conducted research, long-term investors:

- more often benefit from co-financing for modernization of heating systems and thermal insulation,
- tend to spend larger amounts on environmental protection equipment,
- pay more attention to the eco-label when doing shopping,
- are more interested in receiving an environmental subsidy.

Also income significantly differentiates the obtained groups in a similar manner. As could be expected, long-term pro-ecological investments are more often made by people with higher incomes. Nevertheless, in both groups, a similar percentage of respondents is not willing to pay more for green energy. The ecological awareness of poles in 2018 was much lower than today. The conducted research confirms this conclusion; the environmental motivation of the households in the Silesian Province, which have mostly been classified among the short-term investors sector, is much lower. Moreover, the higher percentage of respondents from the SI sector have not used or heard about the prosument program. Owing to the fact that the attitudes of poles regarding electricity consumption and environmental protection are dynamically changing, it is important to continue research and observe the developing trends. In 2018, the public was to a small extent ready for long-term pro-ecological investments, but it can be hypothesized that the percentage of respondents who could be classified as long-term investors has currently increased. This issue, nonetheless, requires further, in-depth research.

Acknowledgments This work is supported by European Union's Horizon 2020 research and innovation program: "Reducing energy consumption and carbon footprint by smart and sustainable use", as a part of the currently implemented project "Personalised ICT-tools for the Active Engagement of Consumers Towards Sustainable Energy. Eco-bot" under grant agreement No. 767625.

References

Accenture (2010) Understanding consumer preferences in energy efficiency. Accenture end-consumer observatory on electricity management 2010. Accenture. https://www.accenture.com/t20160811t002327__w__/us-en/_acnmedia/accenture/next-gen/insight-unlocking-value-of-digital-consumer/pdf/accenture-understanding-consumer-preferences-energy-efficiency-10-0229-mar-11.pdf. Accessed 30 Oct 2020

Albert A, Maasoumy M (2016) Predictive segmentation of energy consumers. Appl Energy 177:435–448

Bator A, Kukuła W (2016) Rola konsumenta w transformacji energetycznej. Fundacja ClientEarth Prawnicy dla Ziemi, Warszawa

Bylok F (2014) Prosumpcja na rynku energii elektrycznej w perspektywie teoretycznej. *Biblioteka Źródłowa Energetyki Prosumenckiej. Klaster 3x20*. www.klaster3x20.pl. Accessed 30 Oct 2020

Cabinet Office Behavioural Insights Team (2011) *Behaviour change and energy use: behavioural insights team paper*. Cabinet Office Behavioural Insights Team. https://www.gov.uk/government/publications/behaviour-change-and-energy-use-behavioural-insights-team-paper. Accessed 30 Oct 2020

Clancy D, O'Loughlin D (2002) Identifying the 'energy champion': a consumer behaviour approach to understanding the home energy conservation market in Ireland. Int J Nonprofit Voluntary Sector Mark 7(3):258–270

Dąbrowska A, Bylok F, Janoś-Kresło M, Kiełczewski D, Ozimek I (2015) Kompetencje konsumentów. Innowacyjne zachowania. Zrównoważona konsumpcja. PWE, Warszawa

de Almeida A, Fonseca P, Schlomann B, Feilberg N (2011) Characterization of the household electricity consumption in the EU, potential energy savings and specific policy recommendations. Energy Build 43(8):1884–1894

European Commission (2006) Communication from the commission—action plan for energy efficiency: realising the potential. European Commission Report. COM (2006), 545 Final

Frederiks E, Stenner K, Hobman E (2015) Household energy use: applying behavioural economics to understand consumer decision-making and behaviour. Renew Sustain Energy Rev 41:1385–1394

Gardner G, Stern P (1996) Environmental problems and human behavior. Allyn & Bacon, Boston

GUS (2020) Bank Danych Lokalnych. Retrieved from https://bdl.stat.gov.pl/BDL/start

Hille S (2016) The myth of the unscrupulous energy user's dilemma: evidence from Switzerland. J Consum Policy 39:327–347

Kiełczewski D (2005) Style konsumpcji jako przejaw zróżnicowania poziomu życia. Gospodarka Narodowa 5–6:87–100

Kiełczewski D (2015) Wpływ ekologizacji konsumpcji na zmiany w zarządzaniu organizacjami. Handel Wewnętrzny 6(359):55–63

Lutzenhiser L (1993) Social and behavioral aspects of energy use. Annu Rev Energy Env 18:247–289

Matel A (2016) Przesłanki ekologizacji konsumpcji z perspektywy zachowań konsumenckich. Reasons for green consumption from the perspective of consumer behavior. Zarządzanie. Teoria i Praktyka 16(2):55–61

Matel A, Poskrobko T, Andrejuk D, Dardzińska M, Kulesza J, Piotrowska B (2018) Różnica między poznawczym a behawioralnym komponentem postaw ekologicznych młodych konsumentów. Handel Wewnętrzny 6(377):404–415

Murawska A, Mrozińska M (2016) Korzystanie z energii elektrycznej w krajach Unii Europejskiej i w Polsce w aspekcie wspierania zrównoważonej konsumpcji [The use of electricity in the European Union and in Poland in terms of promoting sustainable consumption]. Zeszyty Naukowe Szkoły Głównej Gospodarstwa Wiejskiego w Warszawie. Problemy Rolnictwa Światowego 16(XXXI)(2):223–231

Nagaj R (2018) Metody behawioralne w ocenie zachowań konsumentów na rynku energii elektrycznej. Rynek Energii 3(136):3–9

Nakamura E (2016) Electricity saving behavior of households by making efforts, replacing appliances, and renovations: em-pirical analysis using a multivariate ordered probit model. Int J Consum Stud 40:675–684

OECD (2002) Towards sustainable household consumption? Trends and policies in OECD countries. From https://doi.org/10.1787/9789264175068-en. Accessed 30 Oct 2020

Pluskwa-Dąbrowski K (2016) Konsument w energetyce – rzut oka w przyszłość. Federacja Konsumentów. https://www.documents.clientearth.org/wp-content/uploads/library/2016-12-07-konsument-w-energetyce-rzut-oka-w-przyszlosc-ext-pl.pdf. Accessed 30 Oct 2020

Polityka gospodarki niskoemisyjnej dla Województwa Śląskiego. Regionalna polityka energetyczna do roku 2030-projekt (2020) Katowice: Urząd Marszałkowski Województwa Śląskiego. https://www.slaskie.pl/content/gospodarka-niskoemisyjna. Accessed 30 Oct 2020

Popczyk J (2014) Energetyka prosumencka. O dynamice interakcji dwóch trajektorii rozwoju w energetyce: pomostowej/zstępującej i nowej/wstępującej. Europejski Kongres Finansowy. Instytut Badań nad Gospodarką Rynkową – Gdańska Akademia Bankowa & Jan Popczyk

Rogers D, Tanimoto T (1960) A computer program for classifying plants. Science 132:1115–1118

Ropuszyńska-Surma E, Węglarz M (2018a) Profiling end user of renewable energy sources among residential consumers in Poland. Sustainability 10(12):1–21

Ropuszyńska-Surma E, Węglarz M (2018b) Proekologiczne i prooszczędnościowe zachowania gospodarstw domowych jako konsumentów energii. Ekonomia Wroclaw Econ Rev 24(3). Acta Universitatis Wratislaviensis 3881:3–39

SECC (2019) Consumer pulse and market segmentation. Wave 7. Smart Energy Consumer Collaborative. https://smartenergycc.org/consumer-pulse-and-market-segmentation-wave-7-report/. Accessed 30 Oct 2020

Seidl R, Moser C, Blumer Y (2017) Navigating behavioral energy sufficiency. Results from a survey in Swiss cities on potential behavior change. PLoS ONE 12(10):e0185963. From https://doi.org/10.1371/journal.pone.0185963. Accessed 30 Oct 2020

Słupik S (2015) Świadomy konsument energii w województwie śląskim w świetle badań ankietowych. Studia ekonomiczne. Zeszyty Naukowe Uniwersytetu Ekonomicznego w Katowicach 232:215–236

Słupik S, Kos J, Trzęsiok J (2019) Report on findings from consultations and online-survey. Personalised ICT-tools for the active engagement of consumers towards sustainable energy. Eco-bot. Project. www.eco-bot.eu. Accessed 30 Oct 2020

Steg L, Dreijerink L, Abrahamse W (2005) Factors influencing the acceptability of energy policies: a test of VBN theory. J Environ Psychol 25:415–425

Walesiak M (2011) Uogólniona miara odległości GDM w statystycznej analizie wielowymiarowej z wykorzystaniem programu R. Uniwersytet Ekonomiczny we Wrocławiu, Wrocław

Zhou K, Yang S (2016) Understanding household energy consumption behavior: the contribution of energy big data analytics. Renew Sustain Energy Rev 56:810–819

The Use of the Spatial Taxonomic Measure of Development to Assess the Tourist Attractiveness of Districts of the Lesser Poland Province

Jacek Wolak

Abstract Lesser Poland is one of the most diversified regions in Poland in terms of landscape features. Despite a relatively small area, it has many environmental and cultural values. These advantages make this region very popular among tourists. The aim of the study is to build a ranking of the Lesser Poland districts in terms of tourist attractiveness with the use of linear ordering techniques. The presented results show that the techniques considering spatial relationships have only a slight impact on the rankings obtained during the study. The most attractive regions are the city of Kraków and the district of Wieliczka, located in the central part of the province, and the mountain districts in the southern part of the region. The areas with relatively the lowest level of tourist attractiveness are the districts of the northeastern and northern Lesser Poland Province.

Keywords Tourist attractiveness · Linear ordering · Taxonomy spatial measure

1 Introduction

Among the topics of research undertaken in the field of tourism sciences, the research focused on tourist attractiveness in the context of a given place, area, or administrative unit is quite popular. In the literature, there are terms of "tourist value" or "landscape value" which are easier to measure. They, in fact, determine the level of tourist attractiveness of a given place or administrative unit. The assessment of tourist attractiveness can be carried out, among others, by methods of statistical multivariate analysis, and more precisely by linear ordering algorithms.

In the context of analyzes on this subject in selected regions of Poland, one can mention papers on the seaside districts (Oleńczuk-Paszel and Nowak 2010) and Sudeten districts (Gryszel and Walesiak 2018). There are exist works dealing with the topic of tourist attractiveness in some regions of Poland (incl. Bąk and

J. Wolak (✉)
Faculty of Management, AGH University of Science and Technology, Kraków, Poland
e-mail: jwolak@agh.edu.pl

K. Jajuga et al. (eds.), *Data Analysis and Classification*, Studies in Classification, Data Analysis, and Knowledge Organization,
https://doi.org/10.1007/978-3-030-75190-6_12

Matlegiewicz 2010; Binderman et al. 2010; Synówka-Bejenka 2017) or individual province (incl. Puciato 2010; Gryszel and Walesiak 2014; Bąk 2014; Stec 2015; Wolak 2020).

The common feature of the above studies is the fact that the composite measure, which is the basis for the ranking of administrative units, is built on classic models that do not consider the spatial relationships in the diagnostic variables.

It should be noted, however, that when choosing a destination, tourists do not follow the administrative units, but the attractions they have easy access to. Therefore, it seems reasonable to consider the situation described above and introduce spatial dependencies into a composite measure.

In the literature, there are several approaches to order objects based on values not only in the studied objects but also in their neighborhood. There are two types of approaches to the input of spatial information. The proposals presented in Antczak (2013) and Pietrzak (2014) works focus on modifying the values of variables characterized by statistically significant spatial autocorrelation (based on the results of the Moran's test). The approach presented by Sobolewski et al. (2014) and Łysoń et al. (2016) consists of implementing information about the neighborhood only in the final stage of calculations when the values of the composite measure determined in the classic way are known.

The aim of this study is to use selected linear ordering algorithms to build a ranking of districts in the Lesser Poland Province in terms of tourist attractiveness using techniques considering potential spatial relationships. The results will not only complement the conclusions described in the paper of Wolak (2020) but may also be helpful for travel agencies and individual tourists looking for an interesting place to visit. It seems that the construction of a reliable ranking may also be a planning support for local authorities, which will thus be able to plan investments more consciously in the tourist development of the region.

2 Methods

Tourist attractiveness is a complex concept. A great influence on its quantity may have both specific (tourist and landscape) values and appropriate infrastructure. This means that making a reliable ranking is not easy and requires the use of multivariate data analysis techniques.

2.1 *Linear Ordering*

The ordering of objects is based on the value of a composite variable, using linear ordering algorithms. Although the synthetic variable is hidden in nature, its realization is determined by observations of diagnostic variables that are measurable and directly affect the intensity of the examined feature. Diagnostic variables may have one of three properties: stimulant, destimulant, and nominant (for some values it is a stimulant and for the other is destimulant).

2.2 Hellwig's Method

In this study, the starting technique of linear ordering is one of the oldest and still popular in applications (see Balicki 2009; Dębkowska and Jarocka 2013; Stec 2015; Bąk 2016)—method presented by Hellwig (1968). The concept of Hellwig's measure is based on the ordering of multidimensional objects with respect to a normalized distance from a hypothetical pattern. The drawback of this method is the fact that, although it usually takes values from 0 to 1, sometimes (in about 4% of cases) for objects far away from the pattern, it can take negative values. When defining ordering techniques using information about neighboring units, we will use the Hellwig method, so for the sake of consistency, we will relate to the ranking building algorithm using it.

The starting point for its application is the conversion of all variables into stimulants. Then, according to formula (1), standardization of the dataset should be performed

$$Z_{ij} = \frac{X_{ij} - \bar{X}_j}{s_j} \tag{1}$$

where $\bar{X}_j$ and s_j are the arithmetic mean and standard deviation, respectively, for variable X_j.

In the second step, a pattern $Z_0 = (Z_{01}, Z_{02}, \ldots, Z_{0m})$ is built using formula (2)

$$Z_{0j} = \max_{i=1,\ldots,n} \{Z_{ij}\} \tag{2}$$

and the distance between i object and pattern Z_0 is calculated

$$d_{i0} = \sqrt{\sum_{j=1}^{m} (Z_{ij} - Z_{0j})^2}. \tag{3}$$

Finally, the taxonomy measure, according to formula (4), is constructed

$$\mu_i = 1 - \frac{d_{i0}}{\bar{d}_{i0} + 2 \cdot s(d_{i0})} \tag{4}$$

where $\bar{d}_{i0}$ and $s(d_{i0})$ are the arithmetic mean and standard deviation of vector d_{i0}, respectively.

2.3 Spatial Taxonomy Measure

Considering the spatial factor to build a taxonomic measure of development was first discussed in Antczak (2013). The idea behind this proposal is to modify the values of diagnostic variables characterized with a statistically significant spatial autocorrelation (based on the results of the Moran's test).

The author suggests taking the values of their spatial delay instead of the original values of these variables, i.e., $x_{ij}^{*} = Wx_{ij}$, where W is the neighborhood matrix adopted by the researcher. For such data, the standard procedure of Hellwig (1)–(4) introduced in the last subsection, is used.

A different approach was proposed by Pietrzak (2014), who said that diagnostic variables may be characterized by a different intensity of spatial interactions. He proposed to estimate, only for variables X_j with the existence of spatial autocorrelation, the SAR model (spatial autoregressive model) of the form

$$X_j = \rho W X_j + \varepsilon_j \tag{5}$$

Then, for the estimated value of ρ, it should finally transform its values according to the formula (6)

$$Z_j = \begin{cases} (I - \rho W)^{-1} X_j, X_j \text{is spatilly dependent} \\ X_j, X_j \text{ is not spatilly dependent} \end{cases} \tag{6}$$

For the new dataset, the classic Hellwig procedure (1)–(4) is used and therefore the final values are obtained.

Another way to solve this problem occurred in the work (Sobolewski et al. 2014). The authors propose that the spatial element should be introduced into the model at the stage of a designated composite measure. In practice, after calculating the Hellwig coefficient μ_i, for the given weight α ($\alpha = 0.6$ was proposed in the original work) and for the given matrix of neighborhoods $W = [w_{ik}]$ use the formula

$$(\mu_{SSM})_i = \alpha\mu_i + (1-\alpha)\sum_{i \neq k} w_{ik}\mu_k. \tag{7}$$

A very similar definition was proposed in (Łysoń et al. 2016). The authors are focused on the assumption that with the increasement of the distance between the administrative units, the influence of the environment on the examined object decreases in a linear manner. The final form of the formula determining the spatial taxonomic measure of development is

$$(\mu_{LSM})_i = \alpha\mu_i + (1-\alpha)\sum_{i \neq k} d_{ik}\mu_k \tag{8}$$

where μ_i is the value of the synthetic measure for the i unit, $d_{ik} = \max\left\{0; 1 - \frac{d_{ik}^*}{d}\right\}$ is the spatial dependence coefficient, where d_{ik}^* is the distance between the units, d is the predetermined maximum distance (in the original work $d = 50$km), and α the set weight (in original paper $\alpha = 0.25$).

3 Dataset and Results

Lesser Poland, inhabited by below 3.4 million people, is one of the sixteen provinces of Poland. There are 22 districts in this region, including three urban districts: the city of Kraków, the city of Tarnów, and the city of Nowy Sącz.

Despite a relatively small area (approx. 15,000 km^2), Lesser Poland has many unique, even on a worldwide scale, cultural values. There are eight groups in its area, out of 16 located in the country, entered as cultural and natural heritage on the UNESCO list,[1] as well as 11 groups of monuments recognized as Monuments of History[2] (this is about 10% of all objects of this type in Poland).

In terms of topography (uplands and mountains constitute 53% of the region's area), it is the most diverse province in Poland in terms of nature and landscape, extremely rich in environmental values. Six out of 23 Polish national parks are located on its territory. Moreover, there are 11 landscape parks, 10 protected landscape areas, over 80 nature reserves, and almost 2,200 nature monuments.[3]

3.1 *Dataset*

In the first stage of the empirical study, a preliminary, substantive, and formal analysis was performed for 21 diagnostic variables characterizing the tourist attractiveness of the districts of the Lesser Poland Province. The data was obtained from the Local Data Bank of the Central Statistical Office (data for 2019) and the Register of Monuments (data for July 2020). Due to the fact that a frequent practical problem in linear ordering is a strong positive asymmetry of selected diagnostic features (Głowicka-Wołoszyn and Wysocki 2020), it was decided to limit high values to the value of the upper whisker, i.e., $Q_3 + 1.5 \cdot (Q_3 - Q_1)$.

[1] UNESCO World Heritage List. https://whc.unesco.org/en/list. Accessed December 1, 2020.

[2] Obiekty wpisane przez Prezydenta RP na listę Pomników Historii. https://www.prezydent.pl/aktualnosci/pomniki-historii/obiekty-wpisane-na-liste-pomnikow-historii/ Accessed December 1, 2020.

[3] Małopolska. https://www.malopolska.pl/publikacje/srodowisko-i-geologia/malopolska-parki-narodowe-i-krajobrazowe-rezerwaty-przyrody. Accessed December 1, 2020.

Table 1 Diagnostic variables used in study

Variable	Unit	Type
Environmental values		
X_1—the share of protected areas	Part of district area	Stimulant
X_2—number of natural monuments	Pcs. per 100 km^2 of district area	Stimulant
X_3—emission of pollutants	Vol. per 1 km^2 of district area	Destimulant
X_4—afforestation	Part of district area	Stimulant
Cultural values		
X_5—number of monuments	Pcs. per 100 km^2 of the district area	Stimulant
X_6—number of museum visitors	Person per 100 km^2 of the district area	Stimulant
X_7—number of screenings in cinemas	Pcs. per 100 km^2 of the district area	Stimulant
X_8—organized mass events	Pcs. per 100 km^2 of the district area	Stimulant
Infrastructure and security		
X_9—number of bed places	Pcs. per 100 km^2 of the district area	Stimulant
X_{10}—number of crimes	Pcs. per 1000 of the district pop.	Destimulant
X_{11}—length of bicycle paths	Km per 100 km^2 of the district area	Stimulant
X_{12}—network of municipal/district roads	Km per 100 km^2 of the district area	Stimulant

Source Own calculations

As a result of the formal analysis (it is assumed that the level of correlation between variables cannot be greater than 90), twelve diagnostic features were obtained. It was decided to divide them into three main categories (Tables 1 and 2).

3.2 *Empirical Study*

The aim of the empirical study is to obtain the ranking of the Lesser Poland districts using three spatial taxonomy measures, proposed by Pietrzak (2014), Sobolewski et al. (2014) and Łysoń et al. (2016). In this paper, these methods will be called: Pietrzak, SMM, and LSW methods.

Firstly, a spatial autocorrelation has been checked by Moran I test (spatial contiguity weight matrix W is considered). The results of which are presented in the second and third column of Table 3. As can be seen, at the significance level 0.05, in the case of four variables, we can talk about the statistically significant of a spatial relationship. These are: afforestation, the number of visitors to museums, the number of beds, and the road network.

To use the algorithm proposed in Pietrzak (2014), for each variable indicating the existence of spatial relationships, it was estimated the SAR model using formula (5). The estimates of the ρ parameter are presented in the last column of Table 3.

Table 2 Descriptive statistics for the diagnostic variables used in the study

Variable	Mean	Sd	Median	Min	Max	Skew	Kurtosis
Protected areas	40.8	33.3	31.9	0.1	92.9	0.23	−1.67
Natural monuments	15.8	11.5	13.3	2.8	37.1	0.78	−0.83
Emission of pollutants	578.1	917.7	44.8	4.3	2534.6	1.27	−0.02
Afforestation	24.6	15.3	22.7	1.5	48.1	0.05	−1.59
Monuments	33.9	20.0	28.7	10.4	70.6	0.73	−0.74
Museum visitors	9584.4	11438.3	5181.0	391.0	33911.6	1.22	−0.01
Screenings in cinemas	1661.9	1201.6	1446.0	0.0	3768.0	0.58	−0.78
Organized mass events	2.5	2.5	1.2	0.2	6.7	0.93	−0.88
Bed places	623.4	502.9	491.9	42.8	1541.2	0.80	−0.74
Crimes	12.1	6.8	9.1	4.5	25.9	0.85	−0.64
Bicycle paths	2.6	2.2	1.9	0.05	6.4	0.76	−0.95
Road network	166.2	42.5	168.5	113.5	240.3	0.28	−1.13

Source Own calculations

Table 3 Results of Moran I test and parameter ρ estimate in SAR model

Diagnostic variable	Moran I	*p*-value	Rho
Protected areas	0.189	0.064	–
Natural monuments	0.089	0.184	–
Emission of pollutants	0.104	0.153	–
Afforestation	0.279*	0.018	0.500
Monuments	0.038	0.286	–
Museum visitors	0.219*	0.036	0.521
Screenings in cinemas	−0.019	0.426	–
Organized mass events	0.107	0.155	–
Bed places	0.228*	0.035	0.312
Crimes	−0.224	0.878	–
Bicycle paths	−0.065	0.545	–
Road network	0.258*	0.023	0.572

Source Own calculations
* denotes statistical significance at significance level 0.05

Finding the ρ value allows for the transformation of the output variables according to formula (6) and after standardization it is enough to carry out steps (2)–(4) in the Hellwig algorithm. As a result, the Pietrzak measure is obtained (Table 4).

In the next phase, the LSW and SMM measures are determined. Both require an initial calculation of a synthetic measure. It will be determined based on Hellwig's algorithm. Now, for $\alpha = 0.6$ and for the spatial contiguity weight matrix W, the SMM measure can be derived. Its values are also presented in Table 4.

For the calculation of the LSW measure, the value of $\alpha = 0.25$ proposed by the authors was adopted, and $d = 40$ km was assumed as the border distance of the

Table 4 Results of the analysis of the diversity of tourist attractiveness of districts in the Lesser Poland Province based on Hellwig, Pietrzak, LSW, and SMM methods

District	Hellwig method		Pietrzak method		LSW method		SMM method	
	Value	Place	Value	Place	Value	Place	Value	Place
The city of Kraków	0.413	1	0.419	1	0.373	1	0.368	1
Wieliczka district	0.390	2	0.377	2	0.345	4	0.330	4
Nowy Sącz district	0.379	3	0.366	3	0.348	3	0.310	5
Oświęcim district	0.364	4	0.353	5	0.315	7	0.300	6
Tatry district	0.361	5	0.361	4	0.354	2	0.350	3
The city of Nowy Sącz	0.338	6	0.345	6	0.332	6	0.353	2
Nowy Targ district	0.335	7	0.333	7	0.340	5	0.289	7
Wadowice district	0.286	8	0.278	8	0.255	8	0.246	8
Bochnia district	0.272	9	0.260	9	0.239	9	0.227	9
Tarnów district	0.229	10	0.218	11	0.205	13	0.223	10
The city of Tarnów	0.217	11	0.225	10	0.210	12	0.222	12
Kraków district	0.212	12	0.205	12	0.237	10	0.217	13
Gorlice district	0.175	13	0.176	13	0.221	11	0.223	11
Chrzanów district	0.168	14	0.163	14	0.186	14	0.194	14
Brzesko district	0.168	15	0.163	15	0.180	16	0.177	16
Miechów district	0.154	16	0.145	17	0.137	19	0.164	19
Myślenice district	0.138	17	0.148	16	0.186	15	0.185	15
Limanowa district	0.131	18	0.120	19	0.170	17	0.173	17
Olkusz district	0.127	19	0.129	18	0.137	20	0.146	20
Sucha district	0.114	20	0.118	20	0.148	18	0.172	18
Dąbrowa district	0.053	21	0.056	21	0.098	21	0.121	21
Proszowice district	0.024	22	0.024	22	0.086	22	0.092	22

Source Own calculations

influence of neighbors on the examined administrative unit. The obtained results are also presented in Table 4.

As can be seen (Table 4), the ranking only slightly depends on the used technique of introducing information about the tourist attractiveness. The results achieved by Pietrzak's method are very similar to the classical Hellwig's method. As it turns out, the spatial relationship has a minimal effect on the ranking. It seems that the reason for such a situation is the fact that among the twelve diagnostic variables, only for four (see Table 3), Moran's I test indicated a statistically significant spatial autocorrelation.

Changes in the ranking caused by considering spatial influences, obtained with the use of LSW and SMM techniques, although still small, are quite more visible. In this case, the obtained result with a certain weight depends on the assessment of tourist attractiveness in the neighboring administrative units. Taking this fact into account contributes to a certain advance in the ranking of districts: Tatry, Gorlice, Myślenice, and Sucha, and decrease in the Oświęcim and the Miechów districts.

Table 5 τ-Kendall rank correlation coefficients for various linear ordering techniques

Method	Hellwig	Pietrzak	LSW	SMM
Hellwig (1968)	1.000	0.965	0.853	0.879
Pietrzak (2014)	0.965	1.000	0.870	0.879
Łysoń et al. (2016)	0.853	0.870	1.000	0.905
Sobolewski et al. (2014)	0.879	0.879	0.905	1.000

Source Own calculations

It should be noted that these changes are small, which is influenced by a relatively spatially consistent ranking obtained by the Hellwig method (units with very high tourist attractiveness are rather rarely neighbors adjacent to the districts least attractive for tourists). The analysis of the τ-Kendall rank correlation coefficients (see Table 5) confirms the remarks concerning the high similarity of the obtained results.

In the next stage, based on the results of spatial linear ordering, districts of the Lesser Poland Province will be divided into four groups in terms of tourist attractiveness (Bąk et al. 2018). To achieve this goal, the three-means method will be considered:

- group I contains the most attractive districts, i.e.,

$$MS_i > \overline{MS} + \mathrm{sd}(MS) \tag{9}$$

- group II contains districts with above average tourist attractiveness, i.e.,

$$\overline{MS} < MS_i \leq \overline{MS} + \mathrm{sd}(MS) \tag{10}$$

- group III contains districts with average tourist attractiveness, i.e.,

$$\overline{MS} - \mathrm{sd}(MS) < MS_i \leq \overline{MS} \tag{11}$$

- group IV contains districts with low tourist attractiveness, i.e.,

$$MS_i < \overline{MS} - \mathrm{sd}(MS) \tag{12}$$

where $\overline{MS}$ and $\mathrm{sd}(MS)$ are arithmetic mean and standard deviation of vector MS_i, respectively.

The results of this division are presented in Table 6 and Figs. 1, 2 and 3. As can be seen, for all rankings, among the most attractive districts there are: the city of Kraków and the district of Wieliczka located in the center of the region, as well as the Tatry, Nowy Sącz districts, and the city of Nowy Sącz located in the southern part of the region.

Table 6 Results of grouping districts of Lesser Poland in terms of tourist attractiveness

Group	Pietrzak measure	LSW measure	SMM measure
I	The city of Kraków, Wieliczka, Nowy Sącz, Tatry, Oświęcim, the city of Nowy Sącz	The city of Kraków, Tatra, Nowy Sącz, Wieliczka, Nowy Targ, the city of Nowy Sącz	The city Kraków, the city of Nowy Sącz, Tatry, wielicki, Nowy Sącz
II	Nowy Targ, Wadowice, Bochnia	Oświęcim, Wadowice, Bochnia, Kraków	Oświęcim, Nowy Targ, Wadowice
III	The city of Tarnów, Tarnów, Kraków, Gorlice, Chrzanów, Brzesko, Myślenice, Miechów, Olkusz, Limanowa, Sucha	Gorlice, the city of Tarnów, Tarnów, Chrzanów, Myślenice, Brzesko, Limanowa, Sucha	Bochnia, Tarnów, Gorlice, the city of Tarnów, Kraków, Chrzanów, Myślenice, Brzesko, Limanowa, Sucha, Miechów
IV	Dąbrowa, Proszowice	Miechów, Olkusz, Dąbrowa, Proszowice	Olkusz, Dąbrowa, Proszowice

Source Own calculations

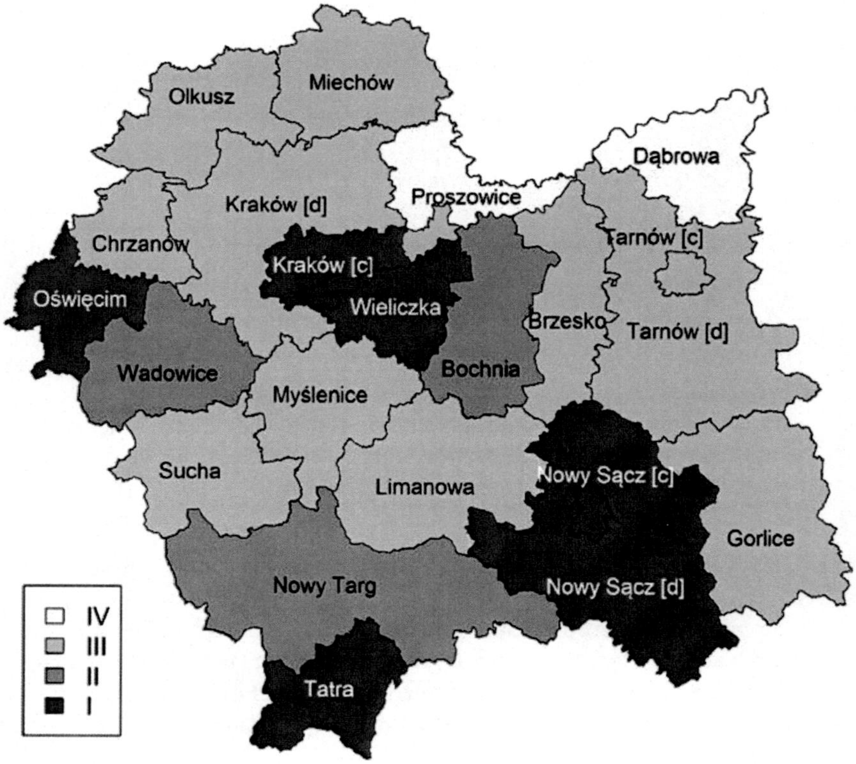

Fig. 1 Spatial diversity of districts in the Lesser Poland Province due to Pietrzak measure. *Source* Own calculations

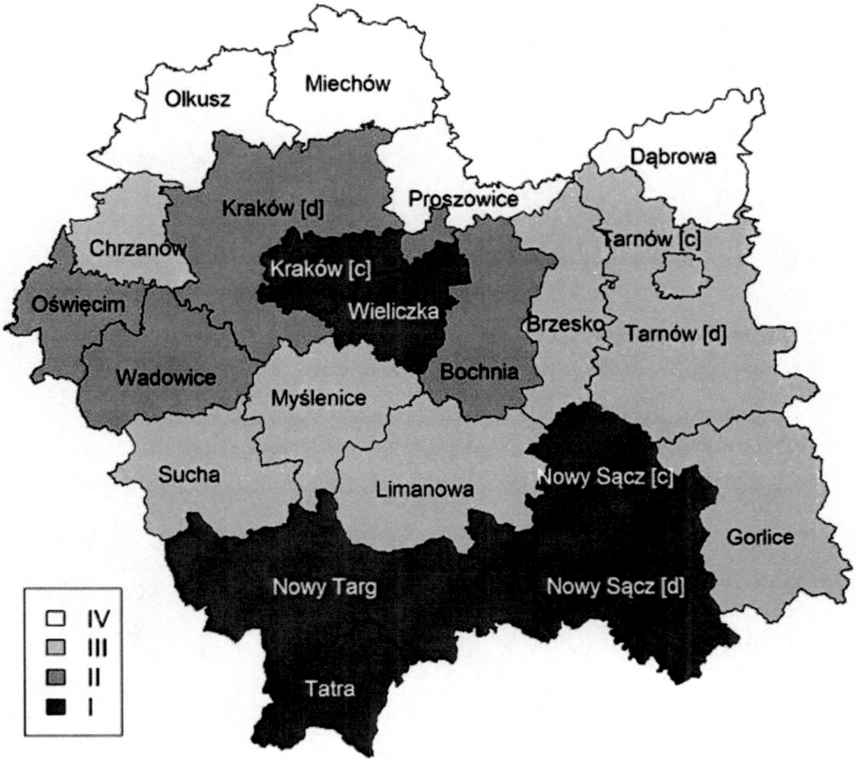

Fig. 2 Spatial diversity of districts in the Lesser Poland Province due to LSW measure. *Source* Own calculations

Among the units with low tourist attractiveness, there are districts located only in the northern part of the province, and regardless of the method used, group IV includes the district of Dąbrowa and the district of Proszowice.

Radar charts are presented in Fig. 4 to illustrate the key factors influencing the attractiveness of the most and least attractive districts. As can be seen, the most attractive districts of the Lesser Poland Province have different tourist values. Districts located in the center of the region (the city of Kraków and Wieliczka district) are characterized by high cultural values and easy access to tourist infrastructure. On the other side, the mountain districts (Tatry and Nowy Sącz), apart from excellent accommodation facilities, encourage potential tourists with their environmental values.

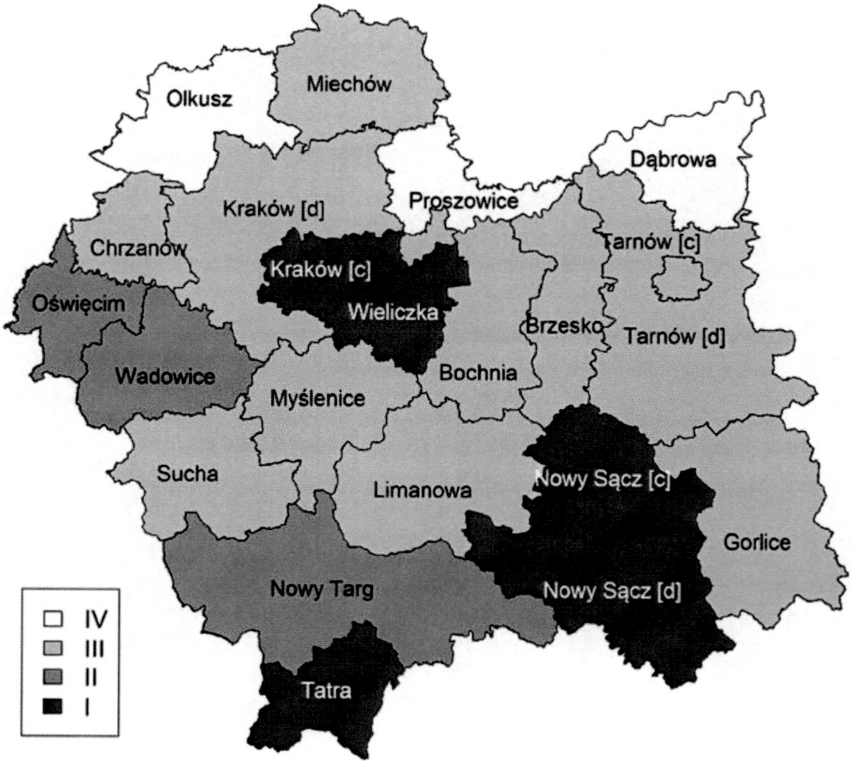

Fig. 3 Spatial diversity of districts in the Lesser Poland Province due to SMM measure. *Source* Own calculations

In the case of the least attractive districts (Fig. 5), the strong advantages of the potential tourist offer include: clean air, low crime rate, and a relatively good road network.

4 Conclusions

The aim of the work was to use spatial techniques to build a ranking of districts in the Lesser Poland Province in terms of the level of tourist attractiveness. As part of the empirical research, appropriate techniques are used, and the obtained results indicate their influence on the obtained form of ranking.

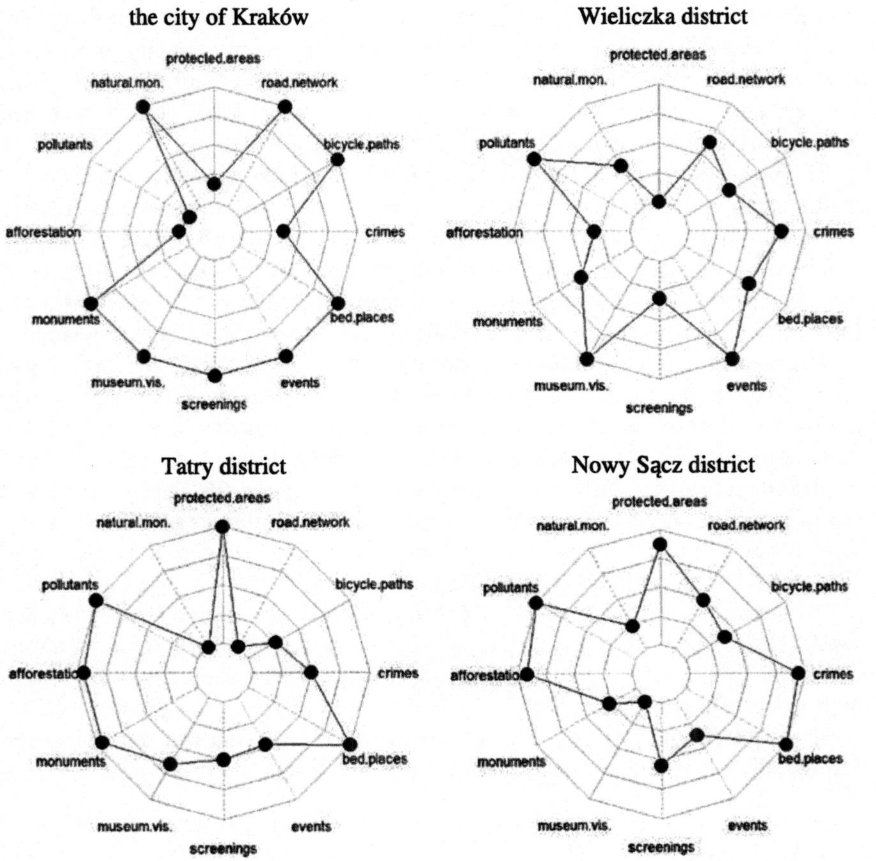

Fig. 4 Radar chart for the most attractive administrative units of the Lesser Poland Province (*Note* Pollution and crime variables have been converted to stimulants). *Source* Own calculations

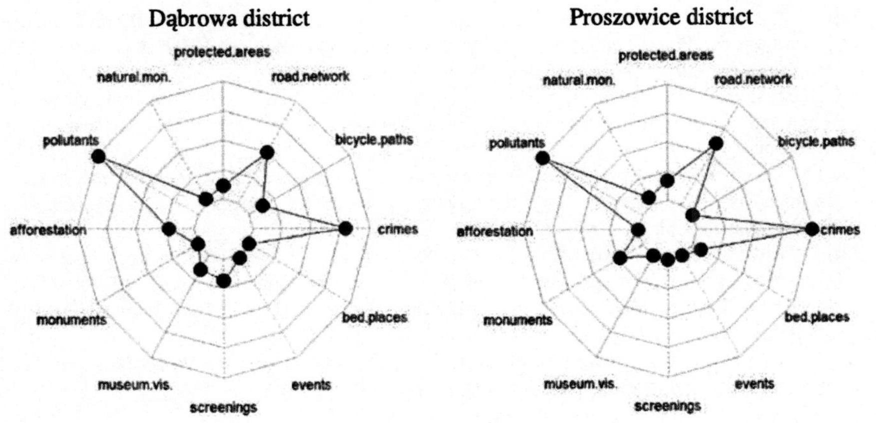

Fig. 5 Radar chart for the least attractive administrative units of the Lesser Poland Province (*Note* Pollution and crime variables have been converted to stimulants). *Source* Own calculations

The results show that the five districts are characterized with the highest level of tourist attractiveness for each of the methods used (these are: the city of Kraków, the city of Nowy Sącz, and Wieliczka, Nowy Sącz, and Tatry districts). Two northern east districts of the Lesser Poland Province are on a low level. It is related to Dąbrowa and Proszowice districts.

The presented results show that the algorithms considering tourist attractiveness in the neighboring districts do not have a significant impact on obtained results. There are two reasons for this. The similarity of the results with the use of Hellwig and Pietrzak measures (administrative units recorded changes in the ranking by at most one place) derived from a small number of diagnostic variables (only four of the twelve considered) indicated a statistically significant spatial autocorrelation.

The impact of the information about the neighbors, although still small, was more visible in the case of using the LSW and SMM methods. In this case, it was not the nature of the considered variables to change the position in the ranking, but the average tourist attractiveness of the neighboring districts (weighted with a spatial weight matrix). As a result, the place in the ranking for some districts was changed. For example, in the ranking, Sucha district was promoted, which is nearby to the much richer in tourist attractions administrative units (the district of Wadowice and the district of Nowy Targ).

The reason the use of LSW and SMM techniques does not give more diversified results (compared to the classic Hellwig method) is the fact that in terms of tourist attractiveness in the Lesser Poland Province there are no hot spots (i.e., an area with high value of the considered feature, which is surrounded by neighbors with low levels of this feature) or cold pots (i.e., areas with low values of the considered feature adjacent to objects with high levels of this feature).

References

Antczak E (2013) Przestrzenny taksonomiczny miernik rozwoju. Wiadomości Statystyczne 58: 37–53

Bąk A (2016) Porządkowanie liniowe obiektów metodą Hellwiga i TOPSIS–analiza porównawcza. Prace Naukowe Uniwersytetu Ekonomicznego we Wrocławiu 426:22–31. https://doi.org/10.15611/pn.2016.426.02

Bąk I (2014) Porównanie jakości grupowań powiatów województwa zachodniopomorskiego pod względem atrakcyjności turystycznej. Prace Naukowe Uniwersytetu Ekonomicznego we Wrocławiu 328:177–185

Bąk I, Matlegiewicz M (2010) Przestrzenne zróżnicowanie atrakcyjności turystycznej województw w Polsce w 2008 roku. Zeszyty Naukowe Uniwersytetu Szczecińskiego. Ekonomiczne Problemy Usług 52:57–67

Bąk I, Wawrzyniak K, Sobolewski A (2018) Przestrzenne zróżnicowanie sytuacji społeczno-gospodarczej a efektywność zatrudnieniowa w powiatowych urzędach pracy w Polsce. Folia Pomeranae Universitatis Technologiae Stetinensis. Oeconomica 93:17–28. https://doi.org/10.21005/oe2018.93.4.02

Balicki A (2009) Statystyczna analiza wielowymiarowa i jej zastosowania społeczno-ekonomiczne. Wydawnictwo Uniwersytetu Gdańskiego, Gdańsk

Binderman Z, Borkowski B, Szczęsny W (2010) Regionalne zróżnicowanie gospodarki turystycznej w Polsce w latach 2002–2008. Acta Scientiarum Polonorum 9(4):71–82

Dębkowska K, Jarocka M (2013) The impact of the methods of the data normalization on the result of linear ordering. Acta Universitatis Lodziensis. Folia Oeconomica 286:181–188

Głowicka-Wołoszyn R, Wysocki F (2020) Right-skewed distribution of features and the identification problem of the financial autonomy of local administrative units. In: Jajuga K, Batóg J, Walesiak M (eds) Classification and data analysis. SKAD 2019. Studies in classification, data analysis, and knowledge organization. Springer, Cham. https://doi.org/10.1007/978-3-030-52348-0_16

Gryszel P, Walesiak M (2014) Zastosowanie uogólnionej miary odległości GDM w ocenie atrakcyjności turystycznej powiatów Dolnego Śląska. Folia Turistica 31:127–147

Gryszel P, Walesiak M (2018) The application of selected multivariate statistical methods for the evaluation of tourism competitiveness of the Sudety Communes. Argumenta Oeconomica 1 (40):147–166. https://doi.org/10.15611/aoe.2018.1.06

Hellwig Z (1968) Zastosowanie metody taksonomicznej do typologicznego podziału krajów ze względu na poziom ich rozwoju oraz zasoby i strukturę wykwalifikowanych kadr. Przegląd Statystyczny 4:307–326

Łysoń P, Szymkowiak M, Wawrowski Ł (2016) Badania porównawcze atrakcyjności turystycznej powiatów z uwzględnieniem ich otoczenia. Wiadomości Statystyczne 12:45–57

Oleńczuk-Paszel A, Nowak MJ (2010) Turystyka w rozwoju społeczno-gospodarczym gmin nadmorskich w Polsce. Folia Pomeranae Universitatis Technologiae Stetinensis. Oeconomica 61:69–78

Pietrzak MB (2014) Taksonomiczny miernik rozwoju (TMR) z uwzględnieniem zależności przestrzennych. Przegląd Statystyczny 61:181–201

Puciato D (2010) Wybrane elementy atrakcyjności turystycznej powiatów województwa opolskiego. Infrastruktura i ekologia terenów wiejskich 1:187–195

Sobolewski M, Migała-Warchoł A, Mentel G (2014) Ranking poziomu życia w powiatach w latach 2003–2012 z uwzględnieniem korelacji przestrzennych. Acta Universitatis Lodziensis. Folia Oeconomica 6:147–159

Stec A (2015) Zastosowanie metody Hellwiga do określenia atrakcyjności turystycznej gmin na przykładzie województwa podkarpackiego. Metody Ilościowe w Badaniach Ekonomicznych 16(4):117–126

Synówka-Bejenka E (2017) Potencjał turystyczny województw Polski. Wiadomości Statystyczne 7:78–92

Wolak J (2020) An analysis of tourist attractiveness of Poviats of the Lesser Poland Voivodeship. Folia Oeconomica Stetinensia 20(1):506–518. https://doi.org/10.2478/foli-2020-0029

Application in Social Issues

Models of Competing Events in Assessing the Effects of the Transition of Unemployed People Between the States of Registration and De-Registration

Beata Bieszk-Stolorz

Abstract Labour market research analyses the transition between the states of economically inactive and active. They show that the characteristics of the unemployed can significantly influence the probability and intensity of transition from one state to another one. Some researchers stress that it is worthwhile to differentiate between the states of the jobless persons, for example the unemployed and those not participating in the labour market. The presented study fits into the scope of multi-state labour market models. Its aim is to assess the gender impact of the unemployed on the duration of registered unemployment and on the duration of staying out of the labour office, taking into account the various reasons for de-registration. Due to their diversity, they are divided into three groups: taking up job, removal and other reasons. The flow of unemployed people in two directions was studied. The probability and intensity of exiting and re-registering in total and according to gender was analysed. In both cases, the reason for de-registration was considered. It applied methods of survival analysis from the area of competing risk models. The study was based on data from the Poviat Labour Office in Szczecin (Poland). Despite the restrictions imposed by the office on persons being de-registered, a frequent reason for registration is still the desire to have health insurance and the desire to receive pre-retirement benefit/assistance allowance.

Keywords Survival analysis · Competing events · Duration in and out of unemployment

B. Bieszk-Stolorz (✉)
University of Szczecin, Institute of Economics and Finance, Szczecin, Poland
e-mail: beata.bieszk-stolorz@usz.edu.pl

K. Jajuga et al. (eds.), *Data Analysis and Classification*, Studies in Classification, Data Analysis, and Knowledge Organization,
https://doi.org/10.1007/978-3-030-75190-6_13

1 Introduction

The changes taking place in the modern labour market are causing the situation of women to change gradually in recent years. The spread of flexible forms of employment, adjustment of education to the needs of labour market, as well as social changes, such as the transition from the traditional family model to the partner one, can improve the situation of women. However, it is still more difficult than the situation of men, as a result of the dual role of the mother and carer and the active person.

The statistics in Poland show that the situation of women compared to men is characterised by a lower activity rate for people aged 15 and over (in 2019: 69.4% for men, 53.7% for women), lower employment rate (in 2019: 67.3% for men, 51.8% for women), and higher unemployment rate (in 2019: 3% for men, 3.6% for women). Women predominate among the economically inactive and make up the majority of those registered in labour offices. Characteristically, the average time spent looking for work by women and men in Poland has been similar in recent years (in 2019: 8.5 months for men, 9 months for women). This situation also occurred in the local labour market in Szczecin (Batóg and Batóg 2016; Bieszk-Stolorz 2013). Women more often than men benefited from subsidised forms of activisation, participate in pro-activity frameworks and increasingly decide to start a business. Numerous studies also pointed to the gender pay gap and the low presence of women in senior positions and in company boards 40 (Kompa and Witkowska 2018).

The aim of the study is to assess the impact of gender of unemployed people on the duration of registered unemployment and on the duration of staying out of the office's register, taking into account different reasons for de-registration. Due to their diversity, they are divided into three groups: taking up job, removal (through the fault of the unemployed person) and other reasons. These three types of causes are the three different states. They can be considered as three different competing events. Particularly undesirable from the point of view of labour market policy is resignation from cooperation with the office (removal). After taking up job, it is the second largest reason for de-registration in Poland. Regulations introduced in offices impose sanctions on such persons. A penalty is imposed on a person who leaves the register without giving a reason. Such a person loses the status of an unemployed person for 180 days and cannot re-register at the office, and consequently cannot benefit from any benefits for 120 days. If the resignation is repeated, the grace period is 180 days. In the case of the third or subsequent resignation, the withdrawal period is extended to 270 days. Competing risk is defined as an event, which occurrence precludes the occurrence of another event or fundamentally changes the probability of that other event occurring. It is assumed that the events are independent of each other, i.e. the occurrence of a certain type of event has no impact on the probability of occurrence of any other event. The observed entity is exposed to different risks at the same time. However, it is assumed that a possible event is due to only one of those factors, which is called the "cause of failure". Due to censored observations, i.e. observations not completed with an event in the analysed period, it has been decided to use selected methods of survival analysis.

2 Literature Review

From the 1970s onwards, articles began to appear in which methods of survival analysis were applied to labour market research. Most often these studies focused on returning to the labour market, and a very important extension of them is the analysis of history of the individuals' participation in the labour market. This review of the literature has been limited to more important research conducted in the 1970s and 1980s. In a sense, their authors can be called pioneers in this field. Due to the phenomenon under study, four research groups can be distinguished (Devine and Kiefer 1991).

The first group is linked to the movement of employees between two states employment and unemployment or non-employment. This group includes studies on the use of Cox proportional hazard models to analyse the risk of transition from employment to either unemployment or non-employment. An example of such analyses is the research by Burdett et al. (1985) of male family heads in the USA. Probability of dismissal decreased along with duration of their employment. Jensen (1987) took an approach like that of Burdett et al. (1985) to study employment and unemployment duration data for young Danish workers. Miller and Volker (1987) also fitted Weibull proportional hazard models in their study of unemployment and employment duration data for Australian youth, aged 15–24, collected in the 1985 Australian Longitudinal Survey. This group of studies also included the ones that used two-state models to solve specific problems. Kiefer (1985) presented some suggestive evidence on the rate-of-return to education—measured in terms of its effects on labour market transitions. Kiefer fitted a non-stochastic constant hazard specification for the two-state model using data for male workers in the USA. Stephenson (1982) studied the employment exit and entry behaviour of young women in USA. His primary concern was the relationship between work during school and employment experience after leaving. Tuma and Robins (1980) studied the effects of a negative income tax on non-employment and employment spell lengths using data from social experiments in the USA. Ridder (1988) attempted to determine the effects of training, recruitment and employment programs on unemployment and employment spell lengths using data for a sample of 337 participants in such programs in 1979 and early 1980 in Rotterdam, the Netherlands. Flinn and Heckman (1983) tested the hypothesis that the classifications "unemployed" and "out of the labour force" were behaviourally meaningless distinctions. This hypothesis was rejected. Distinct behavioural equations governed transitions from out of the labour force to employment and from unemployment to employment.

The second group of studies focused on the movement of workers in three directions—between employment, unemployment and lack of participation in the labour market (inactivity). The studies that were based on three-state models of the labour market, which distinguished unemployment and non-participation, followed. Burdett et al. (1984) considered a dynamic model of an individual's allocation of time among the three labour market states: employment, unemployment and non-participation. Weiner (1984) extended Burdett's et al. empirical model. His analysis focused on the labour market flows underlying the differences in unemployment rates among adult black and white men in the USA. Lundberg (1985) examined the dynamic aspects of the "added worker" hypothesis. He focused on the labour supply response of married women to their husbands' unemployment. Blau and Robins (1986) studied the impact of participation in public aid programmes on labour market changes. Flinn and Heckman (1983) proved the hypothesis that employment and non-participation in state aid schemes were behaviourally separate states.

Studies in the third section focused on transitions in three directions: employment, unemployment and non-participation in the labour market (inactivity), but distinguished unemployment spells following temporary and permanent layoffs. The reason for this was obvious—it was assumed that search behaviour of these workers was distinct from search behaviour following a permanent separation. Ehrenberg and Oaxaca (1976) and Classen (1977, 1979), for example, both noted that the results from duration regressions fit separately for workers on temporary layoff were systematically different from the results for workers on permanent layoff. Ehrenberg and Oaxaca reported a negligible benefit effect for adult males on temporary layoff. Classen's results implied a benefit elasticity that was roughly half the size of the elasticity for workers not on temporary layoff.

In the fourth group of studies, multi-state labour market models were analysed, which did not fall within the groups mentioned above. This research concerned the analysis of transition to jobs offering non-wage benefits in comparison with jobs that did not have them (Khandker 1988). They distinguished, inter alia, between the transition from unemployment to full-time work and alternative destinations (Narendranathan and Stewart 1990) or distinguished between periods of work ending in resignation from job and periods of work ending in dismissal (Farber 1980).

Since the 1990s, a number of studies have been carried out to analyse the likelihood of transition between different states of the individual in the labour market (Böheim and Taylor 2000; Addison and Portugal 2003; Dănăcică and Paliu-Popa 2017; Simon et al. 2017; O'Neill 2019).

Steiner (1997) analysed the influence of specifications of benefit–entitlement on the duration of individual unemployment. The results of the econometric analysis showed that the entitlement to unemployment benefits increased the duration of unemployment for males. For females, benefit–entitlement in general had little effect on the duration of unemployment. The estimation results also showed that, for both males and females, marginal reductions of the income–replacement ratio had very little effect on individual unemployment behaviour.

Van den Berg et al. (2008) simultaneously analysed transitions from unemployment to employment and to non-participation. The individual conditional exit probabilities to employment and non-participation were uncorrelated across individuals, for males as well as females. The exit to employment displayed negative duration dependence after one quarter of unemployment, while the exit to non-participation did not display duration dependence for females, and negative duration dependence during the first quarter for males.

The interesting example of application of analysis of competing risks in the analysis of labour market was presented by Reeuwijk et al. (2017). The study aimed to determine the influence of poor health on competing exit routes from paid employment among older workers in Europe, assess whether these risks are different among welfare state regimes in Europe and evaluate differences in estimates between two different competing risk approaches. Workers with poor health were more likely to leave the labour force than workers with good health. The absolute risks of early retirement and becoming economically inactive were lowest in countries with a Scandinavian welfare state regime. For disability benefit and unemployment, absolute risks were lowest in Southern European welfare state regimes.

Studies carried out in Poland on the impact of gender on the probability of job de-registration indicated that gender differentiated the cause of job de-registration. Taking up a job was the most common reason for de-registration for women throughout their duration in unemployment. The removal occupied the second place. In the case of men, the most likely reason for the de-registration was removal and then taking up job (Bieszk-Stolorz 2017b). Also, in the case of the long-term unemployed, gender was an important classification feature of the unemployed both in case of de-registration and removal (Bieszk-Stolorz and Dmytrów 2018a, b). However, gender was not always a differentiating feature of the employed and unemployed in the labour market. This was the case with the probability of multiple registrations at the labour office (Bieszk-Stolorz 2020). Studies using multiple event models showed that only education and age had a significant impact on subsequent returns to the labour office in Szczecin (Poland). Landmesser (2013) analysed the economic activity of the population in Poland. She examined the transition from employment to unemployment or inactivity. This allowed to identify differences between the transition processes. Men underwent a state of inactivity much later than women. A weaker effect of this type was recorded for exits into unemployment. The study also pointed to the interdependence of women's professional and family careers in the analysed period.

3 Research Methodology

The presented study is part of the multi-state labour market models. In this case, it is the transition between the state of registered unemployment and the three states defined by the causes of de-registration. De-registration due to taking up job is a transition to employment. The resignation from mediation of the office is a transition to an indefinite state because we do not know what the reasons for this were. Maybe an unemployed person has found a job, but has not reported it to the authorities, maybe he has taken up illegal work, or he is still unemployed. The third reason, or the rest, is the transition to a state of non-employment. The two phenomena are analysed. The first one is the transition of an unemployed person from the state of registered unemployment to the state of employment (job), unspecified state (removal) or the state of unemployment (others). The second phenomenon is the return to the state of unemployment from the state of employment (job), unspecified state (removal) and non-employment (other). Schemes of transitions between states are presented in Figs. 1 and 2.

The study used the competing risk models included in the survival analysis. The cumulative incidence function $\mathrm{CIF}_k(t)$ was used to assess the probability of de-registration and registration. The estimation of the intensity of registration and de-registration was made with the help of an empirical hazard model.

Let T and C be continuous random variables, describing, respectively, the time until a certain event occurs and the time until censorship. For K competing risks, pairs (X, δ) are observed, where $X = \min(T, C)$ and $\delta = 0, 1, \ldots, K$. If an observation is censored, then $\delta = 0$ and $\delta = 1, \ldots, K$ for observations ending in an event (one of the competing K). In this context, one of the K events may be considered as a basic event and all others will be considered as competing events.

The cumulative incidence function denoted as $\mathrm{CIF}_k(t)$ is the probability of occurrence of an event due to reason k before time t. It is defined as (Klein and Moeschberger 2003, p. 52):

Fig. 1 Transition scheme for de-registered persons

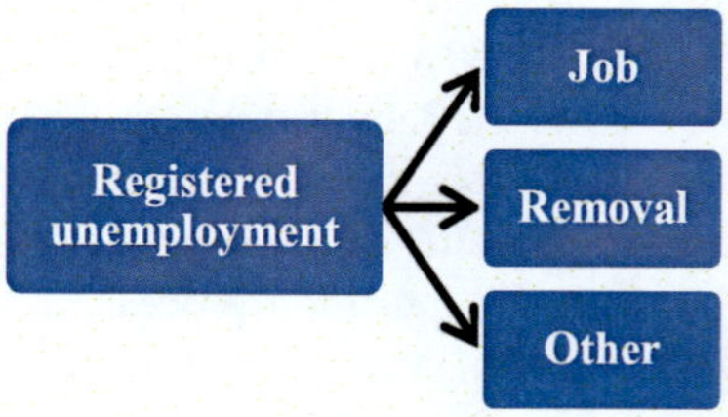

Fig. 2 Transition scheme for re-registered persons

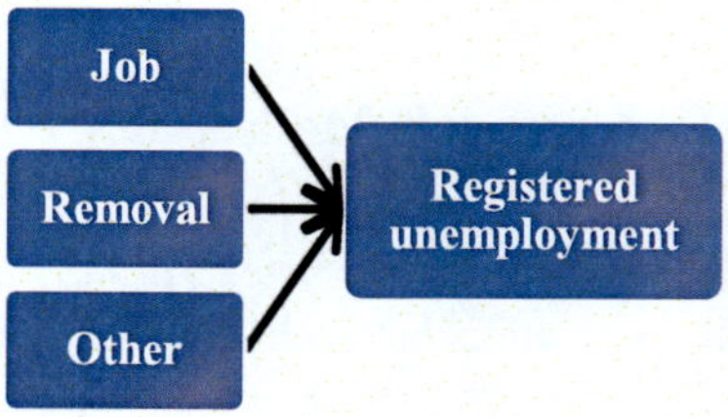

$$\mathrm{CIF}_k(t) = P(t \le T, \delta = k) = \int_0^t S(u)h_k(u)\mathrm{d}u = \int_0^t S(u)\mathrm{d}H_k \quad \text{for } k = 1, 2, 3, \ldots, K \tag{1}$$

where:

T—a random variable describing the time until the event occurs,
K—number of competing risks,
$H_k(t)$—the specific (for the specific k) function of cumulative hazard and $S(t)$ is the survival function.

Let $t_1 < t_2 < \cdots < t_i < \cdots < t_n$ be the moments when events occur. As with the standard cumulated survival analysis function, the cumulated $H_k(t)$ hazard function for cause k can be determined by the Nelson–Aalen estimator (Kleinbaum and Klein 2005):

$$\hat{H}_k(t) = \sum_{j:t_j \le t} \frac{d_{kj}}{n_j} \tag{2}$$

where:

d_{kj}—the number of events due to the occurrence of cause k,
n_j—the number of people at risk in time t_j.

The general survival function $S(t)$ can be determined on the basis of the Kaplan–Meier estimator defined as follows (1958):

$$\hat{S}(t) = \prod_{j:t_j \le t} \left(1 - \frac{d_j}{n_j}\right) \tag{3}$$

After combining these two estimators, the function of cumulative frequency of occurrence due to cause k (Marubini and Valsecchi 1995) can be estimated as:

$$\hat{\mathrm{CIF}}_k(t) = \sum_{j:t_j \le t} \hat{S}(t_{j-1}) \frac{d_{kj}}{n_j} \tag{4}$$

The function of the cumulative frequency of an event can therefore be defined as the cumulative probability of a k-type event occurring before or at time t (Bryant and Dignam 2004). It allows to determine patterns of occurrence of an event due to k and to assess the extent to which each reason contributes to a total failure.

Because $\sum_{k=1}^{K} d_{kj} = d_j$, the following relation is true:

$$\sum_{k=1}^{K} \hat{\mathrm{CIF}}_k(t) = 1 - \hat{S}(t) \tag{5}$$

If there are no competitive events, there is equality:

$$\hat{\mathrm{CIF}}(t) = 1 - \hat{S}(t) \tag{6}$$

Where competitive events occur, sometimes a solution considering the remaining events ending the observation as censored ones is used. However, it should be borne in mind that this reasoning leads to an overestimation of the CIF function (Sherif 2008).

The hazard function at time t is the momentary potential of an emerging event (e.g. death or illness) provided that the observed unit survives until time t. The simplest case is when there is only one risk ($k = 1$). If there are different competing risks ($k \geq 1$), the hazard function is described by the formula (Klein and Moeschberger 2003, p. 50):

$$h_k(t) = \lim_{\Delta t \to 0} \frac{P(t \leq T < t + \Delta t, \delta = k | T \geq t)}{\Delta t} \quad \text{for } k = 1, 2, 3, \ldots, K \tag{7}$$

The hazard function $h_k(t)$ for cause k at the moment t_j can be determined by means of the estimator:

$$\hat{h}_k(t_j) = \frac{d_{kj}}{n_j} \tag{8}$$

where:

d_{kj}—number of events due to the occurrence of cause k,
n_j—number of people at risk at time t_j.

4 Data Used in the Study

In the study, anonymous individual data of persons from the register of the Poviat Labour Office (Polish abbreviation PUP) in Szczecin were used. The data are collected in the Syriusz System, which was implemented in offices in order to support in a comprehensive way their statutory tasks. These data, apart from information about the unemployed, also include the reason for de-registration from the labour office. These reasons are very different, and in total, there are several dozen of them. Since previous research has shown that many of the reasons for de-registration were of marginal importance (Bieszk-Stolorz 2017a), three groups of events ending the observation have been defined: taking up employment, removal and others, which are the competing events. It was decided to combine them into three main groups: job, removal and other reasons (Table 1).

Table 1 Groups of reasons for de-registration from the labour office

State name	Reason for de-registration
Job	Work or other employment
	Subsidised work
	Business
Removal	Refusal to work
	Failure to report to the office within the prescribed time limit
	Refusal or interruption of participation in the form of activisation
	Request for removal
Others	Pension, allowance, retirement pension, pre-retirement benefit, other
	Going abroad
	Others

Two cohorts of persons were analysed. The first of them were the persons registered in the PUP in 2016 and observed until the end of 2016 (19,688 people). The event is de-registration due to taking up work (job), removal or other reasons. The random variable T describes the time from registration (in 2016) to de-registration (by the end of 2016). If the event has not occurred by the end of 2016, such an observation was considered as censored. This cohort is called due to the type of event ending the observation: de-registered persons.

The second cohort were people who were de-registered from office in 2016 due to taking up work (job), removal or other reasons and were observed until the end of 2016. The event is re-registration in the office. The random variable T describes the time from the moment of de-registration (in 2016) to the moment of registration (by the end of 2016). If the event had not occurred by the end of 2016, such an observation was considered as censored. This cohort was named due to the type of event ending the observation: persons registered.

The cohorts together and for the reasons of de-registration and re-registration are shown in Table 2.

Table 2 Number of groups of unemployed persons de-registered and re-registered in the labour office

Status	De-registered persons			Re-registered persons		
	Total	Women	Men	Total	Women	Men
Job	6745	3195	3550	2823	1246	1577
Removal	7610	2999	4611	2143	975	1168
Others	825	372	453	347	139	208
Censored	4508	2128	2380	18421	8623	9798
Total	19688	8694	10994	23734	10983	12751

5 Empirical Results

The probability and intensity of de-registration and re-registration of unemployed people at the labour office was analysed. First, the total unemployed were examined and then in groups by gender: women and men separately. The results are presented in Figs. 3 and 4.

The CIF curves for de-registration are smoother than for re-registration. This can be seen both for the unemployed in general and in groups by gender. This observation is also reflected in the values of empirical hazard. In the case of re-registered persons, these values have characteristic large value spikes. It has been decided to find their causes.

An analysis of all the unemployed de-registered in general leads to the conclusion that the probability of giving up cooperation with the office was higher than the probability of taking up employment (Fig. 3). The third place was the probability of de-registration for other reasons. The intensity of de-registration in case of job and removal was decreasing. In the case of the other reasons, a characteristic jump of values in the fourth and seventh month is noticeable. Analysis of the data indicates that the jump in the fourth month was caused by an increase in de-registrations due to inability to work as a result of illness or being in a closed detoxification centre for a continuous period of 90 days. The jump in the seventh month was caused by granting a pre-retirement benefit/assistance allowance. It is granted provided that a registered unemployed person meets, among other things, the condition that he/she received an unemployment benefit for at least 180 days. The courses of the CIF curves both for males and all the unemployed in general were similar. The difference is that the gap between the probability of being removed and the probability of taking up employment is greater. Starting from the third month, women take up work more often than they give up. The intensity of their resignation increases in the eighth and ninth month after registration.

In the case of re-registration, the course of the CIF curves is slightly different (Fig. 4).

First of all, due to the large number of censored observations, these curves take lower values. From a social point of view, the large number of censored observations is in this case a good information because it shows that many people previously registered do not re-register within the next 12 months. For all the unemployed in general and in groups by gender, the probability of transition from employment to unemployment was greater than the probability of returning to the register after removal. The course of CIF curves in these two cases is not regular. In the case of return after starting work, clear changes can be seen in the third and fourth month after re-registration. In the case of return after removal, these changes occur in the fifth, seventh and tenth month. They are visible in the form of leaps in the value of the hazard function for the unemployed in general, as well as for women and men. In the third and fourth month, those who have completed their participation in public works return to the register. In the fifth, seventh and

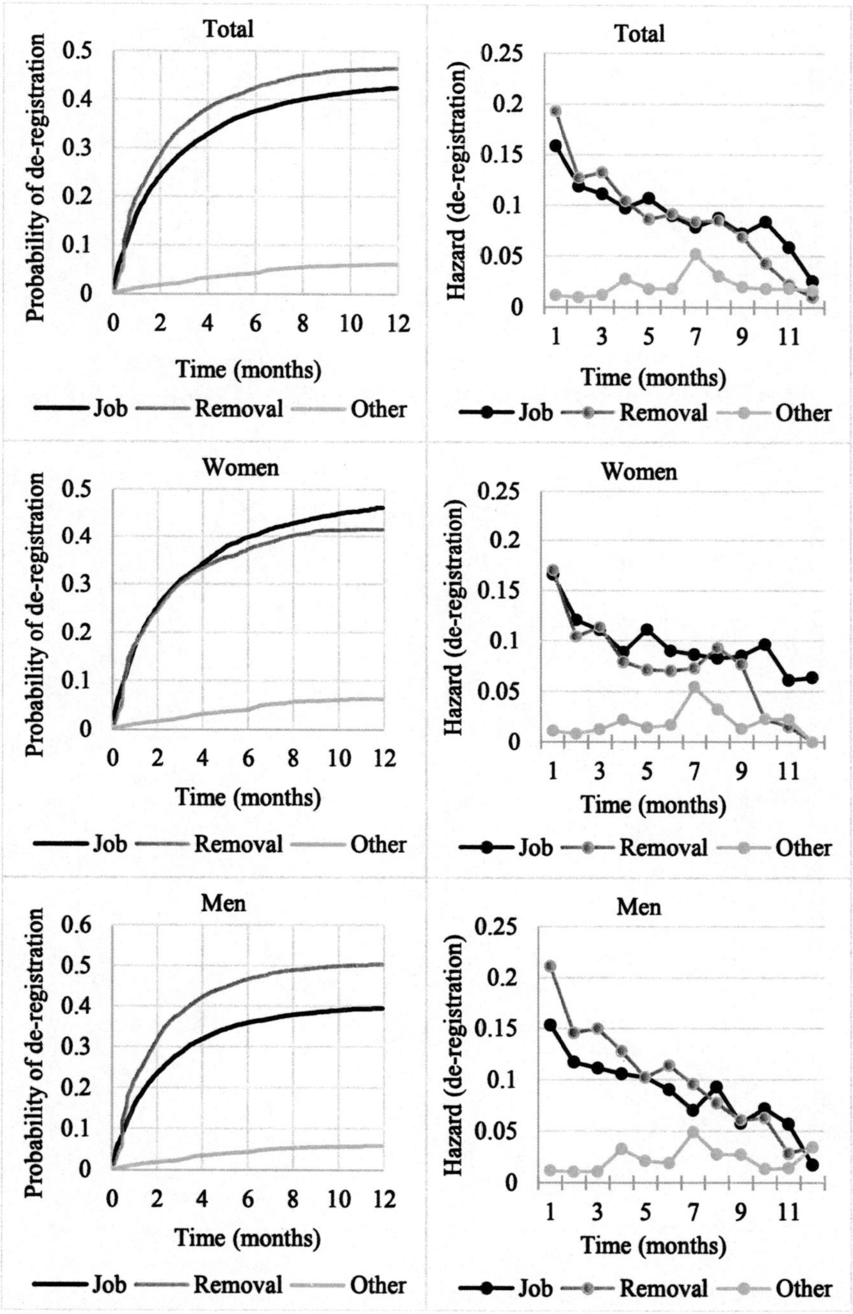

Fig. 3 CIF and hazard functions for the de-registration event total and by gender

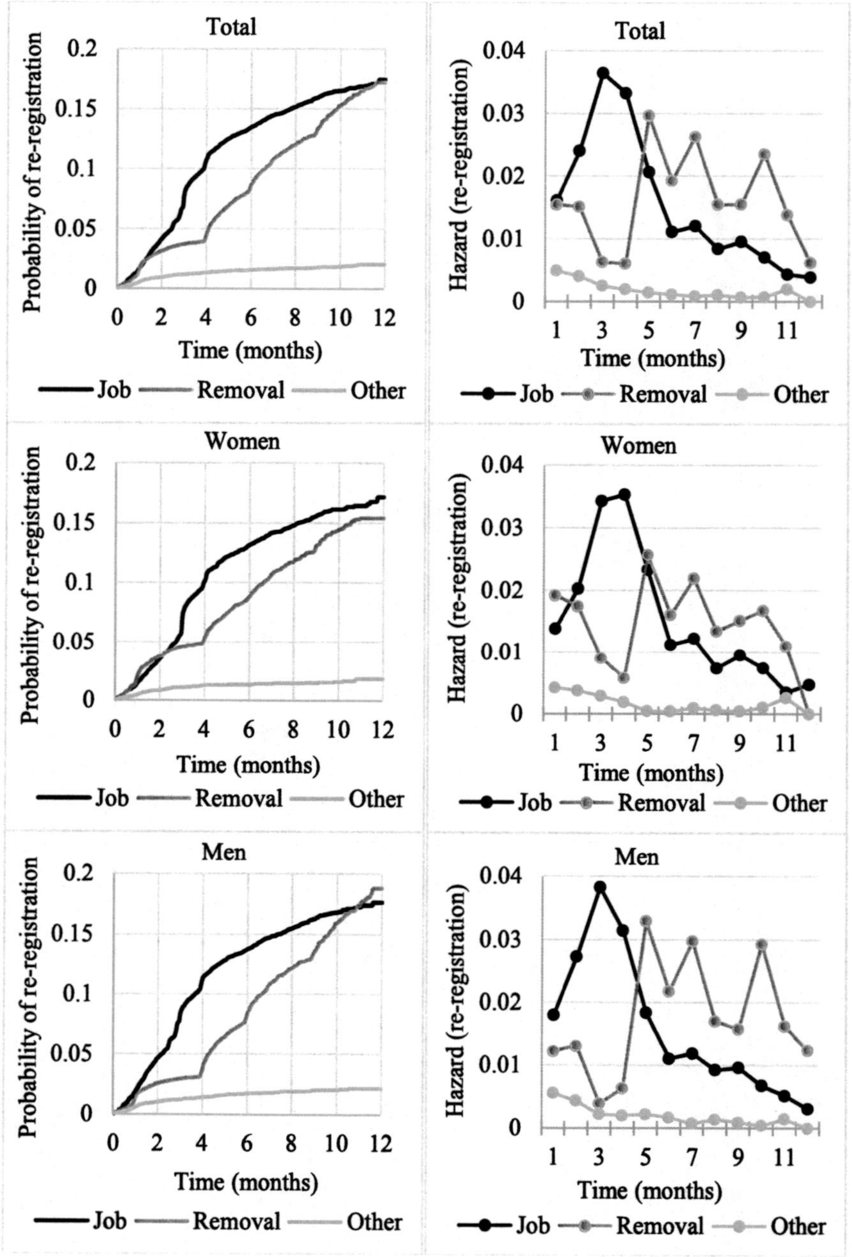

Fig. 4 CIF and hazard functions for the re-registration event at the office total and by gender

tenth months, those who have previously resigned are re-registered. This is linked to the duration of the grace period that has been imposed on them. The jumping value of hazard for men is greater than for women.

6 Conclusions

The study assesses the transition between unemployment and employment and the two states of non-employment, which have been called the resignation of cooperation with the office (removal) and de-registration for other reasons. The assessment was two-way, as the probability and intensity of de-registration and re-registration of unemployed people in total and by gender was analysed. The main objective of the labour offices is to help the unemployed find employment and activate them. However, the purpose of registering in the labour office by unemployed people is not always the will to obtain such assistance. This is evidenced by the large number of removals from the register due to the fault of the unemployed person. Legal solutions and the principles of operation of labour offices cause that one of the main purposes of registration is the will to obtain health insurance. It enables a given person to use public health services free of charge. There are attempts to limit such activities. Restrictions in the form of a grace period imposed on such persons are to serve this purpose. However, even long grace periods (120, 180, 270 days) do not discourage them from even resigning several times. After a fixed period of time, such persons register again, as was shown in the study. These are the characteristic increases in the value of the CIF function and the jumping value of the hazard function for the removal event in the case of re-registration analysis. This may indicate black labour in Poland or due to the location of Szczecin—in Germany. Health insurance is valid in all EU countries. Another purpose of registration visible in the study is the necessity to meet the conditions for receiving pre-retirement benefit. Such a person must be registered and receive unemployment benefit for at least 180 days. The possibility of such a phenomenon occurring is evidenced by the increases in the value of the CIF function and the leaps in the value of the hazard function for the event, the remaining causes in the case of an analysis of persons de-registered from office. One of the activities of the labour office is to organise subsidised work. Among people returning to the office, it can be seen that a large part of them return after the completion of public works. This is visible in the increase in the CIF function and the jump in the hazard function for re-registered persons. Analysing the above problems on the basis of the gender of the unemployed person, there are no significant differences in the behaviour of women and men. The main difference is that women are more often de-registered to work and men more often resign from cooperation with the office. As a consequence of the lower risk of giving up cooperation with the labour office for women, the increase in hazard function has been less intense, as a result of their return after the grace period imposed by the authorities.

References

Addison JT, Portugal P (2003) Unemployment duration: competing and defective risks. J Human Resour 38(1):156–191. https://doi.org/10.2307/1558760

Böheim R; Taylor MP (2000) Unemployment duration and exit states in Britain. ISER Working Paper Series, No. 2000-01, University of Essex, Institute for Social and Economic Research (ISER), Colchester

Batóg J, Batóg B (2016) Application of correspondence analysis to the identification of the influence of features of unemployed persons on the unemployment duration. Econ Bus Rev 16(4):25–44. https://doi.org/10.18559/ebr.2016.4.2

Bieszk-Stolorz B (2013) Analiza historii zdarzeń w badaniu bezrobocia. Volumina.pl Daniel Krzanowski, Szczecin

Bieszk-Stolorz B (2017a) Cumulative incidence function in studies on the duration of the unemployment exit process. Folia Oeconomica Stetinensia 17(1):138–150. https://doi.org/10.1515/foli-2017-0011

Bieszk-Stolorz B (2017b) The impact of gender on routes for registered unemployment exit in Poland. Equilibrium. Q J Econ Economic Policy 12(4):733–749. https://doi.org/10.24136/eq.v12i4.38

Bieszk-Stolorz B (2020) Prentice–Williams–Peterson models in the assessment of the influence of the characteristics of the unemployed on the intensity of subsequent registrations in the labour office. In: Jajuga K, Batóg J, Walesiak M (eds) Classification and data analysis. SKAD 2019. Studies in classification, data analysis, and knowledge organization. Springer, Cham, pp 237–250. https://doi.org/10.1007/978-3-030-52348-0_15

Bieszk-Stolorz B, Dmytrów K (2018a) Application of the survival trees for estimation of the influence of determinants on probability of exit from the registered unemployment. In: Papież M, Śmiech S (eds) Social-economic modelling and forecasting, vol 1. Foundation Cracow Univ Economics, Cracow, pp 30–39. https://doi.org/10.14659/SEMF.2018.01.03

Bieszk-Stolorz B, Dmytrów K (2018b) Application of the survival trees for estimation of the propensity to accepting a job and resignation from the labour office mediation by the long-term unemployed people. In: Nermend K, Łatuszyńska M (eds) Problems, methods and tools in experimental and behavioral economics. CMEE 2017. Springer Proceedings in business and economics. Springer, Cham, pp 141–154. https://doi.org/10.1007/978-3-319-99187-0_11

Blau DM, Robins PK (1986) Job search, wage offers, and unemployment insurance. J Public Econ 29(2):173–197

Bryant J, Dignam JJ (2004) Semiparametric models for cumulative incidence functions. Biometrics 60(1):182–190. https://doi.org/10.1111/j.0006-341X.2004.00149.x

Burdett K, Kiefer N, Mortensen D, Neumann G (1984) Earnings, unemployment, and the allocation of time over time. Rev Econ Stud 51(4):559–578

Burdett K, Kiefer N, Sharma S (1985) Layoffs and duration dependence in a model of turnover. J Econometrics 28(1):51–69

Classen KP (1977) The effect of unemployment insurance on the duration of unemployment and subsequent earnings. Ind Labor Relat Rev 30(8):438–444

Classen KP (1979) Unemployment insurance and job search. In: Lippman S, McCall J (eds) Studies in the economics of search. North-Holland, New York, pp 191–219

Dănăcică DE, Paliu-Popa L (2017) Determinants of unemployment spells and exit destinations in Romania in a competing-risks approach. Econ Res-Ekonomska Istraživanja 30(1):964–984. https://doi.org/10.1080/1331677X.2017.1314825

Devine TJ, Kiefer NM (1991) Empirical labor economics: the search approach. Oxford University Press, New York

Ehrenberg RG, Oaxaca RL (1976) Unemployment insurance, duration of unemployment, and subsequent wage gain. Am Econ Rev 66(5):754–766

Farber H (1980) Are quits and firings actually different events? A competing risk model of job duration, Working Paper, MIT
Flinn C, Heckman J (1983) Are unemployment and out of the labor force behaviorally distinct labor force states? J Labor Econ 1(1):28–42
Jensen P (1987) Testing for unobserved heterogeneity and duration dependence in econometric models of duration. Mimeo, University of Aarhus, Denmark
Kaplan EL, Meier P (1958) Non-parametric estimation from incomplete observations. J Am Stat Assoc 53:457–481
Khandker R (1988) Offer heterogeneity in a two state model of sequential search. Rev Econ Stat 70(2):259–265
Kiefer NM (1985) Evidence on the role of education in labor turnover. J Human Resour 20(3):445–452
Klein JP, Moeschberger ML (2003) Survival analysis: techniques for censored and truncated data, 2nd edn. Springer, New York
Kleinbaum D, Klein M (2005) Survival analysis. A self-learning text. Springer, New York
Kompa K, Witkowska D (2018) Gender diversity in the boardrooms of public companies in Poland: changes and implications. Montenegrin J Econ Economic Laboratory Transition Res (ELIT) 14(1):79–92. https://doi.org/10.14254/1800-5845/2018.14-1.6
Landmesser JM (2013) Wykorzystanie metod analizy czasu trwania do badania aktywności ekonomicznej ludności w Polsce. Rozprawy Naukowe i Monografie. Szkoła Główna Gospodarstwa Wiejskiego w Warszawie, Warszawa
Lundberg SJ (1985) The added worker effect. J Labor Econ 3(1):11–37
Marubini E, Valsecchi M (1995) Analysing survival data from clinical trials and observational studies. Wiley, New York
Miller P, Volker P (1987) The youth labour market in Australia. Econ Rec 63(3):203–296. https://doi.org/10.1111/j.1475-4932.1987.tb00652.x
Narendranathan SW, Stewart MB (1990) An examination of the robustness of models of the probability of finding a job for the unemployed. In: Hartog J, Ridder G, Theewes J (eds) Panel data and labor market studies. North-Holland, Amsterdam
O'Neill D (2019) A new competing risks decomposition: application to the effect of cutting unemployment benefit on unemployment durations. J Roy Stat Soc Ser C (Appl Stat) 68(3):793–807. https://doi.org/10.1111/rssc.12335
Reeuwijk KG, van Klaveren D, van Rijn RM, Burdorf A, Robroek SJ (2017) The influence of poor health on competing exit routes from paid employment among older workers in 11 European countries. Scand J Work, Environ Health 43(1):24–33. https://doi.org/10.5271/sjweh.3601
Ridder G (1988) On generalized accelerated failure time models. Mimeo, Rijksuniversiteit, Groningen
Sherif BN (2008) A comparison of Kaplan-Meier and cumulative incidence estimate in the presence or absence of competing risks in breast cancer data, Master's Thesis. University of Pittsburgh. Available via D-Scholarship. http://d-scholarship.pitt.edu/id/eprint/9986. Accessed on 15 Aug 2020
Simon MSL, Gesine S, Wilke AR (2017) Competing risks copula models for unemployment duration. Application to a German Hartz Reform. J Econometric Methods 6(1). https://doi.org/10.1515/jem-2015-0005
Steiner V (1997) Extended benefit-entitlement periods and the duration of unemployment in West Germany. ZEW Discussion-Paper, no. 97–14. Zentrum für Europäische Wirtschaftsforschung (ZEW), Mannheim
Stephenson SP (1982) A turnover analysis of joblessness for young women. Res Labor Econ 5:279–318
Tuma NB, Robins PK (1980) A dynamic model of employment behavior: an application to the Seattle and Denver income maintenance experiments. Econometrica 48(4):1031–1052

van den Berg GJ, van Lomwel AGC, van Ours JC (2008) Nonparametric estimation of a dependent competing risks model for unemployment durations. Empirical Econ 34(3):477–491. https://doi.org/10.1007/s00181-007-0131-8

Weiner SE (1984) A survival analysis of the adult male black/white unemployment differential. In: Neumann GR, Westergaard-Nielsen N (eds) Studies in labor market dynamics. Springer, Heidelberg, pp 132–157

Direct Adjusted Survival Probabilities in the Analysis of Finding a Job by the Unemployed Depending on Their Individual Characteristics

Wioletta Grzenda

Abstract In survival analysis, nonparametric and parametric models are commonly used to estimate survival functions. An important limitation of nonparametric methods is that in the construction of survival function estimators, variables which can affect the duration of the period an individual remains in a given state are not taken into account. The use of parametric methods also has some limitations. First of all, problems with finding the analytical form of the distribution curve of survival times take place. The second limitation is the fact that time-dependent covariates cannot be included in a model. These restrictions do not apply to the Cox model considered in this study. In most studies, Cox models are used to evaluate the effects of explanatory variables on a hazard rate. The purpose of this paper is to indicate the possibility of using Cox regression model to determine direct adjusted probabilities of finding a job by the unemployed depending on their individual characteristics in the context of long-term unemployment risk. The study is based on LFS data from 2017 to 2018 for Poland.

Keywords Survival analysis · Cox regression · Adjusted survival probability · Unemployment

1 Introduction

Nonparametric models are the basic tool for describing survival processes. There are two main methods in this class of models: the traditional life table method and the Kaplan–Meier method (Kaplan and Meier 1958). They are used when a researcher wants to analyze the shape of the survival functions, density functions, or hazard functions. However, with the use of these methods, it is not possible to study the impact of exogenous variables on the duration of events, except for the estimation of these functions separately for individual categories of given categorical

W. Grzenda (✉)
SGH Warsaw School of Economics, Collegium of Economic Analysis, Warsaw, Poland
e-mail: wgrzend@sgh.waw.pl

K. Jajuga et al. (eds.), *Data Analysis and Classification*, Studies in Classification, Data Analysis, and Knowledge Organization,
https://doi.org/10.1007/978-3-030-75190-6_14

covariates. The analysis of the direction and strength of the impact of explanatory variables on the duration can be performed using parametric and semiparametric models (Blossfeld and Rohwer 1995). Parametric models are used when the analytical form of a density function is defined for a random variable describing duration (survival) (Blossfeld et al. 1989; Kalbfleisch and Prentice 2011). Unfortunately, determining the analytical form of the distribution curve of survival times can be very difficult in some cases. Then, the solution may be to use semiparametric models. These models have another important advantage—they can consider both time-independent and time-dependent variables (Fisher and Lin 1999; Klein and Moeschberger 2005).

Semiparametric survival models are among popular methods in labor market research (Gutiérrez-Domènech 2008; Landmesser 2013; Polemis and Stengos 2015). In most of these studies, they are used to assess the direction and strength of the impact of explanatory variables on the duration of the period an individual remains unemployed or professionally inactive. These models are constructed based on historical data and are used to describe the phenomena related to the labor market. In this paper, attention is focused on the use of information from historical data to estimate probabilities of finding a job by the unemployed. The determination of probability of occurrence of a given event in survival analysis is possible by using a survival function estimated with historical data, most often using the Kaplan–Meier method. In monographs (Bieszk-Stolorz and Markowicz 2019; Grzenda 2019a), semiparametric models were also used to determine survival functions, but without explanatory variables. In this paper, the estimation of the survival function was performed using Cox regression models with covariates. The survival function estimated in this way can also be used to estimate the probability of the occurrence of a given event also for the individuals who were not included in the study, as long as the values of explanatory variables included in the model are known for them.

The chances of finding a job largely depend on how long an individual remains unemployed (Landmesser 2013; Grzenda 2019a). In this paper, we propose to use Cox regression models to calculate direct adjusted survival probabilities of finding a job by jobseekers for a maximum of 12 months from the moment of last employment termination. People who during this time fail to find a job, according to the LFS classification, are classified as long-term unemployed. Long-term unemployment has many negative consequences for both an individual and the economy of the whole country (Nichols et al. 2013; Grzenda 2019b). In addition, it is much more difficult to leave it than to leave short-term or medium-term unemployment.

In 2017–2018, the situation on the labor market in Poland improved compared to previous years. The unemployment rate in 2017 ranged from 4.5 to 5.4%, and in 2018 from 3.6 to 4.2%, while, in the period 2010–2014, the unemployment rate exceeded 8%. In the fourth quarter of 2018, the average duration of job search for men was 9.6 months, while for women 9.4 months. Moreover, in this period higher employment rate was observed in the male population (62.6%) than in the female population (46.1%). At the same time, the unemployment rate was not significantly differentiated by gender, it was 5.7% for men and 5.4% for women.

The duration of being unemployed is influenced by many different factors both at the macro and micro scale. At the micro level, the chances of finding a job depend on the characteristics of individuals, including as pointed out by Uysal and Pohlmeier (2011) on their personality traits. Unfortunately, key surveys of the labor market, including LFS, do not contain data enabling the analysis of the latter characteristics. Regardless of the type of characteristics of the unemployed studied, the sole assessment of their impact on the unemployment duration under the assumption of ceteris paribus may be insufficient. In this study, the probability of finding a job by individuals having specific features was estimated using the semiparametric methods. In addition, the possibility of using these models to determine the probability of finding a job by people who were not included in the study was pointed out.

2 Research Method

The basis of the considerations presented in this paper are Cox models. In models of this class, the hazard function is the product of two elements: the first is an unspecified function of the baseline hazard, the second is the exponential transformation of the linear combination of explanatory variables (Cox and Oakes 1984).

Let $\mathbf{X} = [X_1, \ldots, X_k]^T$ denotes the covariate vector, and $\boldsymbol{\beta} = [\beta_1, \ldots, \beta_k]$ be the vector of estimated model parameters. Then, for the Cox proportional hazard model, the hazard is given by the formula:

$$h(t|\mathbf{X}) = h_0(t)\exp(\mathbf{X}\boldsymbol{\beta}), \tag{1}$$

where $h_0(t)$ is the baseline hazard. It is a non-negative and nonparametric function of a continuous random variable T, for $t \geq 0$. The main assumption of the model under consideration is the assumption of proportional hazards. This assumption in this paper was verified with two methods: a graphic method and by using time-dependent variables (Collett 2014).

The Cox semiparametric model can also be represented in an equivalent form:

$$S(t) = [S_0(t)]^{\exp(\mathbf{X}\boldsymbol{\beta})}, \tag{2}$$

where $S(t)$ denotes the probability of survival of the individual and $S_0(t)$ is the baseline survival function corresponding to the baseline hazard $h_0(t)$. For the estimation of parameters in the Cox model, the partial likelihood function is used (Breslow 1975). Then, the survival function estimator $S(t)$ has the form:

$$\hat{S}(t) = \left[\hat{S}_0(t)\right]^{\exp\left(\mathbf{X}\hat{\boldsymbol{\beta}}\right)}, \tag{3}$$

where $\hat{\boldsymbol{\beta}} = \left[\hat{\beta}_1, \ldots, \hat{\boldsymbol{\beta}}_k\right]$ denotes the estimator of the parameter vector $\boldsymbol{\beta}$ and $\hat{S}_0$ is the estimator of a baseline survival function, which is given by the formula:

$$\hat{S}_0(t) = \prod_{u|t_{(u)} < t} \left(1 - \frac{d_u}{\sum_{l \in R\left(t_{(u)}\right)} \exp\left(\mathbf{X}_l \hat{\boldsymbol{\beta}}\right)}\right), \tag{4}$$

where d_u, $u = 1, 2, \ldots, m$ denotes the numbers of observations, for which the event occurred at the moment $t_{(u)}$, $u = 1, 2, \ldots, m$, and $R\left(t_{(u)}\right)$, $u = 1, 2, \ldots, m$, denotes hazard set. The hazard set includes all individuals for whom the survival or censoring time is greater than $t_{(u)}$.

Taking into account the relationship between the basic functions used to describe the duration, the baseline survival function S_0 can be presented using cumulative hazard function H_0 in the following way:

$$S_0(t) = \exp(-H_0(t)), \tag{5}$$

where $H_0(t) = \int_0^t h_0(u)du$, $t \geq 0$.

Commonly used methods of estimating the survival function do not allow for the inclusion in the assessment of the survival time of other factors potentially differentiating the survival time of individuals from the analyzed groups. However, there exist methods estimating the survival curve following a Cox regression analysis, thanks to which various combinations of variables and their levels can be taken into account when estimating survival curves (Zhang et al. 2007).

Let j denote an individual belonging to the i-th group, and then, the observed values for this individual can be described by $\left\{T_{ij}, D_{ij}, \mathbf{X}_{ij}\right\}$, $i = 1, 2, \ldots, K$, $j = 1, 2, \ldots, n_i$, where T_{ij} denotes observed time, $D_{ij} = 0$, when censoring occurs and $D_{ij} = 1$, otherwise, and $\mathbf{X}_{ij}$ denotes the covariate vectors.

One of the first methods for estimating the survival function taking into account the values of other variables was to replace the covariate vector $\mathbf{X}$ by $\bar{\mathbf{X}}$ denoting the vector of average values of the covariate vectors in the input data (Neuberger et al 1986):

$$\hat{S}_i(t) = \exp\left\{-\hat{H}_{0i}(t) \exp\left(\bar{\mathbf{X}}\hat{\boldsymbol{\beta}}\right)\right\}, \tag{6}$$

where $\hat{H}_{0i}(t)$ is an estimated cumulative hazard function. However, this approach is not the best solution for two reasons. First of all, such an analysis is limited to the continuous variables only. Secondly, the question is whether such an individual with average continuous values exists. Therefore, the desirable approach is to determine the survival function separately for each group of individuals characterized by a specific set of characteristics. With the adopted notation, the survival curve at the moment t, for the individual from the i-th group, with values of variables $\mathbf{x}$, has the form (Zhang et al. 2007):

$$\hat{S}_i(t;\mathbf{x}) = \exp\left\{-\hat{H}_{0i}(t)\exp\left(\mathbf{x}\hat{\boldsymbol{\beta}}\right)\right\} \tag{7}$$

Then, the general formula for the direct adjusted survival curve is:

$$\hat{S}_i(t) = \frac{1}{n}\sum_{l=1}^{n}\exp\left\{-\hat{H}_{0i}(t)\exp\left(\mathbf{X}_l\hat{\boldsymbol{\beta}}\right)\right\} \tag{8}$$

where $n = \sum_{i=1}^{K} n_i$.

Such an approach also enables the use of the Cox regression model to determine the probability of a specific event occurring for individuals for which the model was not trained. In this paper, the survival function estimated in this way was used to estimate the probability of finding a job.

3 Empirical Data and the Estimation of the Models

The calculation of the direct adjusted survival probabilities of finding a job by the unemployed was based on the data from the LFS survey from 2017 to 2018 from Poland. Therefore, in this study, the unemployed are those, who were not employed during the week under consideration, in addition during the last 4 weeks, including as the last week the week under consideration, they were looking for a job and could take up a job in the 2 weeks following the week in question. At the same time, the unemployed did not include persons who were not looking for a job because they already had a job and waited for it to start within no more than 3 months and were ready to take it. Short-term and medium-term unemployment was studied. Therefore, the study included people who were unemployed in 2017 and were observed for a maximum of one year. In addition, the study only included individuals who had previously worked, so in the study characteristics related to previous employment could be included. People aged 18–44 were considered. There were 396 such persons, 152 of whom found a job during the period considered. For the purposes of survival analysis, it was assumed that they were persons for whom the event occurred (152 observations), other individuals were censored observations (244 observations). The time was calculated in months from the moment of termination of employment at the previous place of employment until finding a job or until the end of survey. The set of characteristics included in the study is presented in Table 1.

In the study of transition from the unemployment state to employment state, the Cox semiparametric model was used. Therefore, in the first stage of the study, the assumption of proportional hazards was verified with the graphic method and using

Table 1 Sample characteristics

Variable	Description	Levels	Proportion (%)
Sex	Respondent's gender	0 = woman	44.44
		1 = man	55.56
Age group	Age group of a respondent at the time of the survey	1 = from 18 to 24 years old	15.15
		2 = from 25 to 34 years old	50.00
		3 = from 35 to 44 years old	34.85
Marital status	Marital status of the respondent	0 = unmarried, separated or divorced, a widower, a widow	64.39
		1 = married	35.61
Education status	The level of education of the respondent at the time of the survey	1 = higher/undergraduate or engineering	29.29
		2 = post-secondary or secondary professional	24.24
		3 = secondary general	15.91
		4 = basic vocational	16.41
		5 = primary school, incomplete basic and without education	14.14
Place of residence	Class of place of residence during the survey	1 = city of 20 thousand residents and more	39.39
		2 = city under 20 thousand residents	22.47
		3 = rural areas	38.13
Work experience	Work experience	1 = up to 1 year	19.70
		2 = from 2 to 5 years	29.80
		3 = over 5 years	50.51
Reasons for stopping a job	Reasons for stopping the previous job	1 = dismissal by the employer or the end of the contract	62.37
		2 = employee side cause	24.75
		3 = other	12.88
Type of job	Type of recent job	0 = salaried employee	95.20
		1 = self-employed or helping family member	4.80

time-dependent variables. In order to make it possible, in groups established for individual categorical variables, survival functions were estimated using the Kaplan–Meier method and verified with log-rank test and Wilcoxon test whether there are statistically significant differences between the obtained curves. Then, plots of the negative logarithm were determined from the negative logarithm of the estimated survival function $(-ln(-lnS))$ for the qualitative categories of explanatory variables under comparison, and it was examined whether the obtained curves

Table 2 Results obtained from a semiparametric model with covariates

Parameter		Parameter estimate	Standard error	Chi-square	*p*-value	Hazard ratio
Sex	0	0.111	0.178	0.392	0.531	1.118
Age group	1	0.095	0.318	0.089	0.766	1.099
	2	−0.092	0.222	0.173	0.676	0.912
Marital status	0	−0.331	0.196	2.857	0.091	0.718
Education status	1	0.185	0.285	0.422	0.516	1.203
	2	−0.306	0.302	1.026	0.311	0.737
	3	−0.276	0.329	0.703	0.402	0.759
	4	0.130	0.295	0.193	0.660	1.138
Work experience	1	0.318	0.284	1.261	0.262	1.375
	2	0.424	0.231	3.382	0.066	1.528
Reasons for stopping a job	1	0.870	0.317	7.542	0.006	2.387
	2	1.014	0.340	8.870	0.003	2.756
Type of job	0	1.051	0.520	4.080	0.043	2.860

are parallel. Due to the lack of clear conclusions for some of the variables, the assumption of proportional hazards was also verified using time-dependent variables. As a consequence, it was found that the variables presented in Table 1 can be used to construct the Cox semiparametric model. The results of the estimation of the model best fitted to the empirical data are presented in Table 2.

Based on the analysis, it can be concluded that the respondent's characteristics of previous employment and marital status had the greatest impact on finding a job during one year period. Taking into account work experience, it was received that persons who had work experience from two to five years had the best chance of transition from the unemployment state to employment. In addition, people who had work experience up to a year were more effective in finding a job, compared to people whose work experience was longer than 5 years. Such a result may be due to the fact that unemployment was analyzed in the first year since the employment termination. Moreover, based on the data in Table 1, it can be concluded that some people quit their jobs because they had already found another job. It was obtained that persons who resigned from their jobs had about 175% more chance of transition from the unemployment state to employment than persons who resigned for other reasons than dismissal by the employer or termination of the contract.

People who were dismissed by the employer or whose employment automatically terminated at the end of a fixed term contract were also good at finding a new job. To sum up, it can be stated that people who had difficulties finding a job were the people who terminated their employment, for example, because of taking care of children or the elderly or terminated the employment for reasons other than resignation from work, dismissal by the employer or termination of the contract. It was almost three times harder to find a job for the people who conducted their own business activity or who helped in running a family business than for employees.

Unmarried persons had fewer chances of transition from the unemployment state to employment than married persons.

Based on the results obtained, it can be concluded that there are no longer large differences between genders in terms of the chances of finding a job on the Polish labor market. Interestingly, it was found that women had a slightly higher chance of finding a job than men, but according to other LFS studies (LFS 2018), women are more intensely looking for a job than men, and the study concerns a short period of time. There were no major differences in the transition from the unemployment state to employment in terms of age. In addition, it was received, that both people with higher education and vocational education had better chance of finding a job, compared to people with the least education.

This paper presents only the most important conclusions regarding the impact of the factors studied on the transition from the unemployment state to employment, because the main purpose of this study was the estimation of the probability of finding a job. To determine the direct adjusted survival probabilities, and more specifically, to calculate the probability of the opposite event, the methods described in Chap. 2 of this paper were used.

In the first stage of this part of the study, the survival function was estimated using the average values of features taken into account in the study of these features (Fig. 1—survival for average individual). Based on the survival function estimated in this way, some general conclusions can be drawn about the course of the phenomenon. However, the result obtained applies to an individual with the average values of features, and if there are categorical variables in a model, such an individual does not exist at all. Therefore, the survival function was estimated, which is an average of the survival functions for the individuals that occurred in the considered dataset (Fig. 1—direct adjusted survival curve). In the case of the dataset, no significant differences between the survival function estimated for the average individual and the direct adjusted survival curve were obtained (Fig. 1). Based on the results obtained, it can be concluded that in the analyzed period, the probability of finding a job by an unemployed person within half a year was about 0.2, and within 12 months about 0.4.

In order to verify the earlier finding that there are no significant differences in the transition from the unemployment state to employment due to gender, the direct adjusted survival curve was estimated in the groups determined by this variable (Fig. 2). Comparing the obtained graphs (Fig. 2) to the graph of the direct adjusted survival curve for the whole sample (Fig. 1), it can be seen that all these curves are shaped in a similar way, which confirms the previous finding. It has been obtained that from about second month women had slightly less probability of not finding a job compared to men. At the end of the period considered, this difference was around 0.04.

In the next stage of the study, direct adjusted survival curves were determined and compared for individuals characterized only by selected sets of characteristics. Due to the low percentage of self-employed and helping in a family business, only salaried employees were considered. Figure 3 presents the survival function for single women aged 18–24 with higher education, depending on the reason for

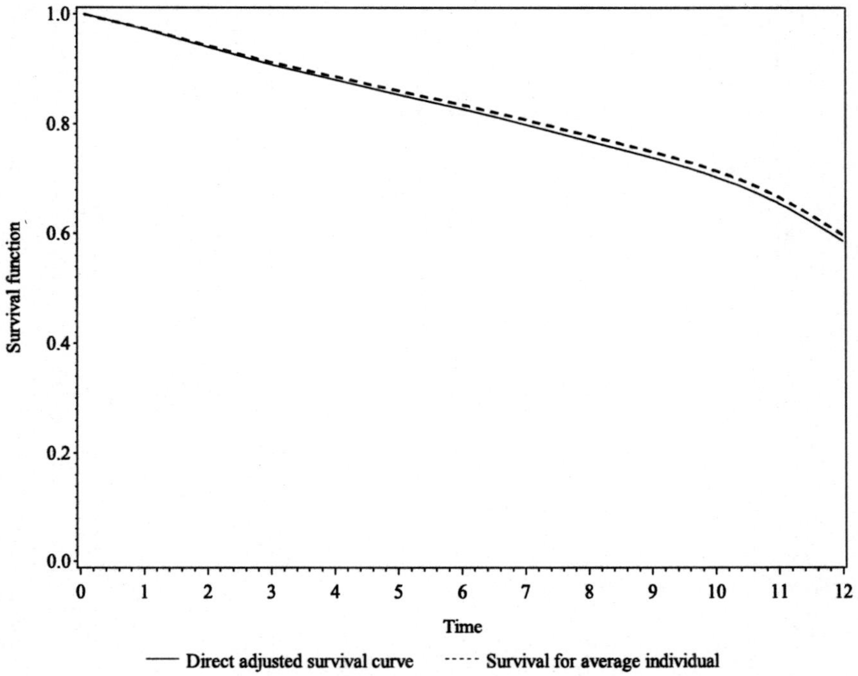

Fig. 1 Survival curve for the average individual and direct adjusted survival curve

leaving a job. It can be seen that young women who terminated an employment contract for reasons other than dismissal by the employer, the end of the contract, or resignation from their job had the least chance of transition from the unemployment state to employment. One year after leaving last job, the probability of finding employment for these women was only 0.38. Women who terminated their employment contract for reasons attributable to the employee, had the best situation on the labor market, for them this probability was 0.65.

Then, the survival functions for married women of the same age with higher or undergraduate or engineering education were also estimated, depending on the reason for leaving a job (Fig. 4). Comparing the curves obtained, to the curves obtained for single women, it can be seen that the probability of finding a job by women who terminated the employment contract for reasons other than dismissal by the employer, the end of the contract or resignation from job was 0.42, and for women who terminated their employment contract for reasons attributable to the employee, this probability was 0.77. It can therefore be concluded that married women fared better in the labor market, but in the case of these women, there were slightly larger differences between women, who terminated their employment contract for reasons attributable to the employee, and those who terminated their employment contract for reasons other than dismissal by the employer, or the end of the contract.

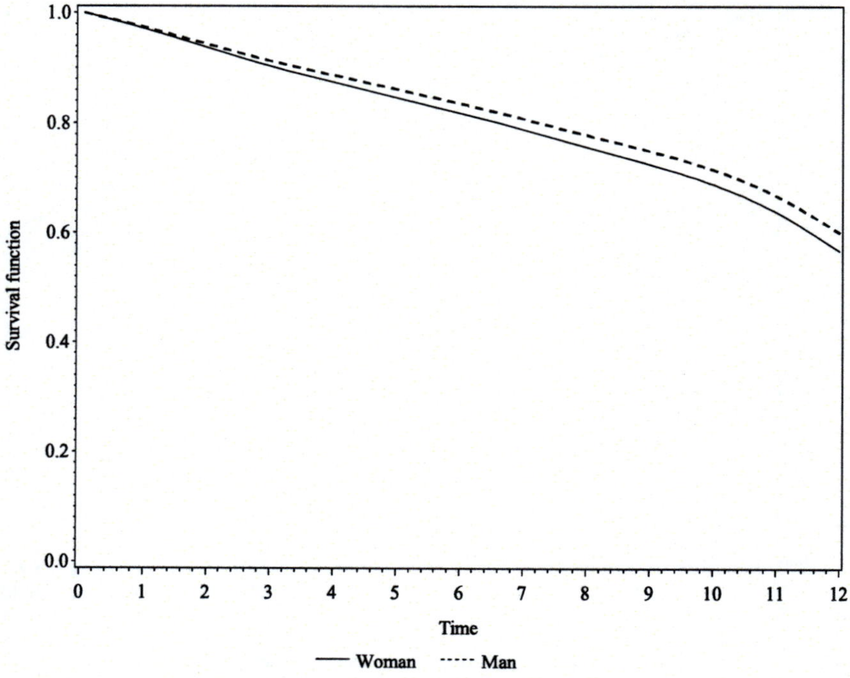

Fig. 2 Direct adjusted survival curves for women and men

In the next stage of the research, the chances of finding a job by people who were dismissed from their job or those who terminated the employment contract and were aged 24–35 years old were analyzed. By estimating survival functions for various combinations of their other characteristics, it is possible to examine which individuals are similar to each other and which differ in terms of the probability of transition from the unemployment state to employment state. People with post-secondary or secondary professional education (Fig. 5) and vocational education (Fig. 6) were considered separately. Comparing survival functions presented in both graphs, it can be concluded that the probability of not finding a job is higher in the case of people with post-secondary and secondary vocational education (Fig. 5) than in the case of people with vocational education (Fig. 6). Analyzing only people with post-secondary and secondary vocational education, it can be seen that married people with seniority from 2 to 5 years have a better chance of finding a job than unmarried people with seniority up to 1 year. Similar relation was obtained for people with vocational education. In addition, it was obtained that the probability of finding a job is at a similar level for people with post-secondary and secondary vocational education, married, with work experience from 2 to 5 years (Fig. 5), as for people with vocational education, unmarried, and professional experience up to 1 year (Fig. 6).

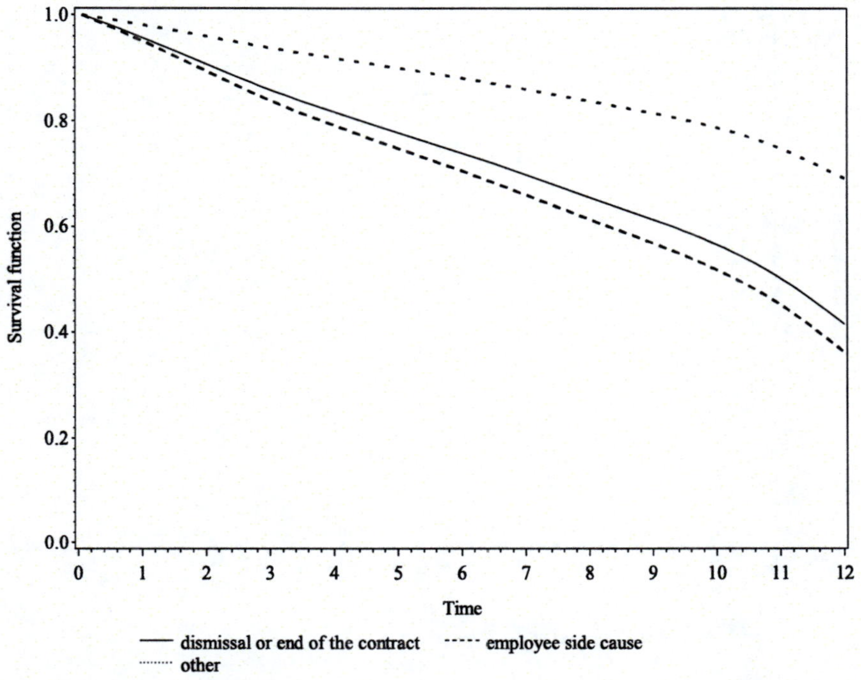

Fig. 3 Direct adjusted survival curves for single women aged 18–24 with higher or undergraduate or engineering education, depending on the reason for leaving a job

According to the report of the Labor Market Department of the Polish Ministry of Labor and Social Policy (2019), the situation of young people on the labor market in Poland, despite the fact that it is constantly improving, remains difficult. Analyzing various survival curves, it was possible to identify a group of people aged 18–24 who have a probability of finding a job within a year of 0.8. They are married women with at least undergraduate and engineering education, with professional experience of 2–5 years, who terminated their employment contract for reasons attributable to the employee, but these reasons did not include taking care of children or the elderly. A similar result was also obtained for married men with at least undergraduate and engineering education, having professional experience of 2–5 years, who also terminated their employment contract for reasons attributable to the employee (Fig. 7). Based on the received survival curves, it can be concluded that it is easier to find a job if employees decide to terminate the employment contract, than when the contract is terminated by employers. This may be due to the fact that in the former case the reason for resignation from work may be the fact of preparing the contract with a new employer.

Table 3 presents the direct adjusted probabilities of not finding a job in individual months since leaving the last job by married women aged 18–24 with at least undergraduate and engineering education, having professional experience from 2 to

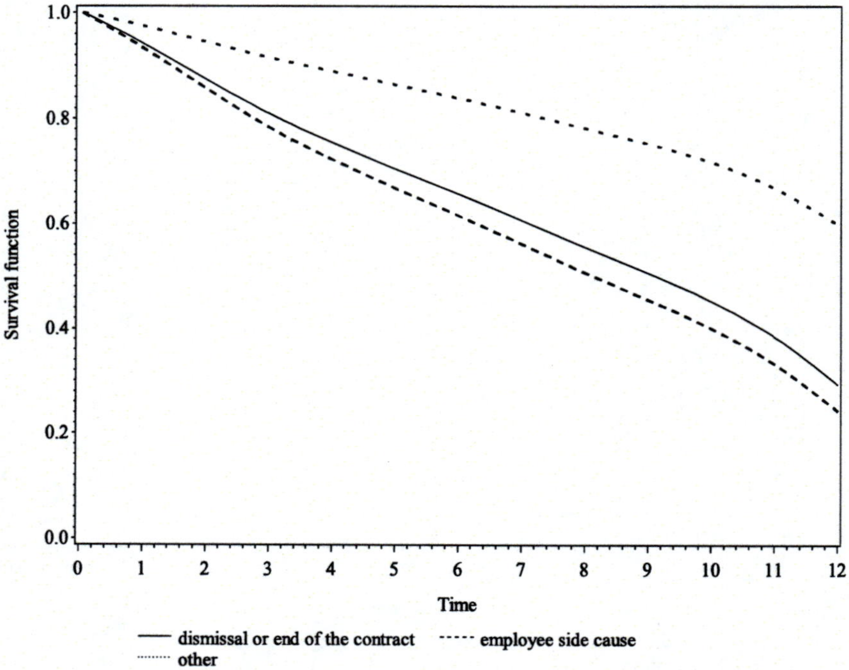

Fig. 4 Direct adjusted survival curves for married women aged 18–24 with higher or undergraduate or engineering education depending on the reason for leaving job

5 years, who terminated their employment contract for reasons attributable to the employee. Similar probabilities can be calculated for each individual for whom the values of features included in the model are known.

4 Conclusions

In this work, the evaluation of the probability of transition from the unemployment state to employment state using the Cox regression model has been discussed. The study considered people who were unemployed for a maximum period of 12 months since the time of leaving last job. In research on the labor market, this is a crucial moment, because people who do not find a job after such a period are considered long-term unemployed. Long-term unemployment is a very undesirable phenomenon that has many negative effects (Nichols et al. 2013). The correct definition of the profile of people at risk of long-term unemployment can be helpful in preventing it.

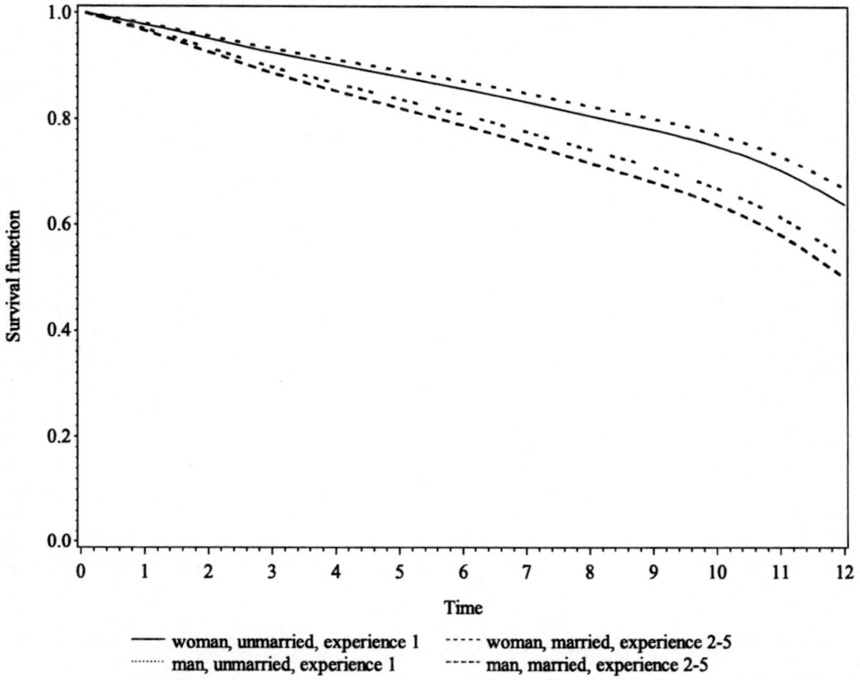

Fig. 5 Direct adjusted survival curves for people who were dismissed from their job or those who terminated the employment contract and were aged 24–35 years with post-secondary or secondary professional education

Based on the presented research, it can be concluded that the probability of not finding a job within one year decreases the fastest in the last two months. On the one hand, it can mean that the unemployed are also aware that after 12 months of being unemployed, it is more difficult to find a job, for example, due to loss of professional skills. On the other hand, it can be assumed that the result obtained is due to the fact that in the regions with high unemployment, unemployment benefit is granted during the first 12 months of unemployment period. However, no changes were observed on the survival curve graph after 6 months, i.e., after the standard period for granting unemployment benefits.

This study found that the characteristics of an individual related to their previous professional experience have the greatest impact on the probability of transition from unemployment to employment. People with work experience of 2–5 years and those who gave up their jobs had the greatest chance of finding a job. However, there were no major differences in the chances of finding a job due to such features as gender, age, education, or place of residence.

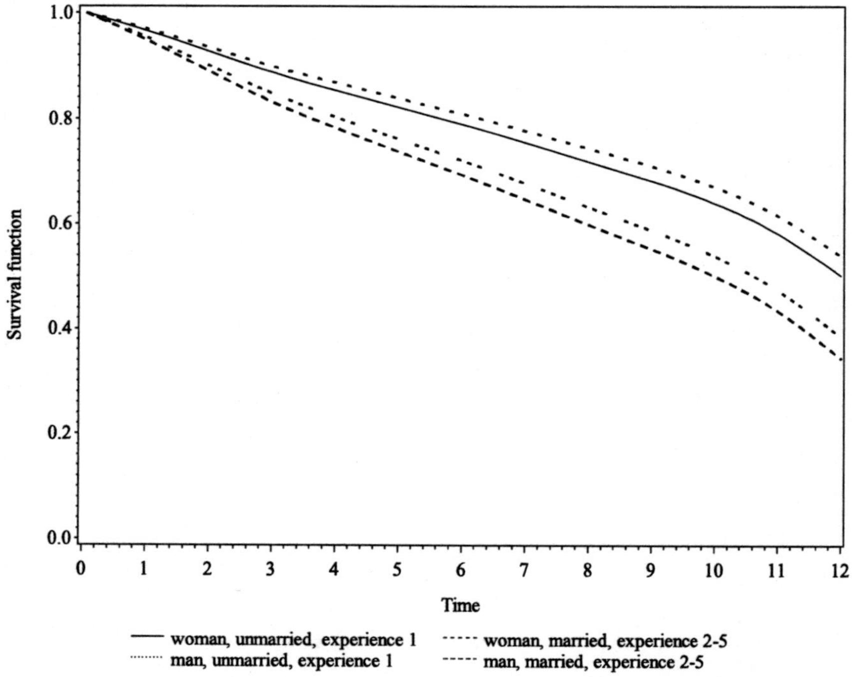

Fig. 6 Direct adjusted survival curves for people who were dismissed from their job or those who terminated the employment contract and were aged 24–35 years with vocational education

Although there were no clear differences in the behavior of women and men on the labor market in Poland in the analyzed period, the approach proposed in this work enabled the assessment of opportunities for women and men to find jobs also due to their other characteristics. Based on the analyzes carried out, it can be concluded that the probability of transition from unemployment to employment in the case of women is strongly conditioned by their other characteristics. It was received that married women who stopped working for reasons related to, among others, caring for children or other persons requiring care had more difficulty in finding a job than women who terminated employment contract for reasons attributable to the employee, due to terminating their contract by their employer or the period of their employment contract ended. At the same time, it was received that married women with at least undergraduate and engineering education, having professional experience of 2–5 years, who terminated employment contract for reasons attributable to an employee have a very good chance of finding a job within a year since the moment of leaving their last job.

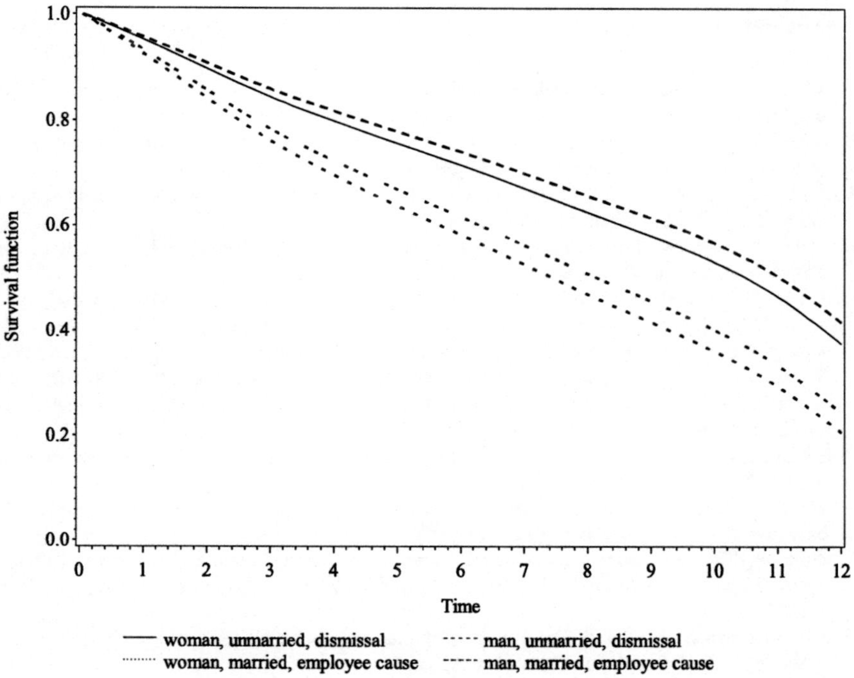

Fig. 7 Direct adjusted survival curves for people aged 18–24 with at least undergraduate and engineering education, with professional experience of 2–5 years

Table 3 Direct adjusted probabilities of not finding a job of individuals having different covariate values

Month											
1	2	3	4	5	6	7	8	9	10	11	12
Survival											
0.94	0.85	0.75	0.70	0.63	0.59	0.53	0.63	0.42	0.36	0.31	0.20

Based on the analyzes carried out in this work, it can be concluded that the probability of transition from unemployment to employment is conditioned by many different factors. Therefore, analyzing social groups in terms of one characteristic, with fixed values of other characteristics, may be in some cases insufficient. The approach presented in this paper enables the estimation of the direct adjusted survival probabilities of finding a job for people characterized by any set of characteristics included in a model. The use of the presented methods by relevant institutions can be helpful in correctly defining target groups for various programs of professional activation of the unemployed.

References

Bieszk-Stolorz B, Markowicz I (2019) Analiza trwania w badaniach ekonomicznych. Modele nieparametryczne i semiparametryczne. CeDeWu, Warszawa

Blossfeld HP, Hamerle A, Mayer K (1989) Event history analysis statistical theory and application in the social sciences. L. Erlbaum, Hillsdale, New York

Blossfeld HP, Rohwer G (1995) Techniques of event history modeling. New approaches to causal analysis. L. Erlbaum, Mahwah, New York

Breslow NE (1975) Analysis of survival data under the proportional hazards model. Int Stat Rev/ Revue Internationale de Statistique 43(1):45–57

Collett D (2014) Modelling survival data in medical research. Chapman and Hall, London

Cox DR, Oakes D (1984) Analysis of survival data. Chapman and Hall, London

Departament Rynku Pracy MRPiPS (2019) Sytuacja na rynku pracy osób młodych w 2018 roku, Warszawa. https://psz.praca.gov.pl/documents/10828/167955/Sytuacja%20na%20rynku%20pracy%20os%C3%B3b%20m%C5%82odych%20w%202018%20roku.PDF/988213a4-02ad-425d-b6b8-7e0145a5d65b?t=1557826527278. Accessed on 2 Sept 2020

Fisher LD, Lin DY (1999) Time-dependent covariates in the cox proportional-hazards regression model. Annu Rev Public Health 20(1):145–157

Grzenda W (2019a) Modelowanie karier zawodowej i rodzinnej z wykorzystaniem podejścia bayesowskiego. wydawnictwo Naukowe PWN, Warszawa

Grzenda W (2019b) Socioeconomic aspects of long-term unemployment in the context of the ageing population of Europe: the case of Poland. Econ Res Ekonomska Istraživanja 32(1): 1561–1582

Gutiérrez-Domènech M (2008) The impact of the labour market on the timing of marriage and births in Spain. J Population Econ 21(1):83–110

Kalbfleisch JD, Prentice RL (2011) The statistical analysis of failure time data. Wiley, USA

Kaplan EL, Meier P (1958) Nonparametric estimation from incomplete observations. J Am Stat Assoc 53:457–481

Klein JP, Moeschberger ML (2005) Survival analysis: techniques for censored and truncated data. Springer, New York

Landmesser J (2013) Wykorzystanie metod analizy czasu trwania do badania aktywności ekonomicznej ludności w Polsce. Wydawnictwo SGGW, Warszawa

LFS (2018) Labour force survey in Poland IV quarter 2018. Statistics Poland, Labour Market Department, Warszawa

Neuberger JAMES, Altman DG, Christensen ERIK, Tygstrup N, Williams R (1986) Use of a prognostic index in evaluation of liver transplantation for primary biliary cirrhosis. Transplant 41(6):713–716

Nichols A, Mitchell J, Lindner S (2013) Consequences of long-term unemployment. The Urban Institute, Washington, DC

Polemis ML, Stengos T (2015) Does market structure affect labour productivity and wages? Evidence from a smooth coefficient semiparametric panel model. Econ Lett 137:182–186

Uysal SD, Pohlmeier W (2011) Unemployment duration and personality. J Econ Psychol 32(6): 980–992

Zhang X, Loberiza FR, Klein JP (2007) A SAS macro for estimation of direct adjusted survival curves based on a stratified Cox regression model. Comput Methods Programs Biomed 88(2):95–101

Europe 2020 Strategy—Objective Evaluation of Realization and Subjective Assessment by Seniors as Beneficiaries of Social Assumptions

Klaudia Przybysz, Agnieszka Stanimir, and Marta Wasiak

Abstract The study attempts to evaluate the implementation of the Europe 2020 Strategy in the area relating to the elderly. Due to the aging of the EU population, a broad analysis of various aspects of the life of seniors is essential. An important element of the Europe 2020 Strategy, highlighted in the Integrated Guidelines for Growth and Employment (Part II/No. 8), is activities for social inclusion, combating poverty and promoting equal opportunities. This guideline also applies to seniors and here it is necessary to implement appropriate actions. In the study, we proposed to use the methods of multidimensional comparative analysis to assess the level of implementation of the Europe 2020 Strategy, indicating areas important for the quality of life of seniors and identifying changes in the assessment of the implementation of this Strategy by this generation. In the conducted research, we determined the taxonomic measures: TMGO and TMMI in relation to Europe 2020 indicators. We identified the characteristics of seniors who assessed the Europe 2020 Strategy positively and negatively and the resulting benefits for this generation. The study showed the existence of a very large diversity of seniors in terms of their life quality and their assessment of the Strategy.

Keywords Europe 2020 · Taxonomic measure of good oldness (TMGO) · Taxonomic measure of main indicators (TMMI) · Seniors' subjective ratings · Hellwig's ordering

K. Przybysz · A. Stanimir (✉)
Wroclaw University of Economics and Business, Wroclaw, Poland
e-mail: agnieszka.stanimir@ue.wroc.pl

K. Przybysz
e-mail: klaudia.przybysz@ue.wroc.pl

M. Wasiak
OptumRx, Dublin, Ireland
e-mail: marta.wasiak@optum.com

K. Jajuga et al. (eds.), *Data Analysis and Classification*, Studies in Classification, Data Analysis, and Knowledge Organization,
https://doi.org/10.1007/978-3-030-75190-6_15

1 Introduction

The period of planned activities under the Europe 2020 Strategy is coming to an end. The project was aimed at providing equal opportunities for the citizens of the European Union Member States. Thus, it is necessary to conduct a thorough evaluation of the implementation of the postulates and the assumptions of this Strategy. The simplest method to do it is by comparing the level of implementation of the assumptions with reference values or by assessing the period of reaching the postulated values and comparing them with the level of economic factors characterizing individual economies of the EU countries.

Our research was aimed at filling the research gap in the area of perception of the assumptions and the Strategy implementation by the beneficiaries of this project. In the study, we focused on a group of seniors. Due to the aging of European societies, they are becoming a large social group. Their opinions, needs, and conclusions are a valuable source of information, necessary while planning social policy instruments as well as making economic decisions. The conducted research concerned both the summary of the implementation of the Europe 2020 Strategy and the analysis of opinions of the Europeans aged 60+ on the implementation of the Strategy's objectives, also embracing demographic characteristics of this social group.

Our study also aims to present a method with similar assumptions to TOPSIS, but published much earlier—this is Hellwig's linear ordering.

2 Research Background and Literature Review

Europe 2020—a Strategy for smart, sustainable, and inclusive growth was endorsed by the European Council on the June 17, 2010. It is a long-term program of the European Union's socio-economic development policy, which has replaced the Lisbon Strategy. While its main goal remains economic development, as in the Lisbon Strategy, more emphasis has been placed on balancing this process. It deals not only with the structural problems of the European economy but also takes into account long-term problems, including globalization, limited resources, and rational use of them, and the aging of societies (Sulmicka 2011; Grosse 2010; Vanhercke et al. 2010). The main goals of the Strategy are:

- G1—more than 75% of the population aged 20–64 years to be employed,
- G2—more than 3% of GDP to be invested in the R&D sector,
- G3—reducing greenhouse gas emissions by 20% compared to 1990,
- G4—increasing the share of renewable energy sources in final energy consumption to 20%,
- G5—improving the energy efficiency by 20%,
- G6—reducing the share of early school leavers to less than 10%,

- G7—at least 40% of 30–34 years old to have completed tertiary or equivalent education,
- G8—at least 20 million people fewer at risk of poverty or social exclusion.

The result of implementing the Europe 2020 Strategy is to be an economy based on knowledge, low-emission, popularizing and favoring environmentally friendly technologies, using resources sparingly, creating new jobs while respecting the natural environment, and maintaining care for social cohesion (http://www.mg.gov.pl). In line with the established priorities, five overarching goals were assigned and attributed indicators, thanks to which it is possible to accurately assess the degree of implementation of the Strategy objectives, both on the basis of the values of these indicators as well as the progress made by individual member countries in the adopted time horizon (Balcerzak 2015; Pasimeni 2012). Many authors have discussed the possibilities of the implementation and the existing threats (Stanickova 2017; Sulmicka 2011; Zalewska and Świetlikowski 2017). In literature, one can also find studies on the planned effects of the implementation, based on probable scenarios (Hobza and Mourre 2010). Numerous studies have also concerned the degree of achievement of the Strategy's goals (Kedaitiene and Kedaitis 2012; Manafi 2012; Megyeri 2018; Młynarzewska-Borowiec 2020; Rappai 2016; Stec and Grzebyk 2018; Walheer 2018) and also the research of the most important indicators of the impact of the Europe 2020 Strategy on economic performance (Daly 2012; Radulescu et al. 2018). Some authors have proposed new methods of measuring the effects of the Strategy (Pasimeni 2012; Talmaciu and Cismas 2016).

However, the issue of implementing the Strategy objectives in relation to how it is assessed by its beneficiaries has not been discussed until now. The social group that is at the center of interest in our study, i.e., seniors, so far has been denied the right to provide their expertise. Most often, such words as seniors, the elderly or the aging society appear in literature as a definition of an economic or social activity. With regard to the Europe 2020 Strategy, older people appear in researches as a reason to take action to secure the participation of younger women in the labor market or as beneficiaries of older-oriented services so that younger ones can get into work (European Commission 2019a). Older people are indicated here as indirect beneficiaries of the elderly services economy, i.e., care, health, and other age-friendly environments services in long-term as well as innovation for active and healthy aging (Begg 2010; Delmas 2015; European Commission 2019a; Fernández-Carro et al. 2015; Fico et al. 2015). In European Platform against Poverty which is one of the Flagship Initiative of European 2020 Strategy elders appear in the context of defining measures addressing the specific circumstances of their particular risk of poverty (European Commission 2010) and as a social group against which protective measures should be taken (European Commission 2019a). Studies on the implementation of the Europe 2020 Strategy also concern active aging (Loureiro and Barbas 2014). In this perspective, this article provides an innovative view of the implementation of the Europe 2020 goals. Taking into account one of the priorities which is social inclusion, and the main goal related to this priority—preventing social exclusion, we have decided to analyze the opinions of the oldest part of the European society—seniors 60+.

The goals of our study were therefore:

1. Using taxonomic measures to compile differences in the living conditions of seniors with differentiation in the implementation of the Strategy goals.
2. Checking whether there are any differences between Europeans by age in the assessment of the Europe 2020 Strategy, and whether the Strategy is clearly well assessed.
3. Verifying the assumptions about the influence of demographic characteristics and knowledge of the Strategy goals on the assessment of their implementation.
4. Examining the relationship between a positive opinion on the Strategy's objectives and an opinion on the EU policy concerning the areas compliant with the Strategy.
5. Comparing Europeans' opinions on EU policies and the rights of the EU citizens on the background of respondents' current satisfaction with their lives.

The result of such research is a potential source of information about seniors' attitudes in various member countries. It can also help to identify the differentiating factors. This, in turn, may translate into a more effective construction of social policy instruments in relation to seniors, who constitute an increasing social group, particularly vulnerable to exclusion.

3 Data and Methods

The study was conducted as a two-stage process. Since, by definition, the implementation of the assumptions of the Strategy is to contribute to the development of the European economy, and thus to raising the standard of living of its inhabitants, we have decided to compare the differences in the living conditions of seniors with the differences in the implementation of the Strategy objectives. Two taxonomic measures have been created for this purpose, at the first stage of this study. One relating to the living conditions of seniors, the other to the values of the main indicators of the Europe 2020 Strategy. Then, the EU countries were assigned to three classes according to the values of each measure. The results of both classifications were compared, assuming that the similarity of both classifications may indicate the impact of the implementation of Strategy's indicators on the living conditions of seniors, despite the lack of a direct indication of such a relationship.

In the first part of our research, we used a grouping method of taxonomic measures determined on the basis of linear ordering by Hellwig. This method was proposed in 1968 (Hellwig 1968). The way it is conducted indicates the place (linear order) of objects in relation to the ideal solution. These assumptions are compatible with the TOPSIS method (Hwang and Yoon 1981). However, due to the publishing possibilities of the authors of both methods, the latter gained popularity.

In order to popularize Hellwig's approach, we present its algorithm in detail (availability of source data is limited and the method is described in Polish). Other presented methods are widely discussed in literature. There are many online resources where you can find detailed descriptions and algorithms. For this reason, we will only indicate literature sources for them.

Linear ordering methods allow ranking objects on the basis of the set of characteristics of the research problem. The linear ordering is the measure of diversity and should characterize how much, on average, one object is better (or worse) than another object due to variable values. We used Hellwig's ordering to construct the *Taxonomic Measure of Good Oldness* (TMGO) and *Taxonomic Measure of Main Indicators* in relation to Europa 2020 indicators (TMMI). The first one was created on the basis of the values of variables relating to various aspects of seniors' lives, such as:

Income aspect:

X_1—AIC per capita (Actual Individual Consumption),
X_2—income inequality,
X_3—percentage of people who cannot afford to buy the Internet,
X_4—declared satisfaction with your material situation,
X_5—the percentage of people who cannot afford to meet once a month with friends or family at a restaurant,

Housing situation:

X_6—housing cost burden ratio,
X_7—percentage of people who do not have a bathroom with a bath or a shower,

Health aspect:

X_8—the percentage of people positively assessing their health condition,
X_9—percentage of people reporting problems with daily activities due to health obstacles,
X_{10}—average healthy life expectancy according to own assessment,

Lifestyle:

X_{11}—percentage of people who are generally satisfied with their lives,
X_{12}—participation in cultural or sports events in the last 12 months,
X_{13}—satisfaction with spending time,
X_{14}—the percentage of people who meet friends several times a month.

All data were sourced from the Eurostat databases and relate to the year 2018. In ordering Hellwig's method, two presteps are needed:

1. All nominant variables must be converted into stimulants.
2. It is necessary to ensure the comparability of variables (normalization).

There were no nominants among the variables in our study.

For variable normalization, we used the following formula:

$$z_{ij} = \frac{(x_{ij} - \bar{x}_j)}{S_j} \tag{1}$$

where z_{ij}—normalized value of jth variable in ith object; x_{ij}—observed value of jth variable in ith object; $\bar{x}_j$—mean of jth variable; S_j—standard deviation of jth variable.

Standardization method (Eq. 1), by which a variable is rescaled, finally led to a value of each variable equal to zero and a standard deviation equal to 1. In Hellwig's ordering (Hellwig 1968) in the next step, we determined an ideal solution (z_{+j}):

$$z_{+j} = \begin{cases} \max_i\{z_{ij}\} \textit{ asociated with stimulant} \\ \min_i\{z_{ij}\} \textit{ associated with destimulant} \end{cases} \tag{2}$$

Then we calculated the Euclidean distance of each object from the ideal solution (pattern):

$$d_{i+} = \sqrt{\sum_{j=1}^{m} (z_{ij} - z_{+j})^2}, \quad i = 1, 2, \ldots, n \tag{3}$$

The smaller this distance is, the more the object is similar to the pattern (ideal object) and the higher the level of the research problem for this object is.

In Hellwig's ordering synthetic measure is defined as follow (Pluta 1986):

$$m_i = 1 - \frac{d_{i+}}{d_0}, \quad i = 1, 2, \ldots, n \tag{4}$$

where d_0 = distance between ideal and non-ideal solution and m_i is a measure of $TMGO_i$ and $TMMI_i$ in our search for *i*th country.

The distance determined in this way refers to the maximum possible distance, which is d_0 between an ideal solution and a non-ideal solution.

The measure of m_i, usually takes values in the range <0;1>. Higher values of these indicators indicate better-rated objects. An additional advantage of this method is the possibility of dividing objects (in our case, EU countries) into classes according to the established measure values:

$$\begin{aligned} &1.\ m_i \geq m_s; \\ &2.\ m_r < m_i < m_s; \\ &3.\ m_i \leq m_r \end{aligned} \tag{5}$$

where $m_s = \overline{m} + S_{mi}$; $m_r = \overline{m} - S_{mi}$; $\overline{m} = \frac{1}{n}\sum_{i=1}^{n} m_i$; $S_{m_i} = \sqrt{\frac{1}{n}\sum_{i=1}^{n} (m_i - \overline{m})^2}$.

The best results correspond to the objects with relation no. 1, the average results refer to the objects whose measure meets the second inequality, and the measures of the objects with the lowest results do not exceed the value defined as m_r.

In the study of the subjective perception of the implementation of the Europe 2020 Strategy by seniors, we used Standard Eurobarometer data from 2020, 2016, and 2019 (European Commission 2019b, 2020; Papacostas 2012). From 2010 to 2016, the standard Eurobarometer surveyed the Europeans' opinions about eight objectives of the Europe 2020 Strategy in the exact wording (listed in Chap. 2, marked as G1–G8). The aim of our analysis was to determine the characteristics of people from different age groups who assessed the key goals of the Europe 2020 Strategy as too ambitious, about right and too modest, and to check whether this assessment differs in subsequent years. Out of eight goals, we chose only socio-economic goals for analysis, i.e., 1 (75% of the population aged 20–64 years to be employed), 6 (share of early school leavers to be reduced to less than 10%), 7 (at least 40% of 30–34 years old to have completed tertiary or equivalent education), and 8 (reducing the risk of poverty or social exclusion). The first one directly applies not only to young people but also to the group of seniors who are able to be professionally active. This goal indirectly prevents exclusion and discrimination at work on the basis of age. As Di Palo (2019, p. 277), pointed out "the consequent increase in the cohorts of elderly, jointly with the shrinking of the working population, has a negative effect on the financing of social security schemes." Karakaya (2009) indicated that financing related to the aging population may not be available in the future. The implementation of Goals 6 and 7 increases the chances of the efficiency of the pension system, while goal 8 is related to the implementation of the three above-mentioned Goals 1, 6, and 7. For the four selected goals, the assessment of the need for the implementation is easy to carry out by most people. As to the remaining goals, although they seem important, it is difficult to imagine their significance in everyday life.

Since 2016, the scope of the Eurobarometer survey on Europe 2020 has changed significantly. In 2016 and 2019, the variables for the freedom of movement within the EU (for employment, work, travel, and life), the awareness of EU citizenship rights and the feeling of being European were introduced into the study. Therefore, in this case, the aim of our study was to verify the knowledge of the benefits of equal opportunities within the EU.

Using the Eurobarometer survey, we chose respondents over 59, divided them into three groups 60–65 (still working seniors), 66–74 (seniors in the transition into retirement phase), and 75+ (retired seniors). Moreover, we used information on the respondents' life satisfaction, place of residence, age at which education was completed, and gender.

In the study of subjective assessments of seniors, we employed hierarchical charts and a correspondence analysis with a hybrid approach including a concatenated and multidimensional contingency table.

Correspondence analysis is widely discussed in literature. This method owes its popularity to the works of Greenacre (1984), in which not only the basic algorithms are discussed in detail, but also many modifications. The indicated approach, called by us a hybrid, was presented by Greenacre at the conference described shortly in Greenacre and Blasius (2006) and in detail presented by Michael Greenacre at the IFCS conference in Salonica in August 2019.

Correspondence analysis in its simplest form allows for a graphical presentation of relationships between the categories of two non-metric variables. This method is based on the construction of the contingency table and uses the singular values decomposition. The result of the analysis is a set of coordinates for each category of variables, thanks to which it is possible to present relationships both in the space that fully reflects the connections and in the space of a low dimension, while maintaining the highest quality of the presentation. The problem becomes more complicated when it is necessary to examine a larger number of variables and additionally indicate the dominant variable whose occurrences are to be explained by the other variables. This is the case in our study. The solution we used is based on the following steps:

- constructing a multidimensional contingency table for assessments of the selected Europe 2020 Strategy goal creating layers from the age groups of respondents; so a new goal/age variable was created as follow: G1TA60–65 (G1—goal 1; TA—too ambitious; 60–65); G1AR60–65 (G1—target 1; AR—about right; 60–65); G1TM60–65 (G1—target 1; TM—too modest; 60–65); …; G1TM75+ (G1—target 1; TM—too modest; 75+);
- then this variable formed four contingency tables with life satisfaction (VS—very satisfied, FS—fairly satisfied, NVS—not very satisfied, NS—not at all satisfied), the place of residence (R/V—rural area or village; S/M small or middle sized town; LT—large town); end of education age (1 to 15 years, 2 to 16–19 years, 3 to 20 and more, 5—no education) and gender (F and M);
- combining multiway contingency tables into a concatenated contingency table with a joint goal/age variable.

The hybrid contingency table built in this way was 12 × 9, because 9 is the number of categories of the goal/age variable and 12 is the number of categories of the four demographic variables. Correspondence analysis is performed for such a table. The full dimensional space of the relationships between the categories of variables is equal to eight $((r-1; c-1) = \min(12-1; 9-1))$.

Similar to the first goal, we built hybrid contingency tables for the remaining goals.

Additionally, when we started to analyze the perception of each of the goals of the Europe 2020 Strategy by seniors, we built hierarchical charts. These charts allow us to determine how the study population is divided into categories of successively added variables. In this analysis, the population was first divided according to the assessment of the Europe 2020 Strategy, then the group of respondents who chose a specific assessment was divided by age; further, in each age group, the division was made according to the age at which education ended and finally to the level of life satisfaction.

Another method used in our research was the hierarchical cluster analysis according to Ward's approach. We used this method in the analysis of the subjective assessment of the benefits of carrying out activities related to the Europe 2020 Strategy in 2016 and 2019. In this case, the variables were the percentages of choices made by respondents in particular countries for the categories of the following variables (broken down by age):

- LUE—The right for EU citizens to live in every Member State of the EU (BT—a bad thing, GT—a good thing, N—neither a good nor a bad thing): LUEBT60–65, LUEGT60–65, LUEN60–65, …, LUEN75+;
- LC—The right for EU citizens to live in *my country* (BT—a bad thing, GT—a good thing, N—neither a good nor a bad thing): LCBT60–65, LCGT60–65, LCN60–65, …, LCN75+;
- WUE—The right for EU citizens to work in every Member State of the EU (BT—a bad thing, GT—a good thing, N—neither a good nor a bad thing): WUEBT60–65, WUEGT60–65, WUEN60–65, …, WUEN75+;
- WC—The right for EU citizens to work in *my country* UE (BT—a bad thing, GT—a good thing, N—neither a good nor a bad thing): WCBT60–65, WCGT60–65, WCN60–65, …, WCN75+;
- CIT—You feel you are a citizen of the EU (YD—yes, definitely, Y—yes, to some extent, NR—no, not really, DN—no, definitely not): CITYD60–65, CITY60–65, CITNR60–65, CITDN60–65,…, CITDN75+;
- RIT—You know what your rights are as a citizen of the EU (YD—yes, definitely, Y—yes, to some extent, NR—no, not really, DN—no, definitely not): RITYD60–65, RITY60–65, RITNR60–65, RITDN60–65,…, RITDN75+.

The normalization of the variables was carried out using Eq. 1. It should be noted, however, that because the countries were compared over time, the arithmetic mean and standard deviation were calculated based on the observations from both periods (Strahl 2006). The correctness of the classification was assessed using the Silhouette index. Additionally, the compliance of the classification was assessed with the Spearman coefficient.

4 Taxonomic Measure of Good Oldness

In Table 1, we presented the countries with assigned group numbers which formed the basis for the maps creation. The results of the classification with regard to the calculated measures are shown in Figs. 1 and 2. To compare the results of both classifications, the Rand index was used. The assessment of similarity of the results of two classifications we carried out using the function comparing. Partitions, clusterSim: comparing. Partitions (cl1, cl2, type = "rand") package of *R* program, where:

- cl1 (cl2)—a vector containing cluster numbers to which objects in the first division have been classified (in the second division),
- type—type of index "rand"—*Rand index*.

The *Rand Index* gives a value between 0 and 1, where 1 means the two clustering outcomes match identically. In our case, the Rand index is 0.753. So since the results of both classifications are similar, then on this basis, a conclusion about the impact of achieving goals in individual European Union countries on the living conditions of seniors can be drawn.

Indicators for measuring the achievement of the goals of the Europe 2020 Strategy are useful when we want to check how individual member countries are implementing their main priorities and objectives. However, in this form, it is not possible to relate the results achieved by individual countries to selected groups of beneficiaries. The proposed research method allowed, through indirect inference, to determine the impact of the implementation of the Strategy goals on the living conditions of seniors.

The obtained results clearly show that even those goals of the Europe 2020 Strategy whose impact on the daily life of European citizens is difficult to perceive, translate into their quality of life.

5 Europe 2020 Strategy in the Seniors' Opinion

In the analysis of the subjective perception of the level of achievement of selected goals (G1, G6–G8) of the Europe 2020 Strategy, it was important to check how these goals are assessed by the entire group of seniors. Hierarchical Figs. 3 and 4 present compositions of the variables goal, age, gender, years of education, and life satisfaction of the respondents in 2010 and 2016. Additionally, Figs. 4 and 5 present the decomposition into three categories of the Goals 6–8 assessments in 2010 and 2016.

Table 1 Classification results based on the value of the created measures

Class number	*TMGO*	*TMMI*
1	Belgium, Denmark, Finland, Ireland, Luxembourg, Sweden	Denmark, Finland, Sweden, Austria, France
2	Austria, Cyprus, Czech Republic, Estonia, France, Germany, Greece, Hungary, Italy, Malta, Netherlands, Poland, Slovakia, Slovenia, Spain, UK	Belgium, Ireland, Luxembourg, Czech Republic, Estonia, Germany, Greece, Hungary, Italy, Netherlands, Poland, Slovakia, Slovenia, Spain, UK, Croatia, Latvia, Lithuania, Portugal
3	Bulgaria, Croatia, Latvia, Lithuania, Portugal, Romania	Cyprus, Malta, Bulgaria, Romania

Source Own elaboration based on Eurostat data

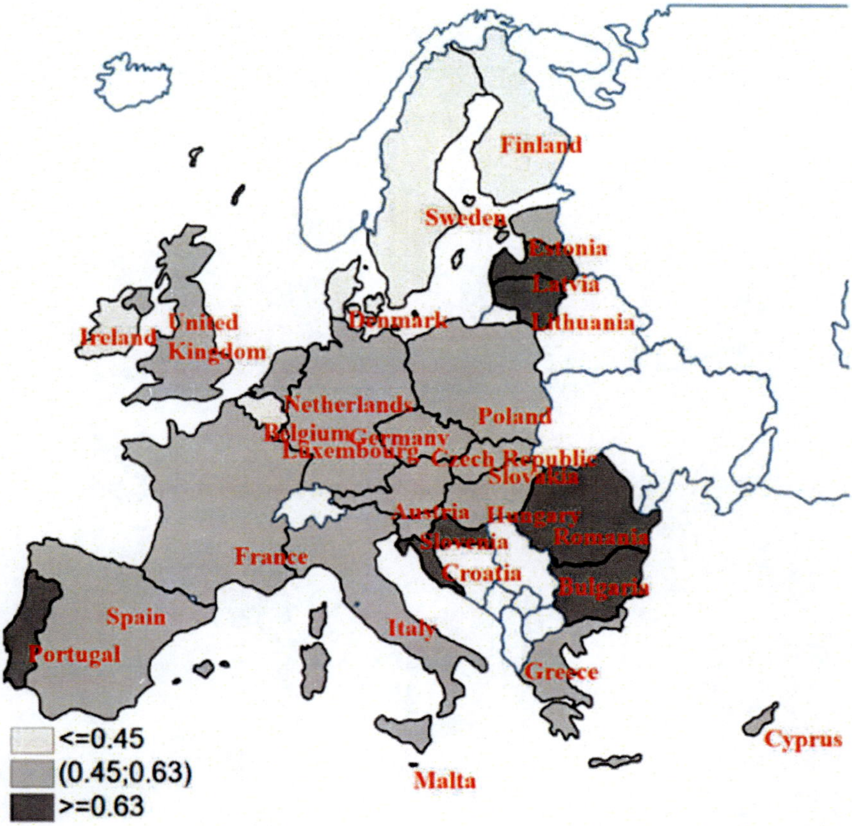

Fig. 1 Value of TMGO in the EU countries. *Source* Own elaboration based on Eurostat data

Observing the choices made by European seniors, we found that the assessments of the four analyzed goals of the Europe 2020 Strategy in 2016, indicated as *about right* (AR), have not changed (Figs. 3, 4, 5 and 6). However, there was a significant change in the shares of the two remaining categories: *too modest* (TM) and *too ambitious* (TA). In 2016, the share of people assessing the social goals of the Strategy as too ambitious increased significantly compared to 2010. Such a change of assessment might have been caused by the fact that people who previously belonged to the younger generation and were professionally and socially active, and paid greater attention to EU activities aimed at equalizing the opportunities for the EU community, entered the 60+ group in 2016. On the other hand, as people grow older, they naturally tend to evaluate their life situation by comparing it to their past, which makes their assessments of the requirements for socio-economic problems more lenient after six years. Therefore, they judge the selected goals of the Strategy as too ambitious. Unfortunately, the Eurobarometer study does not show the real reason for changing the decision.

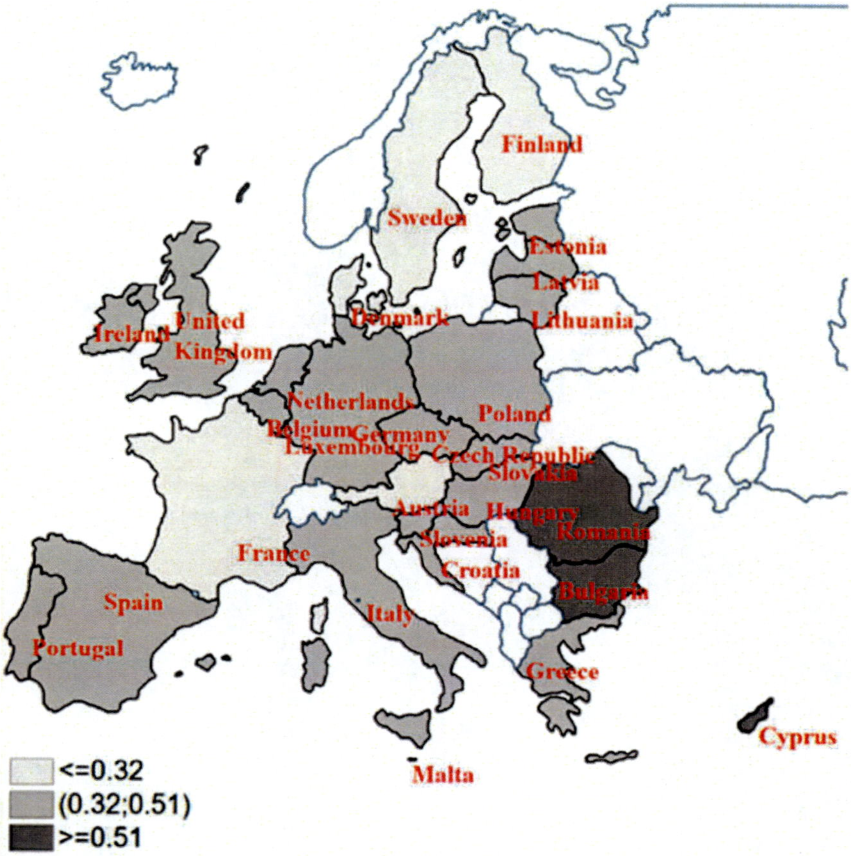

Fig. 2 Value of TMMI in the EU countries. *Source* Own elaboration based on Eurostat data

The full hierarchical chart (for all variables) was presented only for the first goal (Fig. 3), as the distribution of the share of the categories of the demographic variables did not differ in the remaining three objectives. In the chart (Fig. 3), we observed that the shares of the categories of successively added variables, i.e., age, gender, graduation age, life satisfaction, are very similar in each of the three assessments of Goal 1: *too ambitious*, *about right* and *too modest*. Thus, the greatest number of responses for particular categories of the goal assessment was provided by people aged 66–74, more often by women than men, who most often completed their education at the age of 19 and were satisfied with their lives. Since this type of analysis (despite the fact that it is layered, i.e., dividing the population into subgroups) did not give a clear answer about the differentiation of groups of people choosing one of the three categories in assessment of Goal 1, the correspondence analysis was used. In order to compare the changes taking place in the

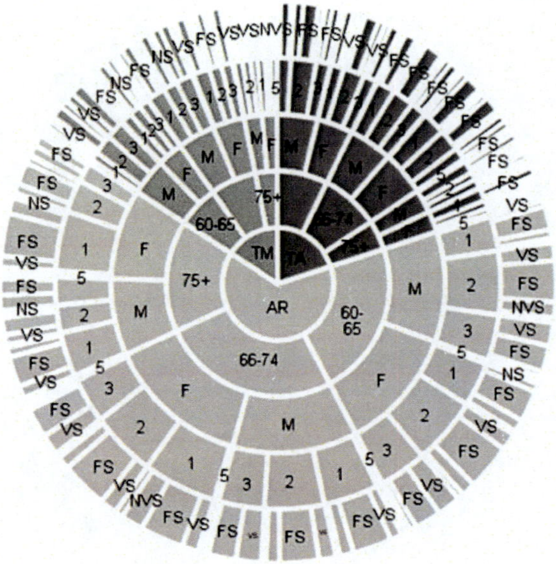

Fig. 3 Goal 1–75% of the population aged 20–64 years to be employed, 2010. *Source* Own elaboration

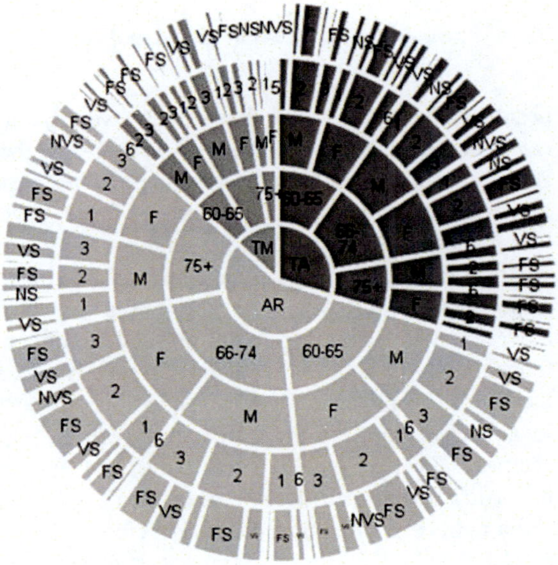

Fig. 4 Goal 1–75% of the population aged 20–64 years to be employed, 2016. *Source* Own elaboration

characteristics of people evaluating Goal 1, we presented the results of the correspondence analysis in separate charts for 2010 and 2016 (Figs. 7 and 8).

Although a very large reduction was made in the scope of the presentation of the results of the analysis (from R^8 to R^2), the quality of the presentation is 81.12% in 2010 and 87.82% in 2016 (Figs. 7 and 8). It means that with two-dimensional

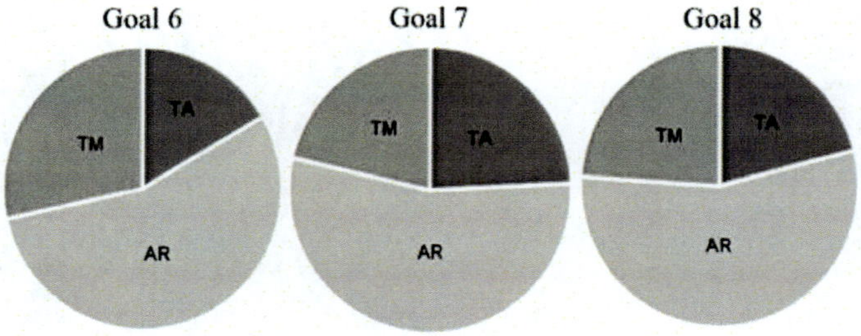

Fig. 5 Assessment of Goals 6–8 by people aged 60+, 2010. *Source* Own elaboration

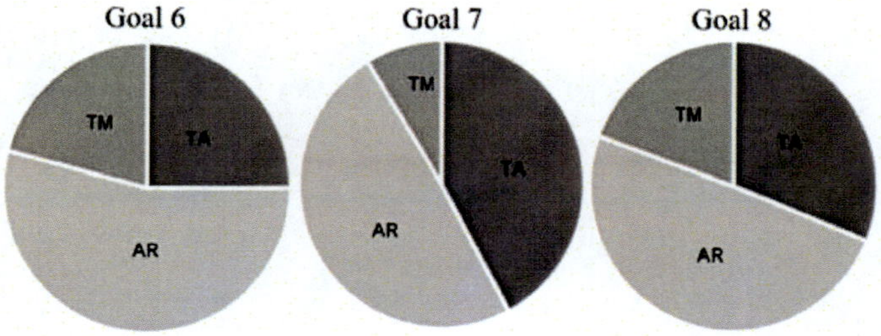

Fig. 6 Assessment of Goals 6–8 by people aged 60+, 2016. *Source* Own elaboration

presentation 81% and 88%, respectively, we explain real relationships between the categories of analyzed variables. While analyzing the scattering of points in Figs. 7 and 8, we noticed that the presentation of the responses along axis 1 is according to age, while axis 2 presents the responses for the level of assessment of the Goal 1 (too modest in the upper part, too ambitious in the lower part). The location of the points in Fig. 7 allowed us to indicate the following three main characteristics of the respondents. People who assess the level of achievement of Goal 1 in 2010 as too modest and are aged 66+ are not satisfied with their lives. This goal is assessed as too ambitious or about right by women aged 66–74, from rural areas, satisfied with their lives. People aged 60–65 consider this goal as too modest, and they have the longest education. For the 2016 ratings, we distinguished other respondent characteristics. People aged 60–65 assessing Goal 1 as about right and too ambitious are not satisfied with their lives Women aged 66–74 who indicate that the first goal is about right or too ambitious live in big cities and are satisfied with their lives. More precise than in 2010 is the description of people aged 75+ who believe that Goal 1 is too modest. They are mostly men living in medium and small towns, who are very satisfied with their lives.

We analyzed the Goals 6, 7, 8 of the Europe 2020 Strategy in a similar way. The obtained characteristics are presented in Table 2.

Table 2 summarizes the characteristics of the respondents obtained after applying the correspondence analysis in 2010 and 2016, taking into account the

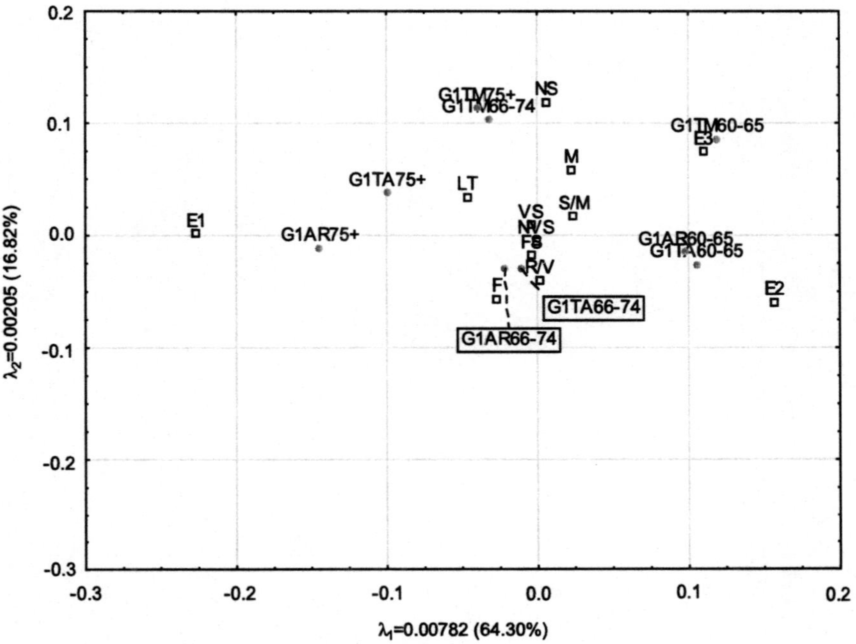

Fig. 7 Characteristics of seniors assessing Goal 1 of the Europe 2020 Strategy in 2010. *Source* Own elaboration

assessment of respective goals made by people in three age groups. We noticed several important characteristics of respondents influencing their assessment of individual goals of the Europe 2020 Strategy. The respondents who have higher requirements regarding the level of goals than those assumed in the Strategy are people aged 60–65 who studied the longest (for all four goals), for Goal 1 they are 66–74-year-olds who are not satisfied with their lives, but for 6, 7, and 8 Goals, they are middle-aged seniors very satisfied with their lives. In 2016, people aged 60–65 who left their education at the latest, assessed Goals 1, 6, 7 as too modest. The Goal 8 in 2016 was assessed as too modest by inhabitants of rural areas. In 2016, middle age seniors who studied the longest, assessed Goal 6 as too modest, and men in this age group assessed Goals 7 and 8 as such. By analyzing the entries in Table 2, we found that respondents aged 60–65 who left their education at the age of 19 consider Goals 1, 6, 8 in both analyzed years as well-defined, and additionally in 2016 this opinion was shared by the youngest seniors who were not satisfied with their lives (NVS, NS). Female seniors who are fairly satisfied with their lives found all four goals about right in both periods. Among the oldest seniors, we indicated only the group of people who participated in education for the shortest time, who assessed as correct Goals 6, 7, and 8 in 2010.

In 2010, for some people aged 60–65 who left school at the age of 19, all goals were too ambitious. A similar opinion about Goals 1, 6, 7 is shared by some women

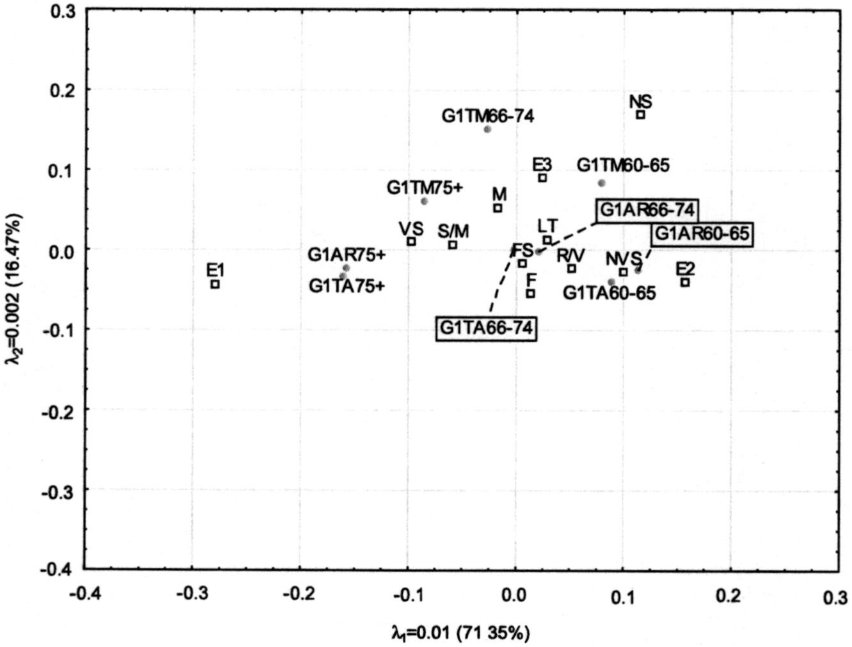

Fig. 8 Characteristics of seniors assessing Goal 1 of the Europe 2020 Strategy in 2016. *Source* Own elaboration

aged 66–74 fairly satisfied with their lives. These two characteristics also appear about right in the goals evaluation. This indicates a large disproportion in the assessments made by the analyzed social groups. In 2010, the oldest women fairly satisfied with their lives and living in small towns or rural areas found Goals 6, 7, 8 as about right, whereas in 2016, such an assessment was made for Goals 7 and 8 by people who were very satisfied with their lives.

Another conclusion is that for the selected socio-economic goals of the Europe 2020 Strategy, it is possible to identify characteristics that appear invariably in both years in each age group and in categories of the Strategy's goal assessment. Most often they relate to the end of education, gender, and life satisfaction. Place of residence is the least constant factor characterizing the respondents. The oldest respondents made the assessments the least homogeneously, which resulted in the lack of indication of detailed attributes. The respondents from the two younger groups of seniors, when assessing the goals as too ambitious or appropriate, were distinguished by a greater number of additional features than those who assessed the goals as too modest. At the same time, the indicated changes that took place between 2010 and 2016 in the characteristics of the groups of respondents divided by age and the selected categories of goals' assessment indicate that the respondents were deeply considering the implementation of particular goals in 2016. Perhaps

Table 2 Profiles of the respondents assessing Goals 1, 6, 7, 8 of the Europe 2020 Strategy in 2010 and 2016

		Evaluation by age groups								
Goals	Years	TA60-65	AR60-65	TM60-65	TA66-74	AR66-74	TM66-74	TA75+	AR75+	TM75+
1	2010	E2	E2	E3	R/V, FS, F	R/V, FS, F	NS			NS
	2016	R/V, NVS	NVS, E2	E3	FS, F, LT	FS, F, LT				
6	2010	E2	E2	E3	F	NS, F, FS, R/V	VS	NS, F, FS, R/V	E1	
	2016	R/V, NVS	NVS, E2	E3, NS	F, FS	F, FS	E3		VS	
7	2010	E2	M, S/M, FS	E3	LT, F, R/V, FS	LT, F, R/V, FS	VS	LT, F, R/V, FS	E1	
	2016	NS, E2	NS	F, E3	FS	F, E3	M	VS		
8	2010	E2	S/M, E2	E3	M, R/V, VS	LT, F, FS	M, R/V, VS	LT, F, FS	E1	
	2016	R/V	NVS, E2	R/V	E3, M	LT, F, FS, E3	M	VS		VS

Source Own elaboration

these goals were more widely commented and promoted publicly, which allowed to reflect on the assessment of the consequences of achieving these goals.

As we have already indicated, in 2016, the scope of the analyzed variables was changed under the Standard Eurobarometer. Measurement of views on energy and migration policy started. The survey raised questions related to the assessment of the possibility of working and living in the EU as well as people from the EU arriving in the respondent's country to work and live. The opinion of seniors on migration within the EU is related to the previously discussed social goals of the Europe 2020 Strategy, especially to Goals 1 and 8, as both research areas concern European policy related to the labor market. For this reason, we conducted an analysis that allowed us to compare the evaluations made by seniors for Goals 1 and 8 with their assessments for intra-EU work migration. The study was conducted with the use of correspondence analysis, presenting approximately 90% of real relationships between the categories of the analyzed variables. In both analyses, in addition to the variables relating to the Europe 2020 goals and work-related migration opportunities, we also added life satisfaction assessment. This factor strongly determines whether an individual decides to change the place of employment and how he/she assesses the risk of poverty and social exclusion.

Analyzing the placement of points in Fig. 9, we found that people assessing Goals 1 and 8 as about right (G1AR, G8AR) are 66–74 years old and believe that the freedom to go to work in another EU country is a good solution (WEUGT66–74), these people are satisfied with their lives. Whereas, people assessing Goals 1 and 8 as too modest (G1TM, G8TM) are 60–65 and 75+ years old and cannot assess whether the freedom to go to work in another EU country is a good or bad solution (WEUN60–65, WEUN60–65). These people are not satisfied with their lives. We also noticed that the opinion that the freedom to work in any EU country is a bad solution is more typical of people who consider Goals 1 and 8 too ambitious than of the other respondents.

In Fig. 10, we present the results of the analysis of the assessments of Goals 1 (employment at the level of 75% for people aged 20–64) and 8 (reducing the number of people at risk of poverty and social exclusion by 25%), together with an assessment of the respondent's possibility of people from other EU countries coming to work. We have noticed that people assessing Goals 1 and 8 as about right (G1AR, G8AR) are 66–74 years old and believe that the freedom to go to work in another EU country is a good solution (WCOGT66–74), they are satisfied with their lives. On the other hand, people aged 66+ who indicated that the arrival of people from the EU to work in their country was a bad solution assessed Goal 1, i.e., increasing employment of people aged 20–64, as too ambitious.

The analyses carried out, the results of which are presented in Figs. 9 and 10, show that migration in order to find employment within the EU is better perceived by people who consider Goals 1 and 8 as well-defined. This attitude is typical of people who are satisfied with their lives. While, people who cannot determine whether a job migration is a good or a bad solution are those who are dissatisfied during their lifetime.

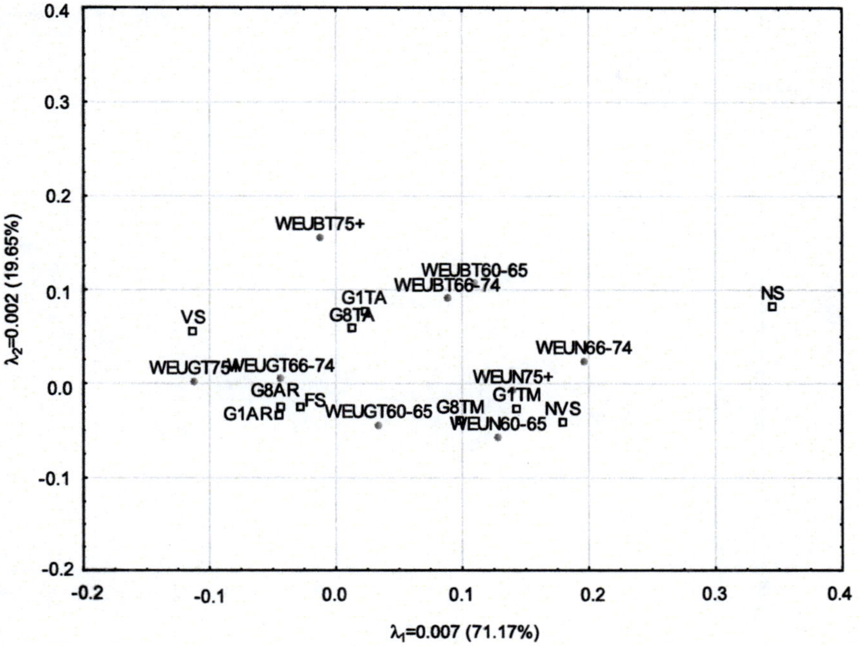

Fig. 9 Linking the assessments of Goals 1 and 8 of the Europe 2020 Strategy with assessments of the possibility of going to work in another EU country, 2016. *Source* Own elaboration

Further analysis concerned only variables related to economic migrations (arrivals to the respondent's country and trips to other countries), freedom of movement (also arrivals to the respondent's country and trips to other countries), knowledge of EU laws, the feeling of being a European and life satisfaction of three groups of seniors. Here, we used a cluster analysis based on the frequency of choices of categories of particular variables. As a result, we have designated, in all age groups and in both years, 11 classes of countries where seniors made similar choices. The results of the cluster analysis are presented in Table 3.

For the classification of the countries presented in Table 3, we determined the Silhouette indices, which were in the range [0.5; 0.6]. So we considered the classifications significant.

We ordered the presented classifications (Table 3) from the best ratings given by respondents in the EU countries to all variables. It is easy to notice that regardless of age of the respondents and the period of the study, seniors from Luxembourg best assess the possibility of free movement within the EU, both for work and tourism purposes, their knowledge of EU rights, they have the highest sense of being a European and assess their lives best. Ireland, Finland and Sweden also fall into the first groups. Furthermore, the latter two appear in the same group in almost every classification, except for the classification of the respondents aged 60–65 in 2019. The respondents from Belgium and Austria were also classified multiple times into

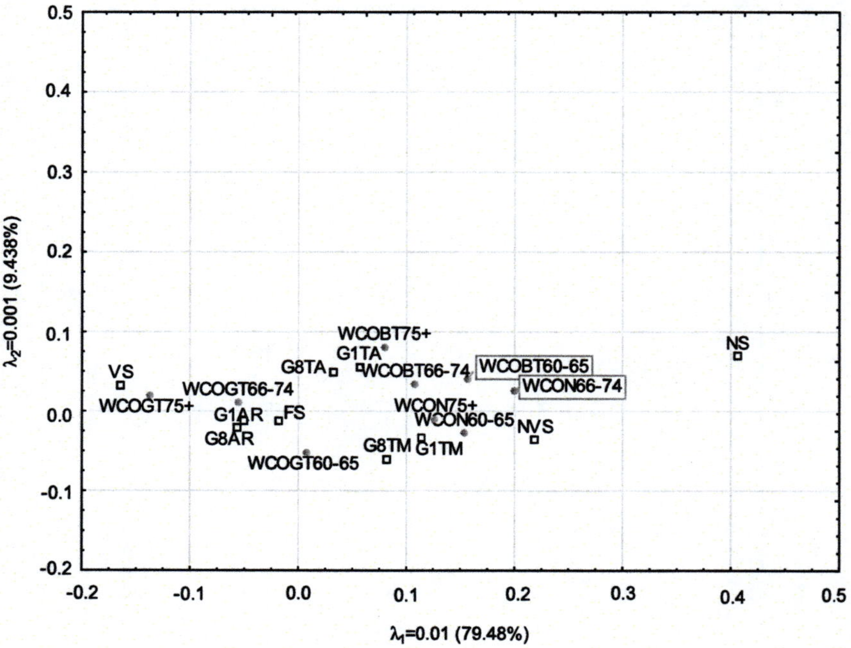

Fig. 10 Linking the assessments of Goals 1 and 8 of the Europe 2020 Strategy with assessments of the arrival possibility for people from EU to the respondent's country to work, 2016. *Source* Own elaboration

the same categories (except for 2019 and the classification for middle-aged and oldest seniors). In the presented classification (Table 3), attention should be paid to people aged 60–65 who assessed the variables in 2019. Contrary to the remaining analyzed groups of respondents, there was a division here, in which only one single-element class was distinguished.

Additionally, in order to check whether the classifications for particular groups of respondents in both analyzed periods are similar, we used the Spearman coefficient (Table 4). Significant values of the coefficients for a *p*-value of 0.5 are pointed out in the table. Based on the correlation coefficients, we found that the similarities between the classifications are not high. The highest rate was recorded for two country classifications in 2016 and respondents aged 60–65 and 66–74. When assessing these two classifications, we noticed that some countries are in the same classes, e.g., Denmark, Finland and Sweden; Germany and Slovenia; Spain and Lithuania; Cyprus; France and Croatia; Belgium and Austria.

Based on the values of Spearman's coefficients, we noticed that there were similarities in the classification of countries for all three age groups of seniors in 2016. On the other hand, the correlations of country classifications made for the three age groups in 2019 have significantly lower values, so they are not similar to each other.

Table 3 Perception of freedom of movement, the feeling of being a European, knowledge of EU citizenship rights, satisfaction with life, 2016, 2019

Group	2016	2019	2016	2019	2016	2019
	60–65	60–65	66–74	66–74	75+	75+
1	LU, IE, ES, HU, LT	LU, IE, FI	LU	LU	NL, IE	DE, LU, IE, FI, SE
2	NL, DK, FI, SE, PL	SE, LT, PL	DE, DK, FI, SE, EE, MT, SI	IE, ES, FI, SE	DE, ES, FI, SE, EE, MT	FR, BE, NL, DK, ES
3	DE, PT, SK, SI	ES, HU	IE, PL	BE, HU, SI	BE, LU, DK, AT	PT, EE, HU, LT, PL
4	EE, LV	DE, PT, EE, LV	ES, LT, SK	NL, DE, DK, CY, EE, LV, LT, PL	LT	CZ
5	CY	BE, AT	BE, AT, NL	FR, PT	PT, CZ, PL	UK, GR, CY
6	FR, IT, RO, HR	NL, DK, CY, SI	CZ, LV	AT	FR, HU, SK, RO, HR	LV
7	BE, AT, MT	FR, BG	FR, PT, HU, HR	MT, SK	LV, SI	MT, SK, SI
8	CZ	MT, SK	CY	UK, CZ, HR	CY	AT, RO
9	BG	GR, CZ, HR	GR, BG, RO	RO	GR, BG	HR
10	UK	UK, RO	IT	GR, BG	UK	BG
11	GR	IT	UK	IT	IT	IT

Austria—AT; Belgium—BE; Bulgaria—BG; Croatia—HR; Cyprus—CY; Czech Republic—CZ; Denmark—DK; Estonia—EE; Finland—FI; France—FR; Germany—DE; Greece—GR; Hungary—HU; Ireland—IE, Italy—IT; Latvia—LV; Lithuania—LT; Luxembourg—LU; Malta—MT; The Netherlands—NL; Poland—PL; Portugal—PT; Romania—RO; Slovakia—SK; Slovenia—SI; Spain—ES; Sweden—SE; United Kingdom—UK
Source Own elaboration

Table 4 Spearman coefficient

	2016 60–65	2019 60–65	2016 66–74	2019 66–74	2016 75+	2019 75+
2016 60–65	1	−0.066	0.530*	−0.056	0.496*	0.099
2019 60–65	−0.066	1	0.300	0.379*	0.122	0.270
2016 66–74	0.530*	0.300	1	0.448*	0.536*	0.164
2019 66–74	−0.056	0.379*	0.448*	1	−0.125	0.371
2016 75+	0.496*	0.122	0.536*	−0.125	1	0.233
2019 75+	0.099	0.270	0.164	0.371	0.233	1

*Significance level = 0.05
Source Own elaboration

6 Conclusions and Discussion

Not all goals of the Europe 2020 Strategy have been achieved, thus far. There is a difference not only in particular EU member countries but also in terms of different areas. It is difficult to clearly define the level of goals achievement in relation to seniors. However, the results of the conducted analysis show that it can be considered a factor in the well-being of seniors in particular EU countries.

In the analysis of objective indicators determining the quality of senior life, we noticed that the best situation is in the Scandinavian countries (Denmark, Finland and Sweden) and the worst in Romania and Bulgaria. We included countries from the first-mentioned group in the *Taxonomic Measure of Good Oldness* (TMGO) and in the *Taxonomic Measure of Main Indicators* in relation to Europa 2020 indicators (TMMI) in the first class. The values of the TMGO and TMMI measures determined for Bulgaria and Romania clearly indicate that the socio-economic conditions of the life of seniors are significantly more difficult than in the Scandinavian countries.

By analyzing the goals of the Europe 2020 Strategy separately, it was possible to indicate that at the European level there are differences in the perception of these guidelines by seniors from different age groups (60+). Unfortunately, an analysis by country was not possible due to the low numbers of respondents in some of them. While analyzing the data from 2010 to 2016, we noticed that there is no tendency for Europeans to judge the objectives of the Strategy as very good or at least appropriate. There are groups of seniors who find these goals too ambitious. Among the youngest people (60–65), additional features dominate, such as graduation at the age of 19, low life satisfaction and a rural place as a place of residence.

The evaluation of the Strategy's goals immediately after its introduction, i.e., in 2010, is assessed differently by people in the earlier distinguished age groups than in 2016, after a six year strive to achieve these goals. This is indicated by changes in the demographic characteristics of the respondents.

In the course of the analysis carried out in 2016, and the change in the scope of the Europe 2020 survey in the Eurobarometer, we have noticed that the EU's efforts to achieve the Strategy goals are assessed according to the perception of the goals themselves. Thus, when they are considered too ambitious, a negative opinion on the accompanying policy follows.

Another important conclusion that we have drawn from the analysis of the senior subjective assessments of the Strategy is their negative opinion on the rights in the EU and the benefits of implementing the Strategy. Citizens of the EU countries aged 60+ may differ greatly in their assessment of the rights as this depends on their membership and also embraces their life satisfaction. In our cluster analysis, it was not possible to distinguish a small number of groups of homogeneous countries, and an increase in the number of groups led to the formation of single-element groups.

We believe that instead of assessing the main objectives of the Strategy, the priorities of the European Energy Union and issues related to migration policy, the introduction into the Eurobarometer should be preceded by examining the existing knowledge of these issues and should be supported by a broad public information campaign about their importance. Otherwise, a subjective opinion of respondents may be difficult to express and may be built only on the basis of media reports.

Seniors are becoming a large social group. Their dismissal will not generate effective tools for social policy. Their respectful treatment is necessary. It is worth remembering that seniors (due to their number) are becoming entities not only in the sphere of social policy, but also sales and services markets. Therefore, their opinions, knowledge about their behaviors and lifestyles are an invaluable source of information indispensable to shape any political and market tools. Therefore, increasing heterogeneity of this group is important. Younger seniors are often participants of the labor market. They willingly use tourist services and all forms of activity. They also use the computer, the Internet, and social media. They perceive reality in different ways. Their needs are therefore different from those of older seniors who struggle with health problems, infirmity, or loneliness. This diversity illustrates a great need for knowledge about each of the subgroups. It is necessary to expand the existing knowledge with the help of various types of research, with particular emphasis on collecting data for different age groups of seniors.

References

Balcerzak AP (2015) Europe 2020 strategy and structural diversity between old and new member states. application of zero unitarization method for dynamic analysis in the years 2004–2013. Econ Sociol 8(2):190–210. https://doi.org/10.14254/2071-789x.2015/8-2/14

Begg I (2010) Europe 2020 and employment. Intereconomics 45:146–151. https://doi.org/10.1007/s10272-010-0332-9

Daly M (2012) Paradigms in EU social policy a critical account of Europe 2020. Transf Eur Rev Labour Res 8(3):273–284

Delmas A (2015) Perspectives for revision of the Europe 2020 strategy. Paris, Economic, Social and Environmental Council. Retrieved from https://www.eesc.europa.eu/sites/default/files/resources/docs/strategieeurope2020.pdf

Di Palo C (2019) Impact of population ageing on the Italian pension expenditure. J Appl Econ Sci XIV 4(66):1203–1215. https://doi.org/10.14505/jaes.v14.4(66).26

European Commission (2010) EUROPE 2020 a strategy for smart, sustainable and inclusive growth. Brussels, European Commission 3.3.2010 COM (2010). Available via https://eur-lex.europa.eu/LexUriServ/LexUriServ.do?uri=COM:2010:2020:FIN:EN:PDF. Accessed 27 Oct 2020

European Commission (2019a) Assessment of the Europe 2020 strategy joint report of the Employment Committee (EMCO) and Social Protection Committee (SPC). Luxembourg, Publications office of the European Union. Available via https://ec.europa.eu/social/main.jsp?langId=en&catId=1063&furtherNews=yes&newsId=9487. Accessed 27 Oct 2020

European Commission (2019b) Directorate General Communication, COMM.A.3, Brussels. Media monitoring and eurobarometer. European Parliament, Directorate-General for Communication, Public Opinion Monitoring Unit (2019b) Eurobarometer 91.5 (2019). GESIS Datenarchiv, Köln. ZA7576 Datenfile Version 1.0.0. https://doi.org/10.4232/1.13393

European Commission (2020) Directorate General Communication COMM.A.1, Brussels. Strategy, corporate communication actions and eurobarometer. Eurobarometer 85.2 (2016). GESIS Datenarchiv, Köln. ZA6694 Datenfile Version 2.0.0. https://doi.org/10.4232/1.13438

Fernández-Carro C, Módenes JA, Spijker J (2015) Living conditions as predictor of elderly residential satisfaction. A cross-European view by poverty status. Eur J Ageing 12:187–202. https://doi.org/10.1007/s10433-015-0338-z

Fico G, Gaeta E, Arredondo MT, Pecchia L (2015) Analytic hierarchy process to define the most important factors and related technologies for empowering elderly people in taking an active role in their health. J Med Syst 39:98. https://doi.org/10.1007/s10916-015-0300-9

Greenacre M (1984) Theory and applications of correspondence analysis. Academic Press, London

Greenacre M, Blasius J (eds) (2006) Multiple correspondence analysis and related methods. Chapman & Hall/CRC, Taylor & Francis Group, Boca Raton

Grosse TG (2010) Doświadczenia Strategii Lizbońskiej—perspektywy Strategii, Europa 2020": o kontynuacji i zmianach w polityce UE. Public Gov 11:5–24

Hellwig Z (1968) Application of the taxonomic method to the typological division of countries due to the level of their development and the structure of qualified personnel. Stat Rev 4:307–327

Hobza A, Mourre G (2010) Quantifying the potential macroeconomic effects of the Europe 2020 strategy: stylised scenarios. European Economy—Economic Papers 2008–2015 424. Directorate General Economic and Financial Affairs (DG ECFIN), European Commission

Hwang CL, Yoon K (1981) Multiple attribute decision-making: methods and applications. Springer, Berlin

Karakaya G (2009) Long-term care: regional disparities in Belgium. J Appl Econ Sci IV 1(7):58–79

Kedaitiene A, Kedaitis V (2012) Macroeconomic effects of the Europe 2020 strategy. Soc Res 4(29):5–19

Loureiro A, Barbas M (2014) Active ageing—enhancing digital literacies in elderly citizens. In: Zaphiris P, Ioannou A (eds) Learning and collaboration technologies. Technology-rich environments for learning and collaboration. LCT 2014. Lecture notes in computer science, 8524. Springer, Cham. https://doi.org/10.1007/978-3-319-07485-6_44

Manafi I (2012) The impact of the Europe 2020 strategy on Romanian labour market. Int J Econ Practices Theor 2(4):247–252

Megyeri E (2018) Old-age poverty and residential property in the EU: an analysis with the EU-SILC 2014 data. In: Eckardt M, Dötsch J, Okruch S (eds) Old-age provision and homeownership—fiscal incentives and other public policy options. Springer, Cham. https://doi.org/10.1007/978-3-319-75211-2_2

Młynarzewska-Borowiec I (2020) Does implementation of the smart growth priority affect per capita income of EU countries? Empirical analysis for the period 2000–2017. J Knowl Econ. https://doi.org/10.1007/s13132-020-00670-0

Papacostas A (2012) Eurobarometer 73.4 (May 2010). GESIS Datenarchiv, Köln. ZA5234 Datenfile Version 2.0.1. https://doi.org/10.4232/1.11479

Pasimeni P (2012) Measuring Europe 2020: a new tool to assess the strategy. Int J Innov Reg Dev 4(5):365–385

Pluta W (1986) Wielowymiarowa analiza porównawcza w modelowaniu ekonometrycznym. PWE, Warszawa

Radulescu M, Fedajev A, Sinisi CI, Popescu C, Iacob SE et al (2018) Europe 2020 implementation as driver of economic performance and competitiveness. Panel analysis of CEE countries. Sustainability 10:566. https://doi.org/10.3390/su10020566

Rappai G (2016) Europe in route to 2020: a new way of evaluating the overall fulfilment of the Europe 2020 strategic goals. Soc Indic Res 129:77–93

Stanickova M (2017) Efficient implementation of the Europe 2020 strategy goals: is social equality achievable reality or myth perhaps? Inst Econ Res. http://hdl.handle.net/10419/219942

Stec M, Grzebyk M (2018) The implementation of the strategy Europe 2020 objectives in European Union countries: the concept analysis and statistical evaluation. Qual Quant 52:119–133. https://doi.org/10.1007/s11135-016-0454-7

Strahl D (ed) (2006) Metody oceny rozwoju regionalnego. Wydawnictwo Akademii Ekonomicznej im. Oskara Langego we Wrocławiu, Wrocław

Sulmicka M (2011) Strategia Europa 2020 – postlizbońska polityka rozwoju Unii Europejskiej, Prace i Materiały Instytutu Rozwoju, Polityka gospodarcza w świetle kryzysowych doświadczeń 85:169–190

Talmaciu AM, Cismas LM (2016) National competitiveness through the Europe 2020 strategy and human development index in CEE countries. A Panel Data Anal. http://ecoforumjournal.ro/index.php/eco/article/view/1104. Accessed 27 Oct 2020

Vanhercke B, Frazer H, Marlier E, Natali D, Van Dam R (2010) Europe 2020: towards a more social EU? Peter Lang, Brussels

Walheer B (2018) Decomposing the Europe 2020 index. Soc Indic Res 140:875–905. https://doi.org/10.1007/s11205-017-1797-8

Zalewska ME, Świetlikowski P (2017) Ocena szans realizacji głównych celów strategii Europa 2020 w krajach Grupy Wyszehradzkiej. Zeszyty Naukowe Politechniki Śląskiej, seria Organizacja i Zarządzanie 104(1979):367–378

Do Seniors Get to the Disco by Bike or in a Taxi?—Classification of Seniors According to Their Preferred Means of Transport

Joanna Kos-Łabędowicz and Joanna Trzęsiok

Abstract Analysis of the preferences of the oldest group of citizens seems especially important and up to date in the context of an ageing Polish society. It is important to determine the mobility of this group by examining the most frequently used means of transport for different travel needs. Studies that define the mobility types of older people can be found, but there are not many and they would require modifications to reflect Polish society specifically. In this paper, an attempt to classify elderly people in Poland in terms of their transport preferences has been made, based on literature research and expert knowledge. At the same time, using primary data from a survey of seniors, a cluster analysis was performed using the Ward's method based on a distance matrix, calculated using the Sokal–Michener metric. The aim of the paper is to test the validity of the obtained classifications. As shown by the results obtained, e.g. the Rand index, there is high similarity, despite the different number of groups in each segmentation; the validity of the proposed expert segmentation is therefore confirmed.

Keywords Seniors · Transport preferences of the elderly · Mobility types of seniors · Cluster analysis · Classification agreement

1 Introduction

Two processes: the ageing society and increasing urbanization mean that more and more seniors live in cities and use the cities' transport systems to meet their needs. Among the means of transport available to seniors, the car was already identified as the most suitable one in the OECD report (2001) on Ageing and Transport. A lot of the research and activities carried out so far have been focused on ensuring the

J. Kos-Łabędowicz · J. Trzęsiok (✉)
University of Economics in Katowice, Katowice, Poland
e-mail: joanna.trzesiok@ue.katowice.pl

J. Kos-Łabędowicz
e-mail: joanna.kos@ue.katowice.pl

K. Jajuga et al. (eds.), *Data Analysis and Classification*, Studies in Classification, Data Analysis, and Knowledge Organization,
https://doi.org/10.1007/978-3-030-75190-6_16

longest possible use of cars by seniors (Coughlin and D'Ambrosio 2012; Haustein et al. 2013). The possible measures to extend the elderly car-use include: driving assistance systems, adapting the road infrastructure to the needs of seniors, changes in regulations regarding the verification of driving licences, special training courses and others. Such a strong focus on the car as the preferred means of transport does not change the fact that not all seniors have access to one and some have to use other available means of transport, usually public or active transport (i.e. walking or cycling). The following issues, differing from one mode of transport to another, can be perceived as barriers to the use of alternatives to cars: service provision, health, safety and personal security, comfort, information and awareness, attitude, affordability, built environment (Luiu et al. 2018).

The issue of transport needs and ensuring mobility of the elderly is not only an interesting and current research problem but also a significant issue from the point of view of various decision-makers responsible for undertaking activities aimed at ensuring the social inclusion of seniors. Most studies dealing with the transport needs of seniors indicate differences depending not only on age or gender, but also on the place of residence, the availability of transport infrastructure, accessibility of various transport means and other factors.

This paper presents two classifications of the elderly in terms of their preferences regarding means of transport:

- one prepared on the basis of literature research and expert knowledge,
- the other with the use of a selected taxonomic method.

The aim of the article is to test the agreement between the obtained classifications and thus to verify the validity of the proposed expert segmentation.

A brief review of the literature on the segmentation of seniors according to their transport behaviour will be presented first. Next, the primary research providing the data for the grouping, both the methodology and the results, will be discussed followed by the methodology of preparing both segmentations. The obtained results, their comparison, and summary, together with the conclusions drawn, will be presented last.

2 Literature Review: Seniors and Their Transport Needs. Attempts at Segmentation

Attempts at segmentation relating to transport behaviour can be performed using different types of variables. The following variables may be used in studies related to transport behaviour: variables describing transport behaviour (such as the means of transport used, frequency of their use, preferences and attitudes towards particular transport options), socio-demographic characteristics of individuals and households (such as age, sex, place of residence, education, household size and others), spatial factors (e.g. spatial availability of individual transport options, quality of transport infrastructure, distribution of the most popular destinations and others), individuals' attitudes (e.g. pro-ecological proclivities, perceived status, independence, etc.) or significant life events (Haustein and Hunecke 2013). Segmentation can be carried out in two ways, preparing a description of segments and a set of rules, based on which individual units will be assigned to a given segment, before conducting the study ("a priori") and "a posteriori", when segmentation is carried on the basis of the obtained data, most often using clustering methods (Sagan 2009).

Studies on the segmentation of seniors in terms of their transport behaviour related to various types of travel can be found in the literature. The Boksberger and Laesser (2008) study on Swiss seniors, which considered Switzerland specifically as a mature market, identified, using cluster analysis, three segments (Grizzled Explorers, Time-honoured Bon Vivants, Retro Travellers) based on the motivation of seniors to undertake travel (and selection of travel means) for touristic purposes. On the one hand, this study indicated some shortcomings of previous segmentations in this area, but also confirmed the existence of previously identified motivations, with the tendency to change depending on the stage of the senior's life. An example of a different approach to such segmentation is the study by Lee and Bowes (2016) who divided seniors according to age into four segments (pre-seniors: younger than 65; the young-old: 65–74; the old-old: 75–84; and the oldest-old: 85+) and investigated the influence of age and transport needs on the perceived barriers in making decisions about tourist travels. Both the above-mentioned studies were aimed at obtaining information that could be helpful in adjusting offers by travel agencies to better suit the preferences of the elderly, who are an important and increasingly more active group of consumers.

Transport behaviour of seniors related to their daily travels, e.g. shopping, commuting, etc., is the main interest for the segmentations prepared and presented in this paper. This type of segmentation can be useful for spatial planning or for the preparation of policies and activities aimed at meeting various transport needs (Elmore-Yalch 1998) (including those of the seniors) (Załoga and Kłos-Adamkiewicz 2019). In the literature on the subject, several examples of such segmentations concerning seniors and their transport behaviour may be found. An overview of the previous segmentations made using both of the above-mentioned approaches ("a priori" and "a posteriori") and based on a number of variables relating to various aspects potentially influencing their behaviour has been presented in Table 1.

Previous attempts at the segmentation of seniors, taking into account their transport behaviour, available in the literature, show a certain similarity in terms of the identified segments. A synthetic review and comparison of the existing segmentations relating to the transport behaviour of seniors carried out by Haustein and Siren (2015) allowed for the grouping of individual segments from the existing segmentations, based on their common aspects (car orientation, activity level, socio-economic economic resources, health and gender), under so-called metasegments (see Table 2).

Table 1 Segmentation studies concerning seniors and their transport choices

Study (project acronym)	Segmentation method	Variables used for segmentation	Segments
Hildebrand (2003)	Cluster analysis	Socio-demographic and household variables	1. Disabled drivers 2. Affluent males 3. Mobile widows 4. Mobility impaired 5. Workers 6. Granny flats
Mollenkopf et al. (2004)—MOBILATE	Cluster analysis	Mobility and satisfaction from mobility	1. Subgroup 1 2. Subgroup 2 3. Subgroup 3 4. Subgroup 4
Haustein et al. (2008)—MOBILANZ	Cluster analysis	Socio-demographic, infrastructure, mobility-related attitudes	1. Restricted mobiles 2. Mobile car-oriented 3. Self-determined mobiles 4. Pragmatic PT-oriented 5. Bike-oriented 6. Eco-friendly PT–oriented
Aigner-Breuss et al. (2010)—MOTION 55+	A priori segmentation	Focus on car use	1. Predominant car user 2. Selective car users 3. Persons without a car
Bell et al. (2010)—SZENAMO	Cluster analysis	Socio-demographic and household variables	1. Mobile person 2. Slightly restricted mobiles 3. Highly restricted mobiles
Haustein (2012)	Cluster analysis	Socio-demographic, infrastructure, mobility-related attitudes	1. Captive car users 2. Affluent mobiles 3. Self-determined mobiles 4. Captive public transport users
Mandl et al. (2013)—GOAL	Cluster analysis	Demographic and health variables	1. Fit as fiddle 2. Happily connected 3. Hole in the heart 4. An oldie but a goodie 5. The car-full
Siren and Haustein (2013)	Cluster analysis	Transport and mobility-related variables	1. Independents 2. Flexibles 3. Restricted

Source Based on Luiu et al. (2018); Haustein and Siren (2015)

The metasegments identified in the study differed significantly from each other in terms of all the included variables, but they indicated some general trends important from the perspective of means of transport used: the high importance of the car (two metasegments: affluent mobile drivers and car-dependent seniors), the perception of public transport as an alternative to the car (transport service-dependent seniors) and large differentiation in relation to the use of other potential means of transport. It should be noted that the segmentation proposed in this paper differs from the ones presented, as it is based on the declared preferences

Table 2 Group of segments/metasegments identified by Haustein and Siren (2015)

Segments	Group of segments	Mobility patterns
Affluent mobiles Mobile car–oriented Workers Affluent males Fit as fiddle Happily connected Independents Fully mobile seniors Subgroup 1	Affluent mobile drivers[1]	Predominant car use, high activity engagement
Captive car users Mobility impaired Hole in the heart Disabled drivers	Car-dependent seniors	Predominant car use, low activity engagement
Subgroup 2 Self–determined mobiles Flexibles Selective car users An oldie but a goodie Self-determined mobiles Bike-oriented Ecology-minded PT–users Slightly impaired seniors	Mobile multi-model seniors[2]	Use of all modes; high/ medium activity engagement
Granny flats Subgroup 3 Persons without a car Highly impaired seniors Mobility impaired The car-full Subgroup 4 Restricted	Transport service-dependent seniors	Walking, public transport and car use as passenger; low activity engagement

[1]Two segments, predominant car users and mobile widows, were classified as transitory ones between metasegments affluent mobile drivers and car-dependent seniors
[2]Two segments, pragmatic PT-oriented and captive PT users, were classified as transitory ones between metasegments mobile multi-model seniors and transport service-dependent seniors
Source Based on Haustein and Siren (2015)

of seniors in relation to individual means of transport, taking into account various travel destinations—that is, it is oriented towards transport behaviour.

3 Methodology: Segmentations Rationale and Data Collection

In this study, primary data collected with the use of an auditorium survey on a sample of 400 students from the Universities of the Third Age (U3A) from the Silesian region will be used. The survey was conducted from December 2018 to

February 2019 in the following cities in Silesia: Bieruń, Dąbrowa Górnicza, Katowice, Rybnik, Sosnowiec. The accessibility of U3A participants and the fact that they are considered active seniors, willing to acquire new skills and expand their knowledge (shown by their enrolling and participation in Long Life Learning —LLL classes), were the main reasons for choosing this demographic for the study. The research addressed two issues: the transport preferences of seniors in relation to various travel destinations and the use of various types of information and communication technologies (ICTs) that could be beneficial for fulfilling the seniors' transport needs. In order to make sure that the questions would be easy to understand and answer, the questionnaire was trialled before starting the research, and during the research an experienced interviewer was present to provided support if needed. The distribution of respondents in terms of their characteristics is presented in Table 3.

The overwhelming majority of respondents participating in the study were female, with secondary or higher education, not working and living in one or two-person households, belonging to the so-called young old age category (60/65–74 years old) (Solecka 2018). More than half of the respondents do not own a car but have a driving licence, the majority declared a monthly income per person in the range of PLN 1500–3000 and assessed their health condition as good. It should be noted that the sample is not representative of the general population of seniors

Table 3 Characteristics of the surveyed group of respondents

Variable	Categories and values of variables						
Gender	Female				Male		
	340				60		
Owns a car	Yes				No		
	164				226		
Has a driving licence	Yes				No		
	238				147		
Education	Primary		Vocational		Secondary		Higher
	2		24		205		154
Employment status	Professionally active		Pensioner		Annuitant		Volunteer
	10		350		28		7
Health	Very good		Good		So–so		Bad
	23		222		103		17
Monthly income per person [PLN]	Up to 1500		1500–3000		3000–5000		Above 5000
	72		252		56		5
Number of people in the household	1		2		3		4 or more
	197		170		16		12
Age	51–55	56–60	61–65	66–70	71–75	76–80	81 and more
	4	18	82	111	93	65	22

(overrepresentation of women and young seniors), but is representative of the total population of Polish U3A students (GUS 2020).

As enrolment at U3A is possible at the age of 50, there was a small group of people younger than 60 years in the sample. Taking into account that different studies vary in defining seniors' age (mostly it is 60+ or 65+, but also at times it can be 50+ or 55+), it was decided to leave in the data on those respondents for the purposes of the analysis. Since some of the questions about respondents' characteristics could relate to sensitive issues (e.g. questions about income, health or the number of people in the household), answers to all questions were voluntary. It turned out to be a good decision: all respondents participating in the survey answered questions about transport preferences, but a small group did not answer particular questions (about 9% did not provide answers regarding the assessment of their health).

Regarding the study of the means of transport preferred by seniors, respondents were asked to indicate the most frequently used means of transport for particular destinations. In order to obtain the most detailed information, it was decided to extend the category of journeys beyond the most frequently used ones, that is necessary (work, school) and optional (other) trips (Hebel 2014), or a more detailed one, where necessary travel also includes shopping and accessing health care. Using that distinction would require either an explanation of the concepts (necessary versus optional travel) or questions about health issues (a potentially sensitive topic). The category depending on fulfilling the mobility needs of different levels (first degree: moving from A to B as quickly, cheaply and effectively as possible; second degree: ensuring a sense of independence, freedom, emphasizing the assumed social roles and status; third degree: enjoying the act of travelling without any particular goal) was also considered, but it was deemed too abstract for the purpose of the study (Curl and Musselwhite 2018). Finally, in order to obtain detailed information about the travel preferences of the elderly, the following seven travel agendas were specified:

- work/school,
- shopping,
- personal errands,
- social meetings,
- recreation/sport,
- culture/entertainment,
- care provided to a family member.

In terms of available means of transport, respondents could choose one of following seven options: car (as a driver), car (as a passenger), public transport, taxi, motorcycle, bicycle, walking. Considering that not every respondent will feel the need to travel in order to fulfil a certain goal (e.g. a given respondent may not provide care services to another family member, or they may live with a family member under their care and do not need to travel for that purpose), it was possible to indicate that the given purpose for travelling did not apply. Additionally, to provide for the possibility of using various means of transport for a given destination due to changing conditions (e.g. everyday small shopping done on foot, but for larger purchases going by car), the respondents were asked to indicate the most frequently used means of travel. Different types of public transport were not included individually (in the case of cities in the Silesian region it is usually a bus or tram), and the motorcycle, despite some doubts, was added following questionnaire feedback.

As was already mentioned, two attempts to classify the elderly in terms of their preferences regarding their means of transport were made. The first (referred to as expert segmentation) is "a priori" segmentation, where the segments have been defined on the basis of a literature review, taking into account both existing segmentations and indicated preferences of seniors towards individual means of transport. The second, "a posteriori" segmentation will be prepared with the use of cluster analysis. Both methods will be presented in the following sections.

3.1 Expert Segmentation

In the case of the first segmentation, the in-depth literature studies on the transport preferences of seniors as well as earlier segmentation of seniors using various indicators were used for describing the transport types of the elderly. The synthetic overview of existing segmentations by Haustein and Siren (2015), that indicated four metasegments (affluent mobile drivers, car-dependent seniors, mobile multi-model seniors, transport service-dependent seniors) was also taken into consideration but with certain modifications. All the segmentations included in that overview were carried out for more developed countries with different socio-cultural background; Polish seniors display certain characteristics that distinguish them from other EU countries (Okólski 2018). Taking into account the expected differences identified in the literature, resulting for example from gender (fewer female drivers, more frequent use of public transport and walking by women) (Böcker et al. 2017), the frequency of using a given means of transport (bicycles generally make a minor contribution to the modal breakdown regardless of age; taxis are used more as a special case or an exception and not as a rule; the same goes for motorcycles, although there is no reference to this mode of transport in the literature) (Hebel and Wyszomirski 2018; Ryan 2020) and after some deliberation and considering several possibilities, the authors decided to include six different segments. The segments were defined on the basis of the preferences declared by the respondents regarding the most frequently used means of transport in relation to various travel destinations, namely:

- active drivers, i.e. those mainly declaring the use of a car as a driver,
- passengers or driver's partners, i.e. those mainly declaring the use of a car as a passenger,

- public transport users, i.e. those declaring that they travel mainly using public transport,
- those who most often declare the use of active transport, i.e. walking or cycling,
- those who declare using both active and public transport, the so-called hybrid segment,
- those using various means of transport, indicating at least three different means without clear preference.

Three of the specified segments are similar to metasegments identified by Haustein and Siren (2015), but it should be emphasized that they were largely based on transport preferences and not on other socio-demographic indicators that were considered in that study. Hence in the proposed segmentation, an active driver is a person who, for the vast majority of travel purposes (destinations), declares the use of a car as a driver without taking into account other factors (high activity, high income or being predominantly male) indicated for the affluent mobile drivers metasegment. The indication of this segment (active drivers) and the second segment referring to the use of the car (as a passenger) results from the great importance of the car for the mobility of the elderly, whether as a driver or a passenger. The segment of PT (Public transport) users results from several factors identified in the literature, such as the inability to use a car (whether due to lack of access, age restrictions for drivers, deteriorating health), or the increasing emphasis on changing the modal division with regard to means of transport used in the city that results in incentives for seniors to use PT (e.g. free rides) and others (Mifsud et al. 2017; Raczyńska-Buława 2017). The segment of mobile seniors using various means of transport depending on their needs and individual travel goals, without indicating a predominant one (e.g. shopping on foot, entertainment by taxi, personal errands by car), was also included. In addition, the authors decided to add two segments, the first one to include seniors using mainly active transport (i.e. walking or cycling), and the second (the hybrid segment) to include respondents using both public and active transport. This was decided partly due to the predominance of women expected in the sample, who use this type of transport (public and active) more often than men (Haustein et al. 2013) and also the fact that most seniors' travels (especially daily ones) are usually confined to areas closest to their place of residence.

3.2 *Segmentation Using Taxonomic Methodology*

Parallel to the segmentation conducted using the expert method, the respondents were classified into groups with statistical methods, in particular, cluster analysis.

Based on the same data concerning the means of transport indicated by the respondents, it was assumed that each destination which the questionnaire enquired about would be represented by a separate variable:

- X_1—work/school,
- X_2—shopping,
- X_3—personal errands,
- X_4—social meetings,
- X_5—recreation/sport,
- X_6—culture/entertainment,
- X_7—care provided to a family member.

Each question concerning the means of transport asked the respondents to choose one of the options: 1—car (as a driver), 2—car (as a passenger), 3—public transport, 4—taxi, 5—motorcycle, 6—bicycle, 7—walking or indicate that a given destination did not apply to them (8). This means that the variables used in cluster analysis are nominal variables, the realizations of which are the respondents' choices coded numerically.

Clustering methods include hierarchical and iterative optimization methods. In this case, we applied one of the hierarchical methods—Ward's method, which is recognized for its strong formal properties, as some researchers posit (Fisher and Van Ness 1973; Trzęsiok 2009). Another argument in favour of the application of the method was the fact that it could be easily implemented to the analysis performed on nominal variables. Ward's method, in particular the ward.D function in **R** used in the study, allows for the use of any metric that can measure the distance between the objects under examination. In this case, the Sokal–Michener metric for nominal variables was used (Rogers and Tanimoto 1960; Walesiak 2011).

The key element of cluster analysis involves determining the optimal number of classes. The Silhouette index was used to validate this number (Kaufman and Rousseeuw 2009). The validation was conducted in the process of dividing the respondents into a few series of clusters, each time the number of clusters was different and the Silhouette index was calculated (I_S). The highest values of the index indicate the best cluster validity. It is also possible to interpret the I_S index in relation to the evaluation of the class structure obtained as a result of clustering. In their study, Kaufman and Rousseeuw (2009) argue that the highest values of $I_S \in (0.7, 1)$ reveal a strong structure of classes, while $I_S \in (0.5, 0.7)$ indicate that a reasonable structure has been found, but $I_S \in (0.25, 0.5)$ mean that the structure is weak and could be artificial. If $I_S \leq 0.25$, no substantial structure has been found.

The final stage of the study involved the comparison of the two segmentations: one performed by the expert and the other obtained as the result of cluster analysis. First, we needed to check whether the two classifications tallied, so we used the Rand index (Rand 1971), which measures the similarity between two clusterings of the same set of objects. The closer the value of the Rand measure is to 1, the more similar the results of two clusterings are. Additionally, we performed the analysis of correspondence to show the relationships between the classes of the two segmentations.

4 Results and Discussion

Based on the survey data on the most frequently used means of transport for different destinations, two segmentations were performed: expert and cluster analysis. The results, i.e. the clustering of respondents into appropriate classes, as well as the characteristics of these classes, are presented further in this part of the paper. In line with the aim of the paper, the two classifications were compared and discussed against the background of the results of research conducted by other researchers.

4.1 Main Findings of the Expert Segmentation

Following the assumptions of the expert segmentation presented earlier in the paper, the respondents were included in a given class according to their preference for a particular means of transport. Based on the means of transport declared as the most frequently used by the respondents for different destinations, the seniors were clustered into six classes:

- active drivers,
- passengers or, in other words, driver's partners,
- public transport users,
- active seniors (travelling mainly on foot or by bike),
- a hybrid segment including the respondents declaring the use of both public and active transport without any clear preference for either,
- the respondents declaring at least three different means of transport, without a clearly defined dominant.

When individual respondents were assigned to predefined segments, it turned out that it was necessary to create one more segment for a group of seniors who chose the answer "not applicable" so often that it prevented their inclusion in any of the original six segments. The absence of information on whether the choice of the "not applicable" option was caused by unrealized (or unconscious) transportation needs or by the lack of need to travel for a specific purpose made it difficult to name a new segment—the final version of the expert segmentation labelled it as "other".

The distribution of the respondents based on their inclusion in the groups proposed in the expert segmentation is presented in Table 4.

Table 4 Distribution of the respondents based on their inclusion in the groups proposed in the expert segmentation

Class	Drivers	Passengers	Public transport	Active seniors	Hybrid segment	Different means of transport	Other
Percentage share	23	9	22.5	5	26.5	3.25	10.75

The largest group (26.5% of respondents) consists of respondents classified in the hybrid segment. They declare the use of public transport as well as active transport. The most numerous subgroup go shopping on foot (56.6%), while using public transport for social meetings (53.8%) and cultural events (66%).

In terms of size, the second largest group includes active drivers. They account for 23% of the respondents. In general, they tend to choose the option of using a car as a driver: when going shopping (87% of the group), for personal errands (87%), for cultural and entertainment events (68.5%), for social meetings (67.4%) or to reach a family member they provide care to (64.1%).

The next group, similar in size (22.5%), includes people who mainly use public transport. They tend to choose this particular means of transport for personal errands (87.8%), cultural and entertainment events (87.8%), social meetings (85.6%) and shopping (73.3%).

In the group of passengers (9% of the total number of respondents), the respondents declared most frequently that they travelled by car as a passenger for shopping (77.8%), social meetings (66.7%), cultural and entertainment events (63.9%) and personal errands (58.3%).

On the other hand, active seniors (5% of the total) generally declare to travel on foot: for personal errands (75% of the respondents in this group), social meetings (75%) and shopping (70%). The respondent seniors did not declare the bicycle as their frequent choice as a means of transport, but the vast majority of respondents who use a bicycle to travel are classified in this group.

As mentioned above, we also identified the segment of the respondents who use different means of transport, without a clear preference for one mode. The means that were indicated most frequently were the car (as a driver or a passenger), public transport and active transport.

As many as 10.75% of the respondents declared no transport needs in the majority of the analysed situations: either for shopping (62.85% of this group), or for personal errands (65.1%), not to mention sport and recreation (93%), culture and entertainment (79.1%) or social meetings (79.1%).

4.2 *Main Findings of the Segmentation Using Taxonomic Methodology*

As mentioned above, we performed cluster analysis on the same data set using Ward's hierarchical method. In this case, it is crucial to determine the optimal number of classes. This was done as a simulation, which involved clustering the respondents into k classes, where the values $k = 2, \ldots, 9$ were checked. Each time the Silhouette index was calculated and the results are presented in Table 5.

The calculated values of the I_S index allowed for the determination of the optimal number of classes. Although the highest value of the index was obtained for $k = 2$, the clustering of respondents into two classes only was considered

Table 5 Silhouette index I_S calculated for clustering the respondents into k classes

k	2	3	4	5	6	7	8	9
I_S	0.255	0.243	0.241	*0.251*	0.184	0.157	0.140	0.135

uninteresting and hindering further research. Preliminary analysis of the data set and the expert segmentation revealed a diversity of behaviours and preferences of elderly respondents when they were asked about their choice of the means of transport. The classification of the seniors into two groups meant that only a group of people travelling by car was separate from all the others. This would significantly reduce the possibility of discovering the real class structure and make it impossible to conduct further research (e.g. analysis of correspondence). Based on expert knowledge, the authors decided that the optimal solution would involve dividing the respondents into five groups, because the value of the Silhouette index for $k = 5$ was only slightly lower from its highest value (it was the second highest value).

Both in this case and in the case of $k = 2$, we can only say that a weak class structure has been found (Kaufman and Rousseeuw 2009), but it is typical of clustering objects described by variables measured on weak scales.

The distribution of the respondents into five groups using Ward's method is presented in Table 6.

The first group consists of 22.5% of the respondents. The analysis of the answers of the respondents in this particular class to the questions used in the segmentation shows that the majority chose travelling by car as a driver (Fig. 1). This applies to most of their destinations: shopping (86.7%), personal errands (85.6%), social meetings (68.9%), visiting a family member they provide care to (67.8%), cultural and entertainment events (66.7%), sport and recreation (56.7%) and work or school (32.2%). Therefore, similarly to the expert segmentation, this group was labelled *active drivers*.

The second group includes only 6.25% of the respondents. The closer analysis of the distributions of their responses to the segmentation questions reveals that they mainly declare travelling by car, but as a passenger (Fig. 2). This way of travelling concerns mainly the following destinations: shopping (84%), social meetings (80%), personal errands (76%), cultural and entertainment events (72%), sport and recreation (48%). In the case of work- or school-related travels as well as when visiting a family member they provide care to, the dominant group of the respondents in this segment declared that those destinations did not apply to them.

The group was labelled *passengers* (*driver's partners*), similarly to the expert segmentation.

Table 6 Distribution of the respondents into five groups created using Ward's method

Class	1	2	3	4	5
Percentage share	22.5	6.25	43.25	13.75	14.25

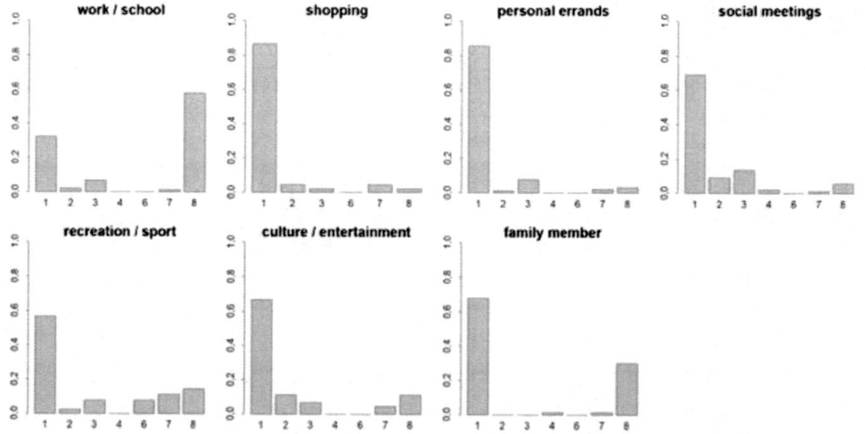

(1 – car (as a driver), 2 – car (as a passenger), 3 – public transport, 4 – taxi, 5 – motorcycle, 6 – bicycle, 7 – walking, 8 – this destination do not apply)

Fig. 1 Distribution of the Class I respondents' answers to segmentation questions

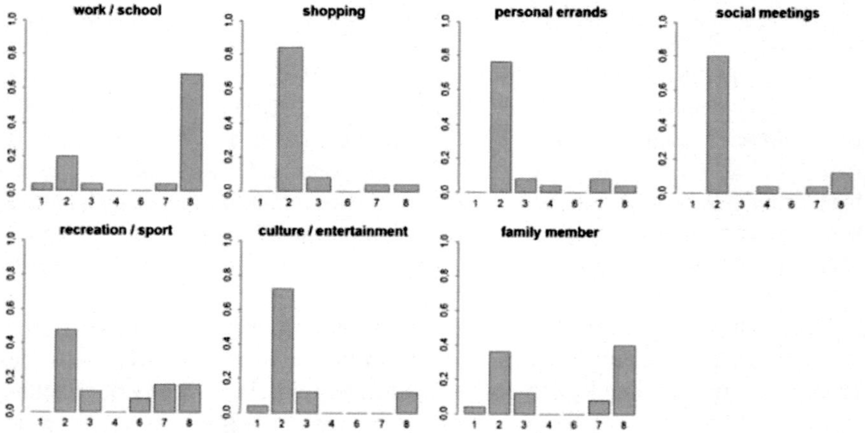

Fig. 2 Distribution of the Class II respondents' answers to segmentation questions

The third largest class includes 43.25% of the seniors participating in the survey. In this group, public transport is declared as the most frequently used means of transport for cultural and entertainment events (85.5%), social meetings (76.3%) and personal errands (68.2%). The respondents in this group usually go shopping by public transport (46.8%) or on foot (35.8%). Similarly, in the case of destinations related to sport and recreation, the most numerous subgroup includes those who declare using public transport (37%), but who also walk (21.4%). Notably, this segment includes a large group of respondents declaring that the following destinations do not apply to them: work or school (52.6%), visits to a family member in need of care (53.8%), or sport and recreation (26%) (Fig. 3).

Given that the means of transport most frequently declared by the respondents classified in this segment is, nonetheless, public transport, we decided to label this group *public transport users*.

The next group includes 13.75% of the respondents. These are the seniors who, to a large extent, use active transport (Fig. 4). They walk mainly to social meetings (67.3%), personal errands (41.8%) or shopping destinations (40%). They declare to travel on foot (52.7%) and by bicycle (27.3%) to sport or recreation destinations.

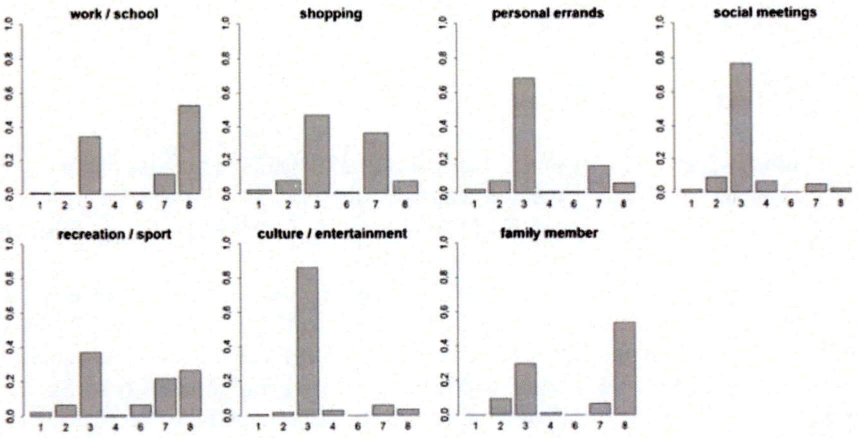

Fig. 3 Distribution of the Class III respondents' answers to segmentation questions

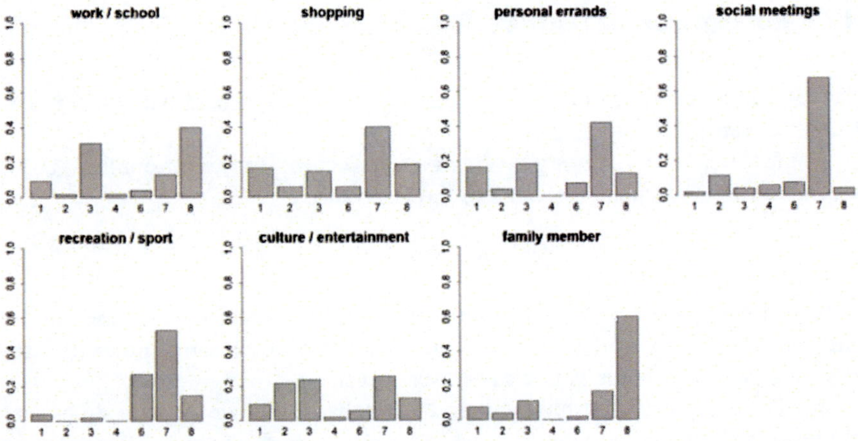

Fig. 4 Distribution of the Class IV respondents' answers to segmentation questions

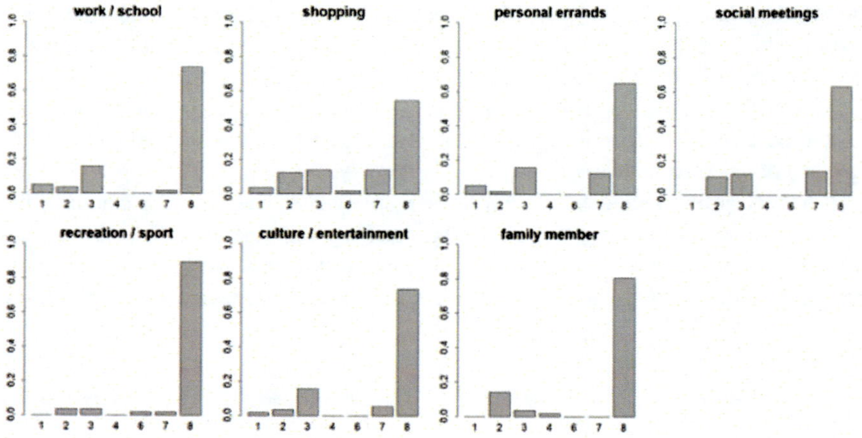

Fig. 5 Distribution of the Class V respondents' answers to segmentation questions

They usually get to work by public transport (30.9%), while in the case cultural and entertainment events, they tend to travel on foot (25.5%), by public transport (23.6%) or as a car passenger (21.8%). Nevertheless, these seniors are labelled *active*, also by analogy to the expert segmentation.

The last group includes 14.25% of the respondents. These are generally the seniors who declared no interest in travels related to sport and recreation (89.5%), visits to a family member they provide care to (80.7%), cultural and entertainment events (73.7%), personal errands (64.9%) or even shopping (54.5%) (Fig. 5). The respondents in this group claim that the destinations listed in the survey usually do not apply to them.

4.3 Comparison of the Two Segmentations

The aim of the analysis performed in the paper was to compare the two segmentations: expert and taxonomic.

The first step involved using the `classAgreement` function from the `e1071` package in **R** to calculate the value of the Rand index

$$R = 0.811. \quad (1)$$

It determines the similarity of two segmentations, which in this case can be interpreted as 81.1% of the pairs of objects classified to the same groups in both classifications. This value is at a satisfactory level.

In addition, the relationship between the results of the two segmentations was examined with the chi-squared test, and then the strength of the relationship was measured using V–Cramer's coefficient (which is a normalized measure). The following results were obtained:

$$\chi^2(24) = 906.87;\ p - \text{value} = 0 \tag{2}$$

and

$$V = 0.753. \tag{3}$$

This means that a significant relationship exists between the classes obtained in the expert segmentation and the classes created by the algorithm in Ward's method. As a result, further examination is performed using analysis of correspondence, which will determine the relationships between particular classes of the two segmentations.

Analysis of correspondence is an exploratory technique used to examine the contingency table. It allows for the creation of the perception map, which illustrates relationships between categories of the variables under examination, which, in this case, are classes obtained as a result of the prior segmentation (Fig. 6).

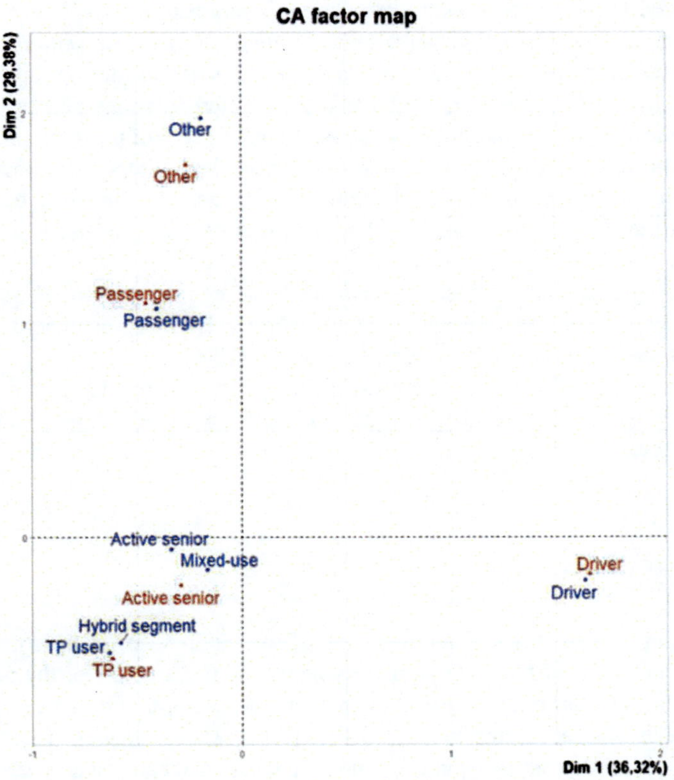

Fig. 6 Perception map illustrating the relationships between classes obtained as a result of the expert segmentation (blue labels) and classes obtained using the Ward's method (red labels)

Objects that are located close to each other on the perception map represent related categories of the variables under examination, in this case segments. It is clear that the classes of drivers are very similar to each other. Thus, it can be inferred that both the expert segmentation and cluster analysis classified the same seniors as drivers. The same applies to passengers (driver's partners). The classes of passengers, created as a result of two different segmentations, also tend to include the same respondents (70%).

The situation is slightly different in the case of the respondents who travel by public transport. The expert segmentation created two groups of respondents: one whose members declared that they chose this particular means of transport on a predominant basis and the other whose members travelled by public transport as often as on foot. The algorithm used in the Ward's method identified the respondents from the two groups as one class; therefore, the points representing these segments are located close to each other on the perception map. This can be treated as a recommendation for expert segmentation to combine those travelling by public transport and those belonging to the hybrid segment. This was considered in the process of analysis, but we decided not to combine the two classes due to the research assumptions that had been adopted.

A similar situation concerns the respondents using active transport. In this case, the perception map shows the links between the classes of the seniors travelling actively, identified as a result of the two segmentations, and the group of seniors classified by the expert as persons travelling by different means of transport. This group is the smallest in size, but it manifests a great variety of transport preferences. Despite the results of analysis of correspondence, the authors argue there are no grounds for including the group of highly mobile seniors, travelling by different means of transport, in the class of the respondents declaring to use mainly public transport.

The two groups including the respondents declaring that most travel destinations in the survey did not apply to them are also similar. Both segmentations classified most of such respondents into these two clusters (75%).

Despite some differences, discussed in the paper, the results of the two segmentations can be regarded as consistent and linked with each other to a relatively strong degree.

5 Conclusions

In the paper, two methods of segmentation of seniors in terms of their preferences for different means of transport were presented. The first called (for the purpose of this study) the expert one was based upon a simple assumption that the respondents should be divided according to the most frequently chosen means of transport for travelling to work, for shopping, pursing personal matters, social meetings and other purposes. Despite its simplicity, this may be considered to be an innovative approach, as the segmentations considered previously used various factors but mostly focused on the car as the preferred means of transport. The second method was made using cluster analysis, using Ward's method and the Sokal–Michener metric, which is based on nominal variables. The aim of the study was to show that the results of both methods are consistent, which would support the validity of the proposed expert segmentation. For this purpose the Rand index was used, and this confirmed the good agreement between the two methods.

The analysis of the correspondence of the classes obtained in both cases showed a clear similarity; in both methods, drivers and passengers are the most separated groups. The study showed that in both approaches used, these two groups are very similar to each other.

Some differences are also noticeable, due to the different number of segments used by each method. Those differences related mainly to the group of people who travelled by public transport. Ward's method identified only one such group, while in the expert approach, people who mainly travel by public transport are separated from those who combine this way of travelling with active transport. The research, the results from which served as an input for the segmentation, was conducted on a sample of U3A students from several cities in the Silesian region. Depending on the spatial planning and development of a given area, not all of the indicated travel destinations may be present in the immediate vicinity of the respondent's place of residence. A compact and mix-used neighbourhood allows for the fulfilment of many needs using only pedestrian (or other types of active) transport, but in the case of less compact or single-use areas (i.e. only residential buildings in the near vicinity), it may be necessary to combine it with another mode of transport like car or public transport. Due to the study's anticipated higher number of women, who, according to the literature on the subject, are less often drivers than men and more often use public transport and walking, it was decided to include such a segment and referred to it in the paper as a hybrid one.

In the case of active transport, despite the different classification of this type of transport in each method (in the expert analysis, both walking and cycling were considered as means of active transport, while for Ward's method these were two separate categories), a certain similarity can be noticed.

The results obtained using Ward's method are closer to the classification in the form of four metasegments used by Haustein and Siren (2015). As already mentioned, according to the authors, this synthetic division is too general, and it would be advisable to break down some of these metagroups. Such an approach may result in potentially greater possibilities of using the proposed segmentation in practice (e.g. in facilitating the preparation of proposals and activities for specific groups of seniors interested in using a given means of transport).

It should be noted that this article is the first approach to the proposed expert segmentation. Expanding the research is possible by taking into account a more representative group of seniors, the most frequently used means of transport and relevant socio-demographic factors.

References

Aigner-Breuss E, Braun E, Schöne ML, Herry M, Steinacher I, Sedlacek N (2010) Mobilitätsszenarienkatalog. Mobilitätszukunft für die Generation 55+. Mobilitätsszenarien für eine aktive Teilnahme am Verk. Retrieved from http://www.kfv.at/fileadmin/webcontent/Bereich_VM/MOTION55_Mobilitaetsszenarienkatalog.pdf

Bell D, Füssl E, Risser R, Braguti I, Oberlader M, Ausserer K (2010) SZENAMO—Szenarien zukünftiger Mobilität älterer Personen. Final project report financed by the Austrian Federal Ministry for Transport, Innovation and Tachnology. Retrieved from http://www2.ffg.at/verkehr/file.php?id=228

Böcker L, van Amen P, Helbich M (2017) Elderly travel frequencies and transport mode choices in Greater Rotterdam, the Netherlands. Transportation 44:831–852

Boksberger P, Laesser C (2008) Segmenting the senior travel market by means of travel motivation—Insights from a mature market (Switzerland). Retrieved from https://ro.uow.edu.au/commpapers/2440

Coughlin JF, D'Ambrosi L (eds) (2012) Ageing America and transportation. Springer Publishing Company, New York

Curl A, Musselwhite C (eds) (2018) Geographies of transport and ageing. Palgrave Macmillan, Cham

Elmore-Yalch R (1998) A handbook: using market segmentation to increase transit ridership. Transportation Research Board, Washington, DC

Fisher L, Van Ness J (1973) Admissible clustering procedures. Biometrika 60:422–424

GUS (2020) Sytuacja osób starszych w Polsce w 2018 roku. Warszawa, Białystok

Haustein S (2012) Mobility behavior of the elderly—an attitude-based segmentation approach for a heterogeneous target group. Transportation 39:1079–1103

Haustein S, Hunecke M (2013) Identifying target groups for environmentally sustainable transport: assessment of different segmentation approaches. Curr Opin Environ Sustain 5:197–204

Haustein S, Siren A (2015) Older people's mobility: segments, factors, trends. Trans Rev 35 (4):466–487

Haustei S, Hunecke M, Kemming H (2008) Mobilität von Senioren. Ein Segmentierungsansatz als Grundlage Zielgruppenspezifischer Angebote. [Seniors mobility. A segmentation approach as basis for target-group specific services]. Internationales Verkehrswesen 60:181–187

Haustein S, Siren A, Framke E, Bell D, Pokriefke E, Alauzet A (2013) Demographic change and transport, commission Europeenne. Retrieved from https://hal.archives-ouvertes.fr/hal-00867029

Hebel K (2014) Podróże obligatoryjne mieszkańców polskich miast na przykładzie Gdyni. Zeszyty Naukowe Uniwersytetu Gdańskiego. Ekonomika Transportu i Logistyka 52:129–145

Hebel K, Wyszomirski O (2018) Transportation preferences and travel behaviour of senior citizens in Gdynia in the light of marketing research. Research journal of the University of Gdańsk. Trans Econ Logist 76:167–177

Hildebrand ED (2003) Dimensions in elderly travel behaviour: a simplified activity-based model using lifestyle clusters. Transportation 30:285–306

Kaufman L, Rousseeuw P (2009) Finding groups in data: an introduction to cluster analysis. John Wiley & Sons, New York

Lee B, Bowes S (2016) A study of older adults' travel barriers by examining age segmentation. J Tour Hosp Manag 4(2):1–16

Luiu C, Tight M, Burrow M (2018) Factors preventing the use of alternative transport modes to the car in later life. Sustainability 10(1982)

Mandl B, Millonig A, Friedl V (2013) The variety of the golden agers: identifying profiles of older people for mobility research. Retrieved from https://www.researchgate.net/publication/234841783

Mifsud D, Attard M, Ison S (2017) To drive or to use the bus? An exploratory study of older people in Malta. J Transp Geogr 64:23–32

Mollenkopf H, Marcellini F, Ruoppila I, Szeman Z, Tacken M, Wahl HW (2004) Social and behavioural science perspectives on out-of-home mobility in later life: findings from the European project MOBILATE. Eur J Ageing 1:45–53

OECD (2001) Ageing and transport: mobility needs and safety issues. OECD Publishing, Paris

Okólski M (ed) (2018) Wyzwania starzejącego się społeczeństwa. Polska dziś i jutro. Wydawnictwo Uniwersytetu Warszawskiego, Warszawa

Raczyńska-Buława E (2017) Mobilność osób starszych. Dlaczego nie transport publiczny? TTS Technika Transportu Szynowego 1–2:24–34

Rand W (1971) Objective criteria for the evaluation of clustering methods. J Am Stat Assoc 66 (336):846–850

Rogers D, Tanimoto T (1960) A computer program for classifying plants. Science 132:1115–1118

Ryan J (2020) Examining the process of modal choice for everyday travel among older people. Int J Environ Res Public Health 17(3):1–19

Sagan A (2009) Podejście modelowe w segemtnacji rynku. Zeszyty Naukowe Uniwersytetu Ekonomicznego w Krakowie 80:21–35

Siren A, Haustein S (2013) Baby boomers' mobility patterns and preferences: What are the implications for future transport? Transp Policy 29:136–144

Solecka K (2018) Potrzeby osób starszych w zakresie mobilności w mieście. Autobusy 6:1252–1259

Trzęsiok M (2009) On some properties of support vector clustering. In: Domański C, Białek J (eds) Acta Universitatis Lodziensis, Folia Oeconomica 228, multivariate statistical analysis—statistical inference, statistical models and applications, pp 221–228

Walesiak M (2011) Uogólniona miara odległości GDM w statystycznej analizie wielowymiarowej z wykorzystaniem programu R. Wydawnictwo Uniwersytetu Ekonomicznego we Wrocławiu, Wrocław

Załoga E, Kłos-Adamkiewicz Z (2019) Potrzeby transportowe starzejącego się społeczeństwa. Transport Miejski i Regionalny 3:14–18

Application with COVID-19 Data

The Impact of the COVID-19 Pandemic on the Economies of European Countries in the Period January–September 2020 Based on Economic Indicators

Ewelina Nojszewska and Agata Sielska

Abstract SARS-CoV-2 coronavirus, causing the COVID-19 disease, reached Europe in early 2020 and quickly spread on the continent. Both the disease itself and the preventive measures introduced by governments lead to a restriction of economic activity. The beginning of the third quarter of 2020 is the time when some of European economies have already experienced the first wave of cases, and many restrictions have already been lifted. On the other hand with the end of this quarter, the number of cases is still growing meaning second wave and further limiting of economic activities. The study analyses the similarities of European countries in terms of the following indicators: Economic Sentiment Indicator (ESI) and Employment Expectations Indicator (EEI) from the beginning of 2020. The obtained results are compared with the course of the pandemic in the studied countries. Ward's method was used in the analysis. Results show that after the collapse in March/April, the values of variables reflecting condition of economies started to increase in most of identified groups of countries. There are very little similarities between the countries with respect to changes in indicators and the course of the pandemic.

Keywords COVID-19 · Coronavirus · Economic indicators · Ward's method

1 Introduction

The first cases of SARS-CoV-2 coronavirus, causing the COVID-19 disease have been noted at the end of 2019 in Wuhan (capital of Hubei province, central China). In early 2020, the virus reached Europe and quickly spread to European Union

E. Nojszewska · A. Sielska (✉)
SGH Warsaw School of Economics, Warsaw, Poland
e-mail: asiels@sgh.waw.pl

E. Nojszewska
e-mail: enojsz@sgh.waw.pl

K. Jajuga et al. (eds.), *Data Analysis and Classification*, Studies in Classification, Data Analysis, and Knowledge Organization,
https://doi.org/10.1007/978-3-030-75190-6_17

(EU) countries. The lack of a vaccine combined with limited resources at the disposal of health care and the rapid spread of the virus resulted in making decisions directly or indirectly leading to a reduction in economic activity.

The aim of the paper is to analyse the relation between the changes in economic indicators and the course of the epidemic in groups of EU countries using clustering approach. Clustering is a method of finding similar objects and grouping them together based on the common characteristics. This approach is also used in the literature in regards to COVID-19. Zarikas et al. (2020) provide clustering using single and complete linkage approach and discuss the algorithm. Kumar analysed the distribution of deaths, confirmed and cured cases in India (2020), Maugeri et al. (2020) focused on Italy, while Shammi et al. (2020) used i.a. hierarchical clustering to study the possible situation in Bangladesh.

This paper is organized as follows. In Sect. 2, we present and discuss the relation between the pandemic and the economic activity. The literature review states Sect. 3, and Sect. 4 is dedicated to the economic indicators used in the study. In the next section, we present the results of the analysis and discuss the common characteristics of countries with respect to the course of the pandemic and changes in economic indicators. The paper ends with conclusions.

2 SARS-CoV-2 and Economies

The economy is a mechanism of interlocking wheels, because everything is related to everything. In the modern economy, these circles also overlap on an international and global scale. That is why economic crises disrupt the functioning of all components of economies. Each of the crises had different causes and transmission mechanisms.

COVID-19 hit the real economy immediately and only then hit the international financial markets and the feedback loops between them. The mechanism of this crisis is completely different from what we know from previous economic collapses.

The basis for the functioning of economies is cooperation in creation in the global reality, and the source of this cooperation is demand. The virus affected all factors influencing the volume of individual and aggregate demand: GDP, disposable income, prices, interest rates, exchange rates, in short—it limited all flows. It appeared general economic domino effect. COVID-19 has caused an economic shock three times worse than the 2008 financial crisis in terms of GDP decline on an annual basis. It has been a crisis like no other, shutting economies and societies, closing borders and putting half of humanity under some form of lockdown during the spring of 2020.

The pandemic proved how important health is, that is, human beings with their human capital and work for the economy. Therefore, the International Labour Organization latest labour market developments (International Labour Organization 2020) inform that:

- After the months of the pandemic so far, restrictions on labour markets have been eased. However, even these new restrictions still have a negative impact on the functioning of these markets.
- The estimated losses in the number of working hours have increased, which confirms the deteriorating situation in the labour markets. It is therefore hard to expect a recovery in the coming months.
- There has been decreases in labour income.

This is the basis for the formulation of effective policies of all kinds.

Apart from the economic changes, COVID-19 affects the society as well. United Nations Committee for the Coordination of Statistical Activities (CCSA 2020) points out that some problems may appear mostly in countries and populations characterized by the low level of income. Economic disturbances in such conditions may increase the risk of undernourishment or violation of human rights.

It is important to note that the disease itself is only one of the factors affecting the economies. Second important cause of disturbances is preventive measures which may be introduced by a government to stop the spreading of the virus. All of those measures have an impact on economic activities and conditions under which markets are functioning. Some of the effects are visible at first glance and basically obvious. Both recommendations issued by governments and most strict measure, i.e. lockdown affects the demand side of the economies as presented in Fig. 1. The supply side is also affected by i.a. by the breakdown of supply chain. In the consequence, the liquidity and income are affected and as a result economic relations are further disrupted.

3 Literature Review

The number of papers on economic impact of COVID-19 pandemic is growing fast. Their authors focus on different aspects of economic relations which are disrupted by either the pandemic itself or remedies.

Infectious diseases may influence economies via various channels (McKibbin and Roshen 2020; Fernandes 2020). Similar topic is discussed by Ozili and Arun (2020). Fernandes (2020) analyses potential costs expressed as percentage of GDP, providing estimates for different countries under different assumptions regarding shutdown. Deb et al. (2020) focus on effects of containment measures on economic activity. Implications of measures are also discussed by Radwan and Radwan (2020) (closure of educational institutions), Kong and Prinz (2020) and Silva et al. (2020). Kong and Prinz analyse effects of six shutdown policies on unemployment. Silva et al. (2020) use agent-based models (ABM) to simulate the effects in seven scenarios which assume different degrees of social distancing from none at all to lockdown. They analyse both epidemiological evolution and economic impact. Despite the fact, that their study is a simulation, the model proposed by authors may be considered a valuable tool for both decision-makers and analysts in their

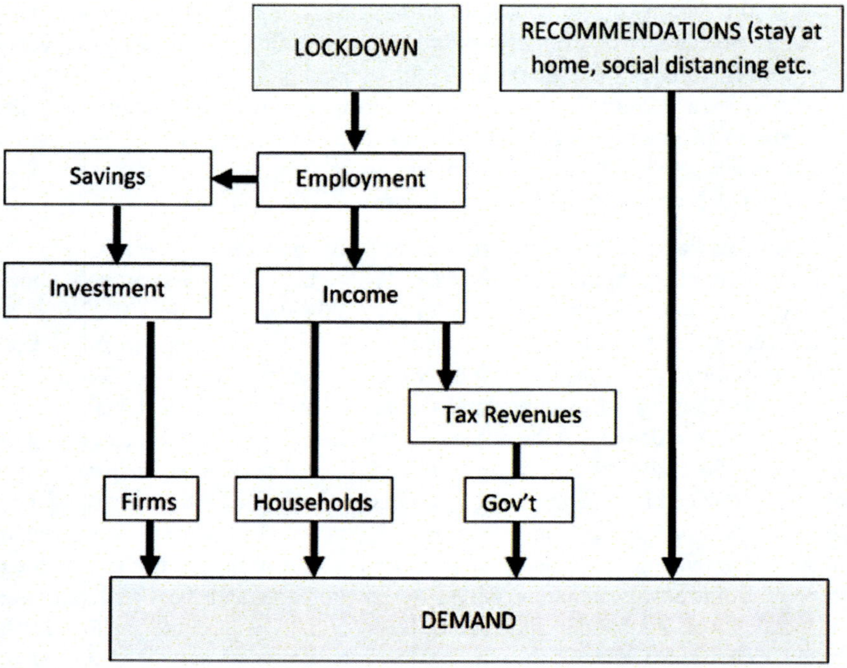

Fig. 1 Chain of economic impacts caused by COVID-19—simplified version. *Source* Own elaboration

predictions. Chetty et al. (2020) discuss the short-run effects on employment and consumer spending. The authors also suggest best interventions to mitigate negative effects. Hall et al. discuss consumption changes (2020), while Norouzi et al. (2020) discuss the impact on the demand on energy resources. It is important to note, that in case of the change in usage patterns, the effect may be long-lasting and multi-level. McKibbin and Roshen (2020) discuss potential macroeconomic impacts and analyse seven different scenarios of how COVID-19 might evolve using a hybrid of equilibrium models. Clemens and Veuger (2020) analyse the impact on tax revenues which may have broader consequences such as limitations in the future availability of public services.

Baker et al. (2020) discuss how COVID-19 changed the uncertainty level in economics and state that the magnitude of uncertainty level caused by current pandemic is similar to the one connected with the crisis of 1929–1933. Uncertain environment is reflected by a situation on stock markets. Topcu and Gulal (2020) analyse the impact on stock markets and find out that on emerging stock markets the effect of the coronavirus was fading. It should be noted, however, that the period taken in the analysis was 10 March 2020–30 April 2020, and in some countries included in the study (e.g. in Poland), the number of cases was growing at that time. Reported results may lead to the conclusion that economies and investors are adapting to the changes in their environment. The second reason may be reduction of the uncertainty by introduction of new preventive measures by local governments, and increasing awareness of the society by education about the pandemic. Also Pak et al. (2020) point out the negative correlations between the number of COVID-19 cases and stocks indices. Ashraf (2020) focuses on the effect of government interventions. Introduction of preventive measures based on social distancing decreases economic activities which result in negative effects on stock market returns. On the other hand, there is a positive effect due to preventing of a number of COVID-19 related cases and deaths. Ashraf also points out on a positive impact of announcing awareness programmes and—which was to be expected—of income support packages (2020).

The different sphere is related to the impact on other fields of health care. There are evidences that rising level of uncertainty in the population, hospitals' focus on COVID-19 cases and preventive measures ordered by governments lead to the decline in diagnostics of other diseases. That could lead to higher mortality in the future and also higher costs (both direct and indirect) to the society. Del Vecchio Blanco et al. (2020) discuss the impact of the pandemic on the colorectal cancer prevention, Farid et al. (2020) provide the percentage rates of change in surgical activity over the course of pandemic. Other aspects are also discussed, such as mental health—Rossi et al. (2020) and inequalities (Palomino et al. 2020; Blundell et al. 2020; Alon et al. 2020).

Because of the multiple levels of impact, we decided to focus on the economic indicators which are supposed to represent the wide spectrum of economic activities in one number.

4 Economic Indicators

In the analysis, we use two indicators published by European Commission (EC): Economic Sentiment Indicator (ESI) and Employment Expectations Indicator (EEI).

Economic Sentiment Indicator (ESI) has been published for 35 years. Its components are confidence indicators computed for individual sectors (industry, services, construction, retail trade and consumers) with weights determined based on the "representativeness" of each sector and performance against the reference variable (European Commission Directorate-General for Economic and Financial Affairs 2020). The components and their respective weights are presented in the right hand side of Fig. 2. Being the aggregate index, ESI reflects the way the economy is viewed as a whole. It is published monthly by EC.

The second index we focus on is the Employment Expectations Indicator (EEI), the values of which are used to represent the situation in the labour market. EEI is calculated based on employment expectations in four business sectors. The components are presented in the left hand side of Fig. 2. It reflects managers' expectations and plans (European Commission Directorate-General for Economic and Financial Affairs 2020).

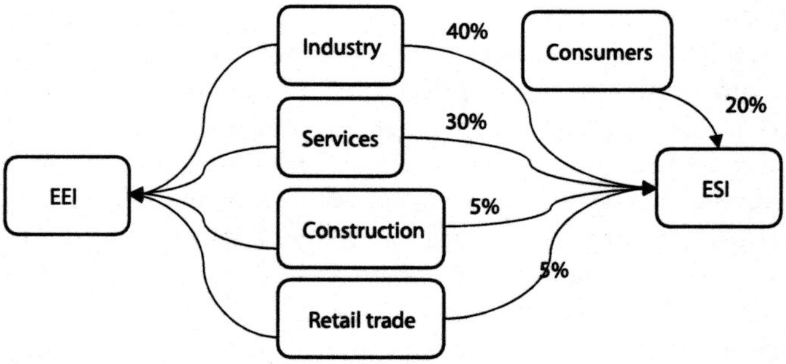

Fig. 2 Components of ESI and EEI—simplified version. *Source* Own elaboration

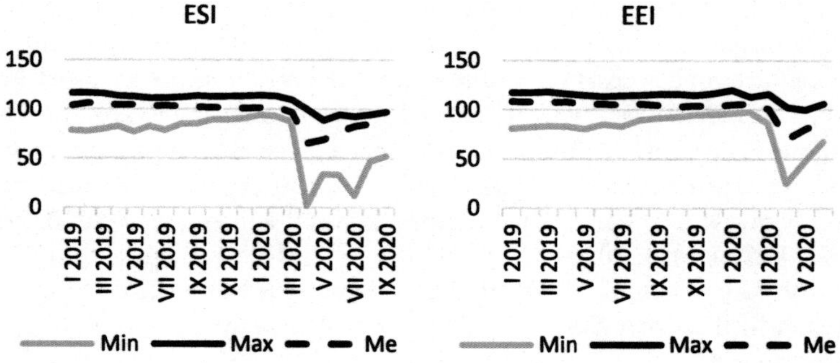

Fig. 3 Changes in ESI and EEI in the European countries due to the outbreak of COVID-19 pandemic. *Source* Own elaboration based on Eurostat data

As can be seen in Fig. 3, the values of indicators decreased strongly in April, which means that the economy reacted to the coronavirus approximately with one month delay. It can be also seen that the differences between countries have risen as the course of the pandemic was different in different territories.

5 Methodology

Analysis covers 31 countries: Albania, Austria, Bulgaria, Croatia, Cyprus, Czech Republic, Denmark, Estonia, Finland, Germany, Greece, Latvia, Lithuania, Malta, Poland, Romania, Slovakia, Slovenia, Belgium, France, Hungary, Italy, Netherlands, Spain, Ireland, Montenegro, North Macedonia, Portugal, Serbia, Turkey and Sweden in the period January–September 2020.

The period has been divided into three subperiods, each corresponding to a quarter of a year. There are several reasons behind this decision. Firstly, due to the fact, that the coronavirus did not appear in the most of the countries in Europe before March, we intend to analyse this first period separately so that the beginning of the pandemic will not affect the overall results disproportionately. Secondly, the beginning of the third quarter of 2020 is the time when some economies have already experienced the first wave of cases and lifted some preventive rules. Moreover in summer season, the tourist traffic was partially restored, and some of the analysed countries are very attractive travel destinations. In such countries like Croatia or Spain, tourism generates a substantial of GDP.

Following variables were selected:

- percentage change in ESI compared to the corresponding month of 2019;
- percentage change in EEI compared to the corresponding month of 2019;
- cumulated COVID-19 cases per capita;
- cumulated deaths per one COVID-19 case.

Percentage changes of ESI and EEI were calculated according to formulas (1–2)

$$y_i^{EEI} = \frac{y_{i,2020}^{EEI}}{y_{i,2019}^{EEI}} - 1, \tag{1}$$

$$y_i^{ESI} = \frac{y_{i,2020}^{ESI}}{y_{i,2019}^{ESI}} - 1, \tag{2}$$

where y_i denotes the value of an index in i-th month.

Close to 0 and negative values of percentage change calculated as above represent small decline in the value of an index as compared with the previous year, while negative and far from 0 values represent greater decline. On the other hand, close to 0 and positive values represent small increase in the value of the index as compared with the previous year, while positive and far from 0 values represent greater increase. Every time a change (increase or decrease) in index is discussed in the text, it should be understood as a change compared to the previous year.

Data on ESI and EEI come from the European Union (1995–2020) Eurostat Database (https://ec.europa.eu/eurostat/), and data on COVID-19 cases come from European Centre for Disease Prevention and Control (ECDC 2020) COVID-19 data. https://www.ecdc.europa.eu/en/COVID-19/).

Due to different ranges of values, the variables used in the study were standardized according to the formula:

$$z_i = \frac{x_i - \min(x_i)}{\max(x_i) - \min(x_i)}, \tag{3}$$

where x_i denotes the value of variable x for i-th object.

Clustering was made using the Ward's method (Ward 1963), and optimal number of clusters was identified using McClain index (McClain and Rao 1975), assuming maximum number of clusters equal to 15 and minimum number of clusters equal to 4. All analyses were done with R Software (R Core Team 2020) and packages NbClust (Charrad et al. 2014) and cluster (Maechler et al. 2019).

6 Economic Situation in European Countries and COVID-19

In this part of the paper, we present both individual results for first three quarters of 2020 and the clustering result for the whole period January–September 2020. Values of McClain index calculated by NbClust were as follows: 0.6847 (first quarter), 0.586 (second quarter), 0.4437 (third quarter) and 1.5016 (whole period). In each case, the minimum number of clusters, i.e. 4, was chosen. Results may indicate the lack of underlying cluster structure. Despite that, the countries were clustered, in order to check whether some similarities can be noticed.

6.1 First Quarter

In the first quarter of first group of countries consists of: Albania, Bulgaria, Cyprus, Czech Republic, Denmark, Greece, Hungary, Ireland, Latvia, Lithuania, Malta, Montenegro, North Macedonia, Poland, Portugal, Romania, Serbia, Slovakia and Slovenia. Second group is made up of: Austria, Croatia, Estonia, Sweden, Finland and Germany. Third group consists of four countries: Belgium, France, Netherlands and Spain. Turkey was the only country in the fourth group.

In January, Turkey stood out due to the values of changes in EEI and ESI, and this will remain unchanged during the whole period. There were no clear differences between the remaining groups in this respect, except for slightly higher values of percentage change in EEI in group 3 than 2 which due to the fact that those values are negative translates into greater size of decline in EEI in group 2 compared to the previous year. Countries in which COVID-19 cases appeared in January were classified into different groups. There were also no fatal cases of the disease yet.

In February, the situation with regard to EEI and ESI was similar. However, some differences could be observed in terms of COVID-19 cases. First group was characterized by rather low morbidity. These values were higher in the second group. There are still virtually no deaths.

In March, third group was characterized by high death rates, while the incidence is similar to the previous month, cases of deaths occurred in all groups, and Turkey has both low morbidity and low death rate. In terms of EEI and ESI, there are no clear differences.

It should be noted that there were two decisive factors in the grouping in this period: Turkey was classified separately due to its percentage changes of EEI and ESI. Death rates in March were crucial in identification of the third group.

As shown in Fig. 4 in terms of trends of the variables, there were no clear differences between the identified groups. In the case of the countries classified in the first group, the values of changes of economic indicators decreased slightly in March, whereas the changes in other groups were more distinct. This was accompanied by the increase in number of cases per capita and in death rates. Countries from other groups experienced similar changes. The changes in mortality in the first three groups were also slightly different.

6.2 *Second Quarter*

In the period April–June, as in the first quarter, the first group consists of the most countries: Albania, Austria, Bulgaria, Croatia, Cyprus, Czech Republic, Denmark, Estonia, Finland, Germany, Greece, Latvia, Lithuania, Malta, Portugal, Romania, Serbia, Slovakia, Slovenia and Turkey. Second group is made up of: Belgium, Ireland, Netherlands, Spain and Sweden. Third group consists of only two countries, i.e. France and Hungary. Montenegro, North Macedonia and Poland are classified into the last, fourth group.

In April, groups 1 and 4 were characterized by relatively low values of both morbidity and death rate. Both of those indicators were high in second group, while the third one was characterized by low morbidity and high death rates. In the countries belonging to the fourth group, relatively larger declines in ESI could be noticed.

In May, characteristics of identified groups of countries with respect to COVID-19 were the same as in April. The results also show that fourth group kept its characteristics.

In June, groups retained their characteristics in terms of values of COVID-related variables. Decline in ESI was small in groups 1–3 and high in 4. There are no clear differences between the countries with regard to the change in the EEI.

Following general changes may be noticed during three analysed months based on Fig. 5. Countries classified into the first group were characterized by death rates fluctuating on a relatively stable level. Declines in economic indicators were getting smaller each month. Situation of the countries from the second group was very similar; however, in this case, death rates were subject to a slight decrease. In the case of the third group, there was a fluctuation of the death rates. Decline in ESI at first grew larger, then decreased noticeably, while the decrease in EEI was getting smaller each month. In the fourth group, the results show increasing morbidity, decreasing death rate, fluctuations of the percentage changes in ESI and an decreasing values of the changes in EEI as compared with 2019.

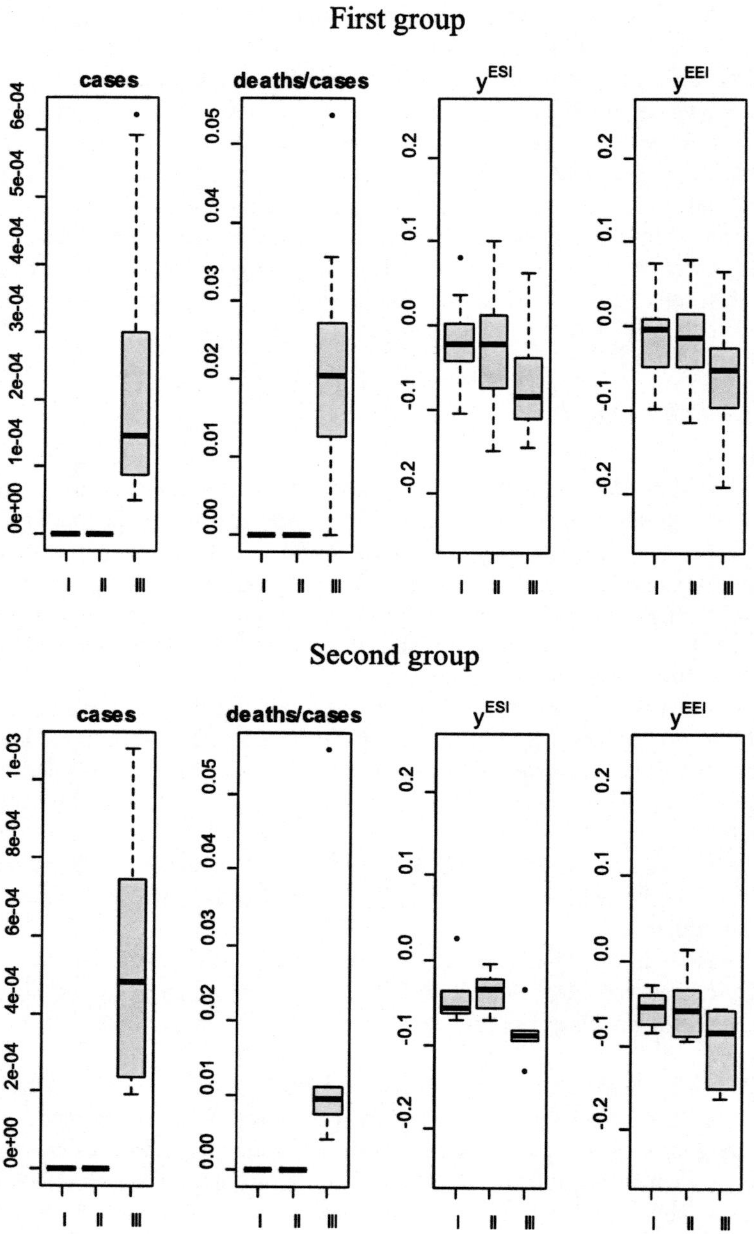

Fig. 4 Evolution of values of variables in the first quarter of 2020. *Source* Authors' calculations based on ECDC and Eurostat data

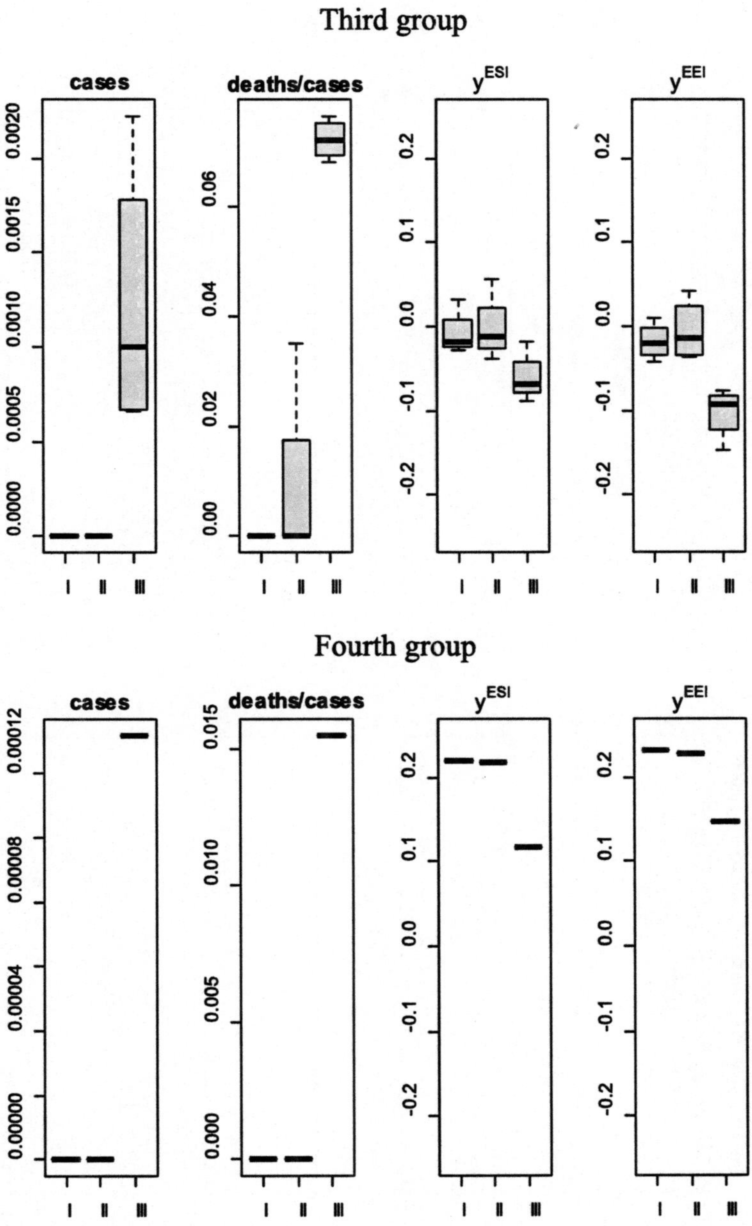

Fig. 4 (continued)

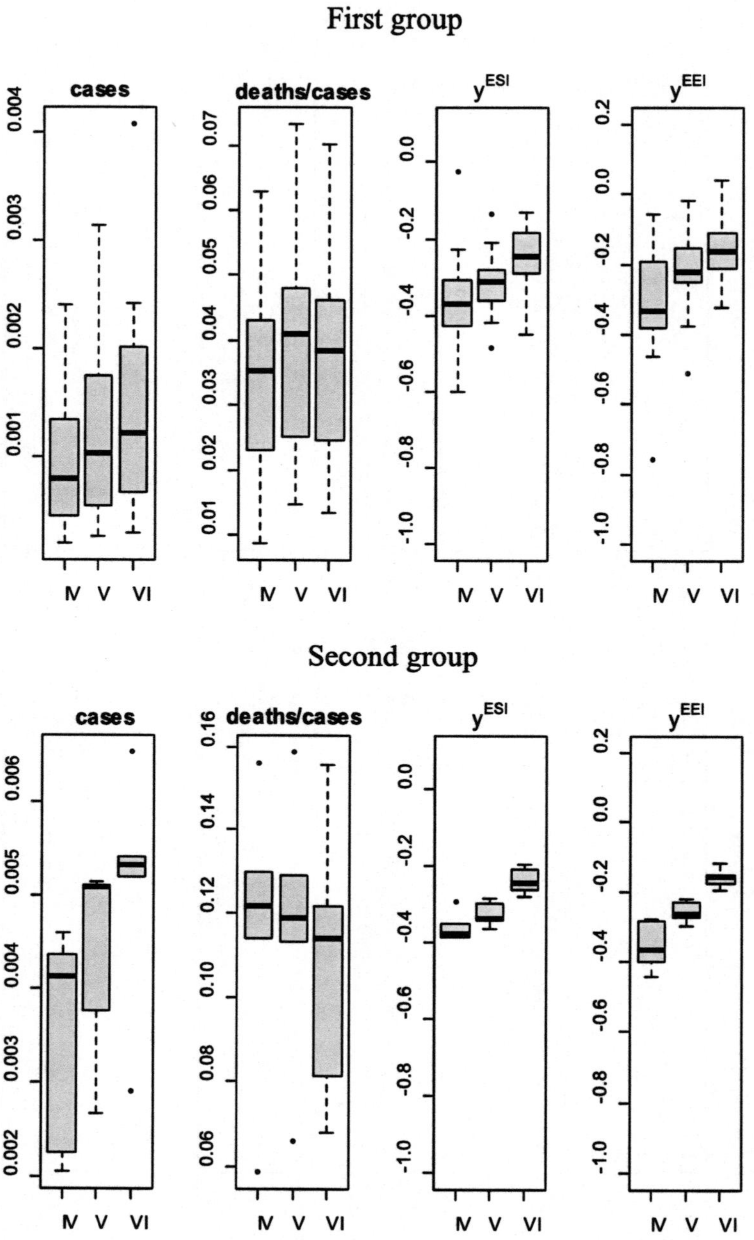

Fig. 5 Evolution of values of variables in the second quarter of 2020. *Source* Authors' calculations based on ECDC and Eurostat data

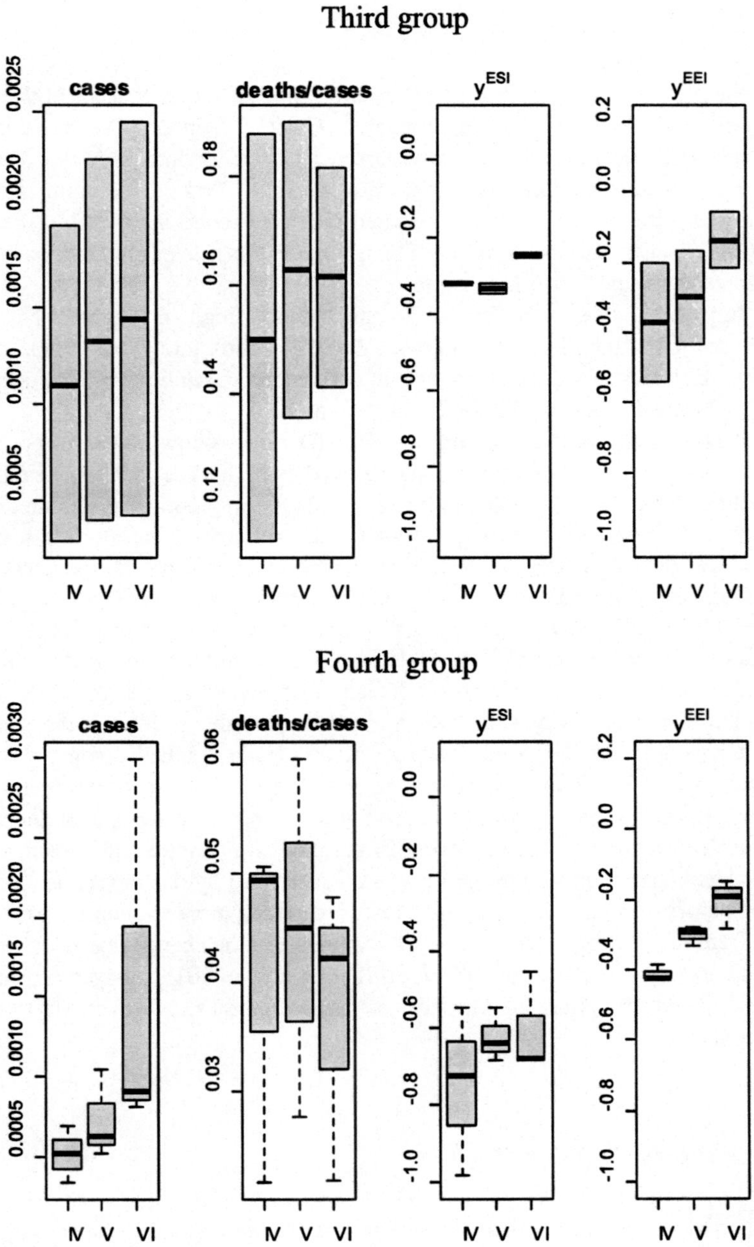

Fig. 5 (continued)

6.3 Third Quarter

As in the previous quarters, in the third quarter, the first group consists of the most countries, i.e.: Albania, Austria, Bulgaria, Croatia, Cyprus, Czech Republic, Denmark, Estonia, Finland, Germany, Greece, Latvia, Lithuania, Malta, Poland, Portugal, Romania, Serbia, Slovakia, Slovenia and Turkey. Belgium, France, Hungary and the Netherlands are classified into the second group, while Ireland, Spain and Sweden into the third one. The last group consists of only two countries, i.e. Montenegro and North Macedonia.

In July, death rates were low in groups 1 and 4, high in group 2. This was accompanied by high morbidity in groups 3 and 4. Fourth group was characterized by large decreases in values of changes in both economic indicators. The decrease in ESI was greater than in the EEI in all groups.

In August, both variables describing COVID-19 took low values in countries from first group. Second group was characterized by highest death rates, third group by high morbidity and moderate death rates, while fourth group—by high morbidity and low death rates. Like in the previous month, countries from the fourth group were characterized by greatest decreases of both economic indicators. Countries' characteristics in September are similar to the previous months.

Results lead to the following general description of changes in values of variables for the identified groups (Fig. 6). In case of the countries belonging to the first group, morbidity is increasing and death rates are decreasing, percentage change of ESI is fairly stable growing little, which translates into the declines in the value of the index getting smaller, while the size of the decline in EEI compared to the previous year is lower in August than in July and remains on a similar level in September. In the second group, the mortality was decreasing, the size of decline in ESI was decreasing, and the decline in EEI got smaller in August and remained on a similar level next month. Third group was characterized by decreasing death rates, and the values of both economic indicators were getting more and more close to the levels from the previous year. In case of both of them, the change was more visible in September. The decrease of EEI and ESI compared to 2019 was getting smaller also in the fourth group. In the case of these countries, however, death rates fluctuated.

6.4 January–September 2020

Again, countries were classified into four groups. The first group included: Albania, Bulgaria, Cyprus, Greece, Hungary, Latvia, Lithuania, Romania, Serbia, Slovakia and Turkey. Second one consisted of: Austria, Croatia, Czech Republic, Denmark, Estonia, Finland, Germany, Ireland, Malta, Poland, Portugal and Slovenia. Belgium, France, Netherlands, Spain and Sweden were classified into the third group, while Montenegro and North Macedonia, into the fourth one.

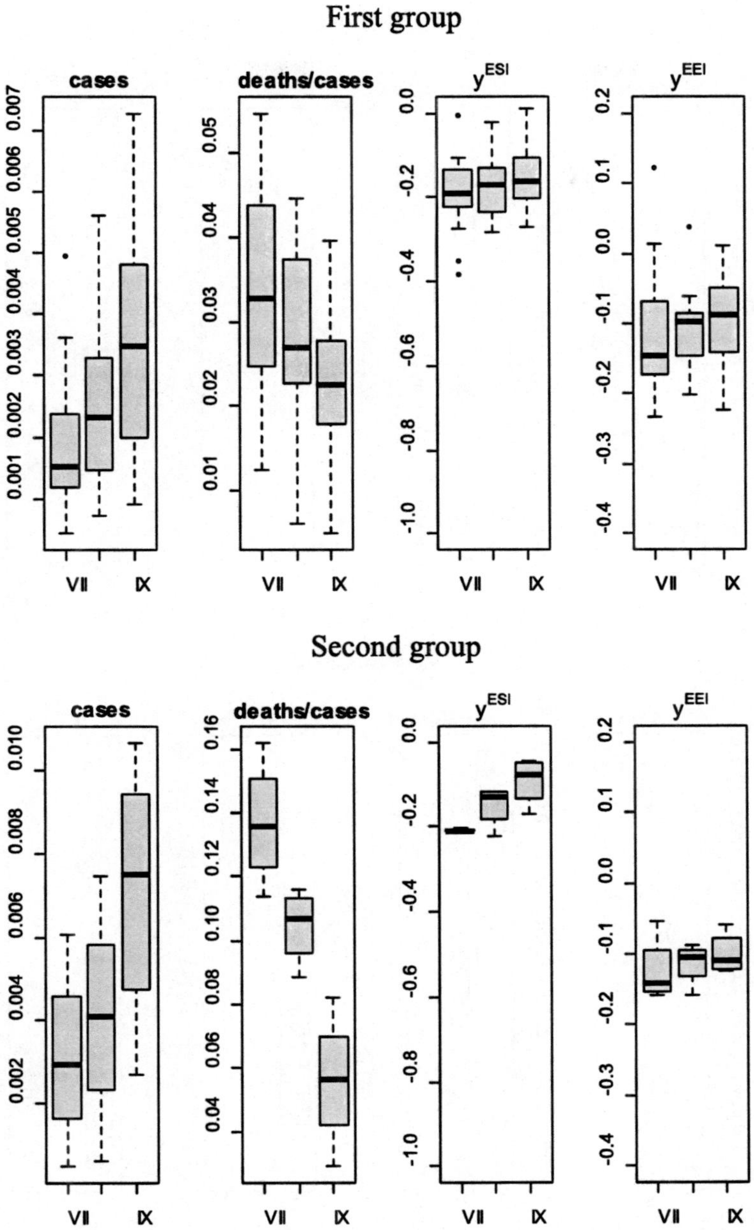

Fig. 6 Evolution of values of variables in the third quarter of 2020. *Source* Authors' calculations based on ECDC and Eurostat data

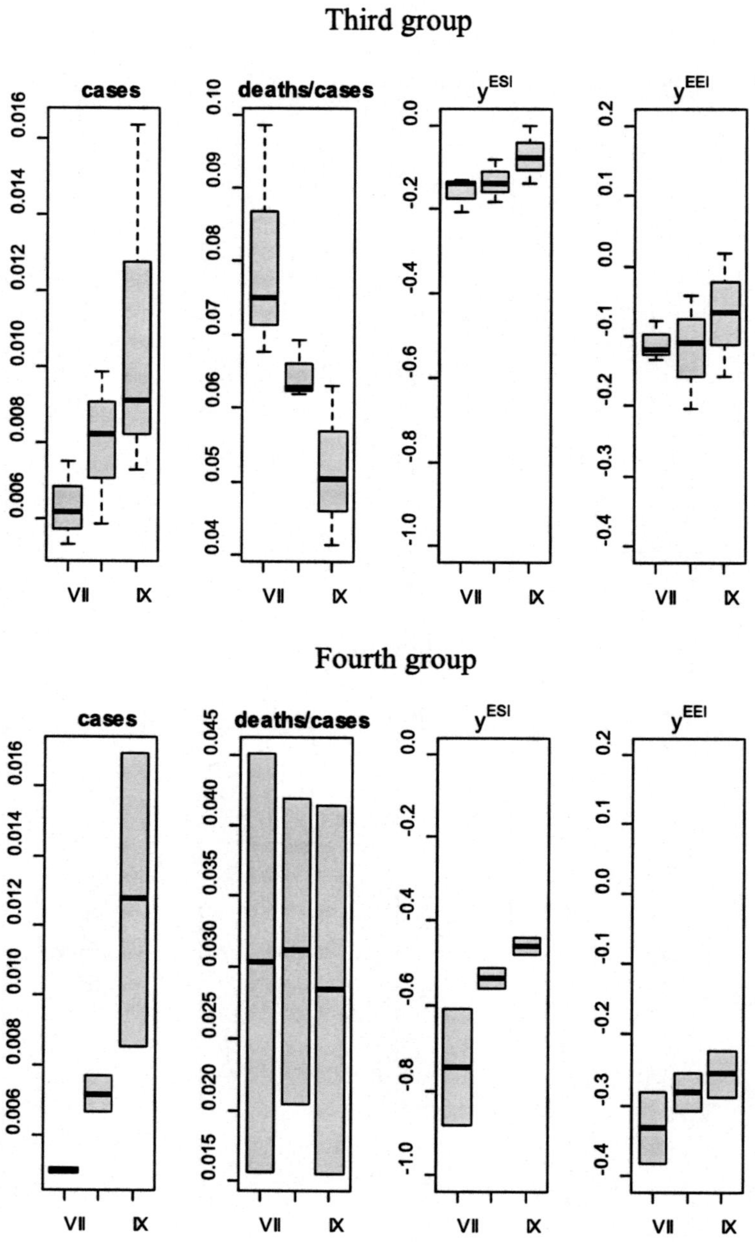

Fig. 6 (continued)

Differences in morbidity appeared in February. In March, group 3 starts to stand out in terms of death rates.

Morbidity in April was low in groups 1 and 4, high in 3. Third group was also characterized by the highest death rates. Fourth group was characterized by relatively large decreases in ESI.

In May, there can be seen low cases and death rates in groups 1, 4. Both of these values are high in the third group, while in the second one death rates are low. Analysing values of economic indicators, we can still see relatively large decreases in ESI in the fourth group and low decreases in EEI in the first group.

In June, morbidity and mortality were low in group 4 and high in group 3. Fourth group was also still characterized by large decreases in ESI while the first one was described by low decreases in both of economic variables.

In July in the first group, morbidity and mortality were low, in the third one high, while the fourth one could be described by high relative number of cases and low death rates. Decreases in both economic indicators were high in the fourth group and low in the first one.

In August, situation was similar: both variables describing COVID-19 situation were relatively low in groups 1 and 2 while high in group 3. Fourth group can be described by high relative number of cases and low death rates. In the fourth group, decreases in both EEI and ESI were large. On contrary, in the first group, they were small. The situation did not change to a great degree in September.

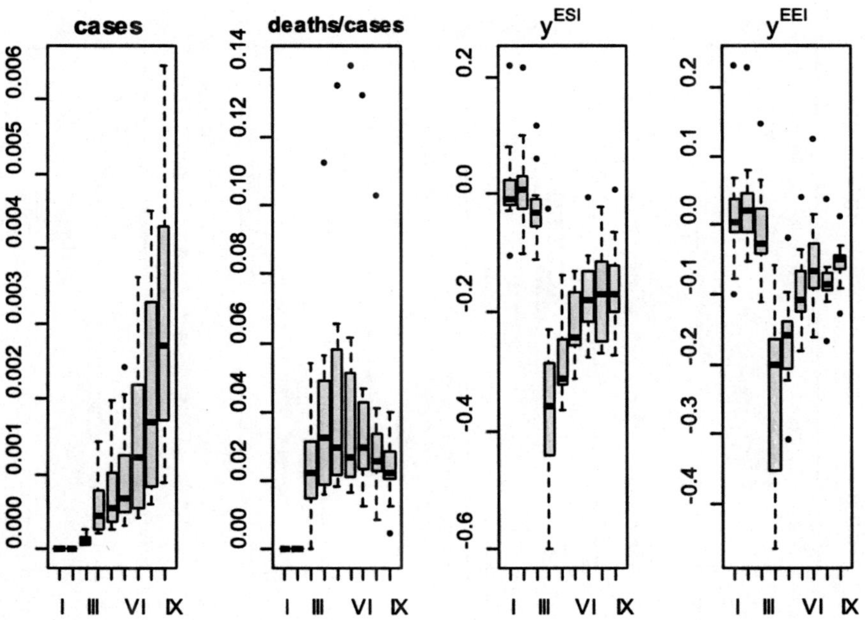

Fig. 7 Evolution of values of variables in the first three quarters of 2020 (first group of countries). *Source* Authors' calculations based on ECDC and Eurostat data

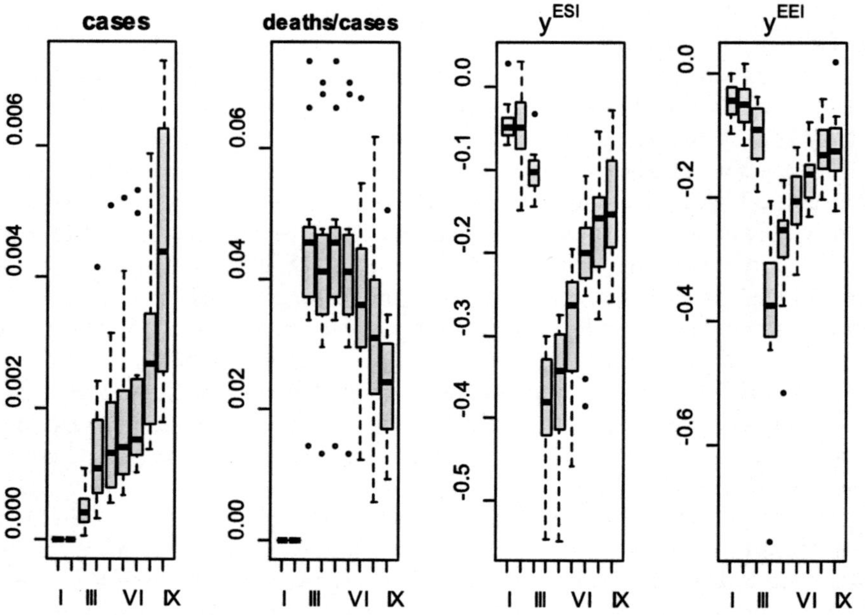

Fig. 8 Evolution of values of variables in the first three quarters of 2020 (second group of countries). *Source* Authors' calculations based on ECDC and Eurostat data

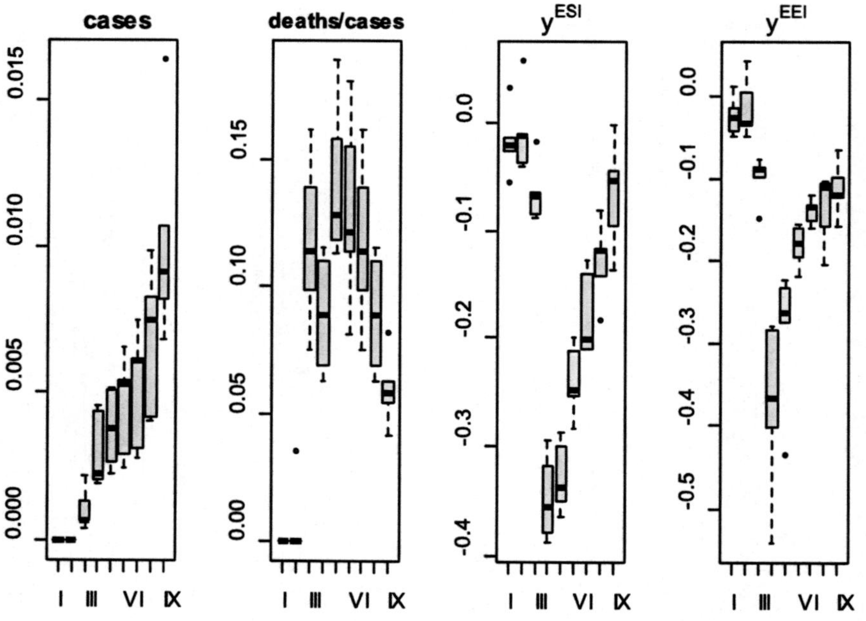

Fig. 9 Evolution of values of variables in the first three quarters of 2020 (third group of countries). *Source* Authors' calculations based on ECDC and Eurostat data

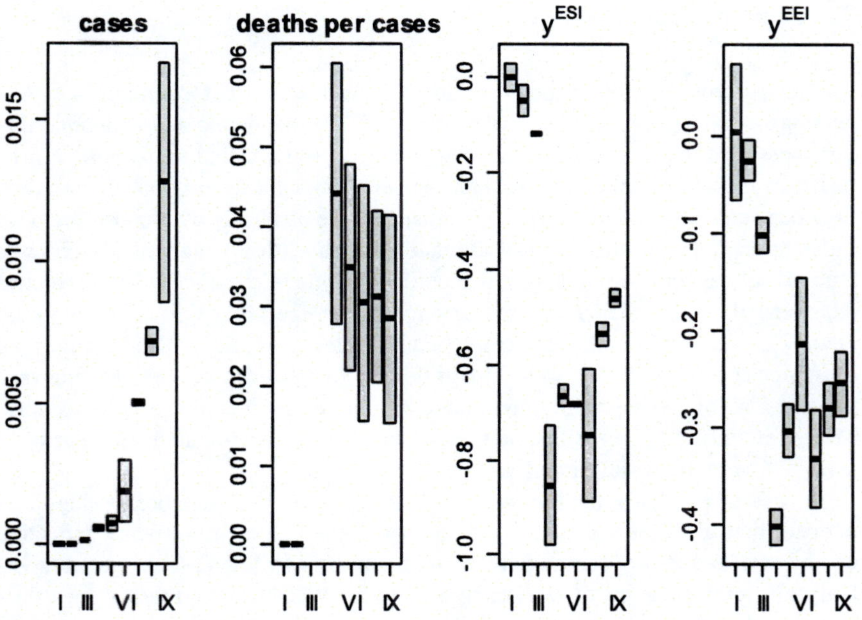

Fig. 10 Evolution of values of variables in the first three quarters of 2020 (fourth group of countries). *Source* Authors' calculations based on ECDC and Eurostat data

Analysing the changes of the variables in given groups, it can be noticed that there are two distinct periods visible for economic indicators. That finding justifies the division of the research period adopted in this paper. The death rate showed a slight decline after April and then fluctuations. After a sharp collapse of EEI and ESI in April, the values of both y^{ESI} and y^{EEI} grew, and in the last analysed quarter, they began to stabilize (y^{ESI}) with fluctuations (y^{EEI}) (Fig. 7).

In case of the second group, the death rates were fluctuating from March to June, and then, a decrease may be noticed. The values of both economic indicators are increasing since April (Fig. 8).

Death rates in countries from the third group fluctuated, but in last months, a decrease may be noticed. The values of both economic variables were rising in this group as well (Fig. 9).

Countries classified into the fourth group may be characterized by decreasing death rates and increasing (with some fluctuations) values of both economic variables (Fig. 10).

7 Conclusions

Results presented in the paper lead to the conclusions that countries are very heterogenous in a way they experienced COVID-19 pandemic and economic disturbances resulting from it. Despite that, some general concluding remarks can be made. In the first quarter, main differences between countries originated from their previous economic situation. Changes caused by the pandemic overlapped previous differences. It seems that a COVID-related variable (death rates in March) was crucial only in case of one group identified by us. Countries from all groups experienced similar changes. In the second quarter, the groups differed from each other due to the cumulated cases, death rates and ESI. The changes of economic-related variables were also different. In the third quarter, all analysed variables seem to have an important contribution in identification of groups. Moreover, except for some fluctuations, values of economic-related variables (y^{ESI} and y^{EEI}) were increasing in all groups.

In case of the last grouping, covering period January–September, it should be noticed, that after the collapse in March/April, the values of variables reflecting condition of economies started to increase in most of identified groups of countries. Only in group made up of Montenegro and North Macedonia, these values fluctuated, but at the end of the analysed period the increase could be noticed. It is important to note that the ratio of deaths to COVID-19 cases also fell down, as the procedures have been worked out in order to treat more effectively, but the trend is not always clear. Even though the economies suffered because of the COVID-induced uncertainty, it seems that the sharp declines in economic indicators resulted from containment measures introduced by various governments.

The presented impact of morbidity and mortality on economic indicators is essential for decision-makers and politicians to make effective economic and social decisions. It will be particularly important to restore jobs, as well as to create new ones in such a way as to return to the lost level of welfare as soon as possible.

Governments have taken up the fight against COVID-19 and the economic crisis it caused. Of course, the most important thing is health and focus on the effective functioning of health care. It has become imperative that the government provides support to households and businesses to survive the pandemic. However, the challenge for governments is to transform economies into post-pandemic situations. Getting people back to work becomes the most important issue.

In this extremely difficult situation, it seems that public investments will play the greatest role. Increasing these investments in developed economies, and especially developing economies, will enable economic recovery, which will allow to improve the functioning of economies, raise the quality of life and restore the sense of security for all people. Public investment can create millions of jobs in the short and long term, which will boost economic growth and development. According to IMF, public investment should increase confidence in expected economic growth, especially if these investments are significant for the economy and of high quality. An additional condition is that public and private debt should not weaken the private sector response to this stimulus. Given these conditions, an increase in public investment by 1% of GDP may increase GDP by 2.7%, private investment by 10% and employment by 1.2% (Gaspar et al. 2020).

References

Alon T, Doepke M, Olmstead-Rumsey J, Tertilt M (2020) The impact of COVID-19 on gender equality. NBER working paper 26947. http://www.nber.org/papers/w26947 or http://dx.doi.org/10.3386/w26947. Accessed 7 Oct 2020

Ashraf BN (2020) Economic impact of government interventions during the COVID-19 pandemic: international evidence from financial markets. J Behav Exp Finance 27: https://doi.org/10.1016/j.jbef.2020.100371

Baker SR, Bloom N, Davis SJ, Terry SJ (2020) COVID-induced economic uncertainty. NBER working paper 26983, April 2020. http://www.nber.org/papers/w26983 or http://dx.doi.org/10.3386/w26983. Accessed 7 Oct 2020

Blundell R, Costa Dias M, Joyce R, Xu X et al (2020) COVID-19 and inequalities. Fisc Stud 41:291–319

Charrad M, Ghazzali N, Boiteau V, Niknafs A (2014) NbClust: an R package for determining the relevant number of clusters in a data set. J Stat Softw 61(6):1–36. https://doi.org/10.18637/jss.v061.i06

Chetty R, Friedman JN, Hendren N, Stepner M, The Opportunity Insights Team (2020) How did COVID-19 and stabilization policies affect spending and employment? A new real-time economic tracker based on private sector data. NBER working paper 27431. http://www.nber.org/papers/w27431 or http://dx.doi.org/10.3386/w27431. Accessed 7 Oct 2020

Clemens J, Veuger S (2020) Implications of the COVID-19 pandemic for state government tax revenues. NBER working paper 27426. http://www.nber.org/papers/w27426 or http://dx.doi.org/10.3386/w27426. Accessed 7 Oct 2020

Deb P, Furceri D, Ostry JD, Tawk N (2020) The economic effects of COVID-19 containment measures (July 2020). CEPR discussion paper DP15087. https://ssrn.com/abstract=3661431. Accessed 7 Oct 2020

Del Vecchio Blanco G, Calabrese E, Biancone L, Monteleone G, Paoluzi OA (2020) The impact of COVID-19 pandemic in the colorectal cancer prevention. Int J Colorectal Dis 35:1951–1954. https://doi.org/10.1007/s00384-020-03635-6

European Centre for Disease Prevention and Control (ECDC) (2020) COVID-19 data. https://www.ecdc.europa.eu/en/COVID-19/data. Accessed 10 Oct 2020

European Commission Directorate-General for Economic and Financial Affairs (2020) The joint harmonised EU programme of business and consumer surveys. User guide (updated February 2020). https://ec.europa.eu/info/sites/info/files/bcs_user_guide_2020_02_en.pdf. Accessed 7 Oct 2020

European Union (1995–2020) Eurostat database. https://ec.europa.eu/eurostat/. Accessed 7 Oct 2020

Farid Y, Schettino M, Kapila AK, Hamdi M, Cuylits N, Wauthy P, Ortiz S (2020) Decrease in surgical activity in the COVID-19 pandemic: an economic crisis. Br J Surg 7:10. https://doi.org/10.1002/bjs.11738

Fernandes N (2020) Economic effects of coronavirus outbreak (COVID-19) on the world economy. https://ssrn.com/abstract=3557504 or http://dx.doi.org/10.2139/ssrn.3557504. Accessed 7 Oct 2020

Gaspar V, Mauro P, Pattillo C, Espinoza R (2020) Public investment for the recovery. https://blogs.imf.org/2020/10/05/public-investment-for-the-recovery/. Accessed 26 Oct 2020

Hall MC, Prayag G, Fieger P, Dyason D (2020) Beyond panic buying: consumption displacement and COVID-19. J Serv Manag. https://doi.org/10.1108/JOSM-05-2020-0151

International Labour Organization (2020) ILO monitor: COVID-19 and the world of work. Sixth edition. Updated estimates and analysis. https://www.ilo.org/wcmsp5/groups/public/—dgreports/—dcomm/documents/briefingnote/wcms_755910.pdf. Accessed 26 Oct 2020

Kong E, Prinz D (2020) Disentangling policy effects using proxy data: which shutdown policies affected unemployment during the COVID-19 pandemic? J Public Econ 189: https://doi.org/10.1016/j.jpubeco.2020.104257

Kumar S (2020) Use of cluster analysis to monitor novel coronavirus-19 infections in Maharashtra, India. Indian J Med Sci 72(2):44–48. https://dx.doi.org/10.25259/IJMS_68_2020

Maechler M, Rousseeuw P, Struyf A, Hubert M, Hornik K (2019) Cluster: cluster analysis basics and extensions. R package version 2.1.0

Maugeri A, Barchitta M, Agodi AA (2020) Clustering approach to classify italian regions and provinces based on prevalence and trend of SARS-CoV-2 cases. Int J Environ Res Public Health 17:5286. https://doi.org/10.3390/ijerph17155286

McClain JO, Rao VR (1975) CLUSTISZ: a program to test for the quality of clustering of a set of objects. J Mark Res 12(4):456–460

McKibbin WJ, Roshen F (2020) The global macroeconomic impacts of COVID-19: seven scenarios. CAMA working paper 19/2020. https://ssrn.com/abstract=3547729 or http://dx.doi.org/10.2139/ssrn.3547729. Accessed 7 Oct 2020

Norouzi N, Zarazua de Rubens G, Choupanpiesheh S, Enevoldsen P (2020) When pandemics impact economies and climate change: exploring the impacts of COVID-19 on oil and electricity demand in China. Energy Res Soc Sci 68:101654. https://doi.org/10.1016/j.erss.2020.101654

Ozili PK, Arun T (2020) Spillover of COVID-19: impact on the global economy (March 27, 2020). Available at SSRN. https://ssrn.com/abstract=3562570 or http://dx.doi.org/10.2139/ssrn.3562570. Accessed 7 Oct 2020

Pak A, Adegboye OA, Adekunle AI, Rahman KM, McBryde ES, Eisen DP (2020) Economic consequences of the COVID-19 outbreak: the need for epidemic preparedness. Front Public Health 8:241. https://doi.org/10.3389/fpubh.2020.00241

Palomino JC, Rodríguez JG, Sebastian R (2020) Wage inequality and poverty effects of lockdown and social distancing in Europe. Eur Econ Rev 29:103564. https://doi.org/10.1016/j.euroecorev.2020.103564

R Core Team (2020) R: a language and environment for statistical computing. R Foundation for Statistical Computing, Vienna, Austria. https://www.R-project.org/. Accessed 3 July 2020

Radwan A, Radwan E (2020) Social and economic impact of school closure during the outbreak of the COVID-19 pandemic: a quick online survey in the Gaza strip. Pedagogical Res 5(4): em0068. https://doi.org/10.29333/pr/8254

Rossi R, Socci V, Talevi D, Mensi S, Niolu C, Pacitti F, Di Marco A, Rossi A, Siracusano A, Di Lorenzo G (2020) COVID-19 pandemic and lockdown measures impact on mental health among the general population in Italy. Front Psychiatry 11:790. https://doi.org/10.3389/fpsyt.2020.00790

Shammi M, Bodrud-Doza M, Islam ARMT, Rahman MM (2020) Strategic assessment of COVID-19 pandemic in Bangladesh: comparative lockdown scenario analysis, public perception, and management for sustainability. Environ Dev Sustain. https://doi.org/10.1007/s10668-020-00867-y

Silva PC, Batista PV, Lima HS, Alves MA, Guimarães FG, Silva RC (2020) COVID-ABS: an agent-based model of COVID-19 epidemic to simulate health and economic effects of social distancing interventions. Chaos Solitons Fract 139: https://doi.org/10.1016/j.chaos.2020.110088

The Committee for the Coordination of Statistical Activities (CCSA) (2020) How COVID-19 is changing the world: a statistical perspective. Available at https://unstats.un.org/unsd/ccsa/documents/COVID19-report-ccsa.pdf. Accessed 6 Sep 2020

Topcu M, Gulal OS (2020) The impact of COVID-19 on emerging stock markets. Finance Res Lett 36: https://doi.org/10.1016/j.frl.2020.101691
Ward JH (1963) Hierarchical grouping to optimize an objective function. J Am Stat 58:236–244
Zarikas V, Poulopoulos SG, Gareiou Z, Zervas E (2020) Clustering analysis of countries using the COVID-19 cases dataset. Data Brief 31: https://doi.org/10.1016/j.dib.2020.105787

Modelling the Risk of Foreign Divestment in the Visegrad Group Countries During the COVID-19 Pandemic

Marcin Salamaga

Abstract The COVID-19 pandemic is an unprecedented global phenomenon posing a risk to the health and life of people around the world. It has forced governments of numerous countries to trigger various activities in order to stop the spread of the virus. Certain sectors of the economy in various countries had to freeze operations. Some industries limited their production or service provision, while domestic and international supply chains were interrupted, which increased economic risk. Such conditions are not favourable to foreign investors and make the odds of foreign divestment real. A lot suggests the pandemic will not subside quickly, and it will probably recur. Therefore, a question arises regarding the influence of the coronacrisis on global foreign investment in the near future. The author of this article is interested in the investment market of the Visegrad Group countries. The article is aimed at analysing the risk of foreign divestment in individual countries of the Visegrad Group using logit models, considering the effects of interactions among independent variables. This will allow for identifying material factors affecting the risk of reduced inflow of FDIs and for comparing the risk of foreign divestment between different countries and different economy sectors.

Keywords COVID-19 · Logistic regression · Foreign divestment · FDI · Visegrad group

1 Introduction

Foreign direct investments (FDIs) are an inherent part of the international economic system, and for many countries they constitute a fundamental element to support economic development. Their strong correlation with an international supply chain, globalisation processes and international distribution of labour has come to be seen

M. Salamaga (✉)
Cracow University of Economics, Kraków, Poland
e-mail: salamaga@uek.krakow.pl

K. Jajuga et al. (eds.), *Data Analysis and Classification*, Studies in Classification, Data Analysis, and Knowledge Organization,
https://doi.org/10.1007/978-3-030-75190-6_18

as a weakness when the COVID-19 pandemic struck the global economy. The interruption of international supply chains and the restrictions imposed on various branches and sectors of the economy have forced many businesses to limit their production and provision of services. Real losses, economic recession and a change of consumption habits have made many investors limit or suspend investments in areas which are more susceptible to the effects of the pandemic. A reaction by investors to the consequences of the COVID-19 pandemic is divestment, which is about restricting the previous scope and scale of operations of a business being the target of direct investment as a result of abandoning part of its operations or a complete transfer of the enterprise by its investor (Borga et al. 2019; Martins and Esteves 2008; Shin 2000). In practical terms, divestment means a change of ownership (co-ownership) rather than the closing of a business. An inclination for such market behaviour usually intensifies in cases of augmented economic or political turbulence causing increased investment uncertainty and thus abandoning all investment. It seems that the coronavirus pandemic has brought new determinants of FDI reduction and might have revalued the importance of previous risk factors associated with this form of investment. For this reason, it is necessary to evaluate the level of risk factors that cause an increased investment risk during a pandemic. The article presents a proposed application of a logit model with effects of interaction for the purpose of analysing the tendency of foreign divestment in the Visegrad Group countries. The purpose of the article is to identify factors and factor interactions which increase or decrease to the greatest extent the odds of divestment in the compared countries amid the global coronavirus pandemic. The modelling of this phenomenon bears certain difficulties due to, for example:

- unprecedented SARS-CoV-2 virus, which has led to economic crises and recession,
- extent of the coronacrisis further exacerbated by globalisation and economic connections as part of international labour distribution (which implies a need to take external factors into account in forecasts),
- the course of the coronacrisis differing from earlier contemporary worldwide economic crises,
- difficulty predicting the direction and intensity of further development of the coronavirus pandemic on a global scale,
- difficulty obtaining up-to-date, complete and official data regarding the inflow of FDIs due to the passing of a short period of time from the start of the pandemic.

Nevertheless, attempts to conduct such research seem necessary in order to monitor tendencies in the investment market on an ongoing basis. They may also be used as support for decision-makers in the management of foreign investments. The analysis has been based on data from a survey among foreign enterprises conducting direct investments in at least one of the four Visegrad Group countries.

2 Review of Literature

There is extensive literature on factors which encourage or discourage foreign direct investment. Research on foreign divestment is less elaborate but has still been conducted for many years. Publications of this type usually become more frequent during local or global economic (financial) crises. Most frequently, researchers analyse causes of divestment and assess their scale and consequences, and indicate the most significant determinants of divestment. Groups of host countries of FDIs are most frequently examined, whereas analyses of economies of individual countries are more rare. Analyses mostly cover microeconomic and macroeconomic factors. The importance of factors typical of the parent company and its affiliates in host countries of FDIs is noted, for example, by Norbäck et al. (2015); Berry (2010); Shimizu and Hitt (2005); Bergh (1997); Hamilton and Chow (1993); Markides (1992); Pashley and Philippatos (1990); Harrigan (1981). Researchers prove that the size of the affiliate or the parent is closely associated with divestment (Norbäck et al. 2015; Berry 2010; Shimizu and Hitt 2005). On the other hand, poor results of the mother company (investor) in host countries of FDIs may also be conducive to divestment (Berry 2010; Markides 1992; Pashley and Philippatos 1990; Harrigan 1981). An unambiguous relation between financial results of affiliates and divestments has not been confirmed (Berry 2013; Hamilton and Chow 1993; Markides 1992; Berry 2010, 2013; Bergh 1997). Researchers have shown that diversification of business operations is also conducive to the withdrawal of FDIs. Furthermore, the level of internationalisation of transnational enterprises, which are the chief providers of capital, is an important diversification factor (Norbäck et al. 2015; Berry 2010, 2013). Borga et al. (2019) prove that the risk of foreign divestment increases along with the intensification of enterprise internationalisation. Norbäck et al. (2015) and Berry (2010, 2013) have proven a positive influence of the presence of other affiliates in the host country of FDIs on the level of divestment, while Borga et al. (2019) argue that this dependence is not unequivocal. Research on divestment takes into account factors at the economic sector level in which branches of transnational businesses operate. Berry (2010); Sembenelli and Vannoni (2000) have demonstrated that the direction of the impact of the condition of the whole sector on the level of divestment is not clear. Jovanovic and MacDonald (1994) prove that technological changes increase the odds of divestment, while Norbäck et al. (2015) and Chatterjee et al. (2003) argue that institutional changes at the sector level increase the chances of divestment. Research shows that GDP, level of economy openness, level of salaries and wages, currency exchange rates, inflation, political stability, membership of a country in economic associations, free trade zones or others belong to macroeconomic factors which may have an influence on divestment. A negative relation between economic growth and divestment has been proven; for example, papers by Norbäck et al. (2015), Blake and Moschieri (2017), Berry (2010). Norbäck et al. (2015) and Borga et al. (2019) have demonstrated that a greater openness of an economy encourages divestment. Higher salaries and wages, and greater employee skills (requiring the appropriate financial reward), in turn, may decrease product competitiveness and incline investors to divest (Berry 2010; Norbäck et al. 2015). The impact of inflation on foreign divestment in OECD countries has not been confirmed in papers by Borga et al. (2019). Norbäck et al. (2015), Berry (2010, 2013) and Borga et al. (2019) do not prove an unequivocal influence of membership in international economic associations or being parties to bilateral trade agreements on divestment. Papers on divestment were written before COVID-19 was declared a pandemic, so researchers did not take into account the risk of divestment resulting from completely new circumstances in the global economy since March 2020. At present, in order to correctly diagnose divestment risk, one needs to consider factors related to the direction of the development of the COVID-19 pandemic and potential economic restrictions imposed by governments of various countries. If the pandemic persists, this may affect risk levels of individual macroeconomic and microeconomic factors which are taken into account in research undertaken in case of "ordinary economic crises". This entails the need to assess the risk of divestment generated both by factors known from earlier research and new factors arising out of the global health crisis.

3 Research Methodology

The logit model containing interactions between independent variables which represent selected macroeconomic factors and economic areas to which FDIs are addressed is used for modelling the risk of foreign divestment in the Visegrad Group countries. Dependent variable Y is a binary variable which takes on a value of 1 if a decision on foreign divestment is made; otherwise, its value is 0. The following independent variables are taken into consideration in the models: ΔGDP_c—investor's expectation as to the change in GDP, ΔGDP—investor's expectation as to the change in GDP in the host country of FDIs, ΔEx—depreciation of the domestic currency in relation to the euro (in Slovakia—in relation to the American dollar), ΔOMI—change in the market openness index, ΔLC—unit labour costs, Δi—change in the inflation level, $\Delta R\&D$—change in domestic expenditures on research and development in relation to the GDP, ΔDC—change in the distance between the capital of the host country of FDIs and the capital of the investor's country, M—industrial sector and the construction industry, S—services sector, IT—IT industry, O—other sectors, $L1$—imposing of minor restrictions on the economy in relation to epidemic risk, $L2$—introduction of moderately burdensome restrictions and a partial freeze of the economy in relation to the development of the COVID-19 pandemic, $L3$—introduction of rigorous restrictions and a considerable economic freeze. The mentioned variables are measured on ordinal scales (ΔGDP_c, ΔGDP, ΔOMI, ΔLC, Δi, $\Delta R\&D$, ΔLC), primarily in accordance with the following variable coding method: 1—change in the value of variable by up to 5%, 2—change by 5–10%, 3—change by 10–15%,[1] etc. or on a nominal zero-one scale (*M, S, IT, O, L1, L2, L3*).

The basic logit model used in the paper takes into account second-order interaction and takes the following form (Jaccard 2001; Harrell 2001):

$$L(p) = \beta_0 + \beta_1 X_1 + \beta_2 X_2 + \cdots + \beta_k X_k + \beta_{k+1} X_1 X_2 + \beta_{k+2} X_1 X_3 + \cdots + \beta_{k+\binom{k}{2}} X_{k-1} X_k \quad (1)$$

where

$X_1, X_2, X_3, \ldots, X_k$—independent variables,
$\beta_1, \beta_2, \ldots \beta_{k+\binom{k}{2}}$—model parameters,
L(p)—logit, where $L(p) = \frac{p}{1-p}$ and $p = P(Y = 1|X_1, X_2, .., X_k)$.

Interactions in this model are entered using the product of independent variables where each is treated as a moderating variable. The presence of interactions in the model causes the odds ratio to be interpreted in a slightly different way based on the parameters next to individual independent variables than in the case of using parameters before products of interacting variables (Jaccard 2001; Harrell 2001). In the former case, the odds ratio calculated with reference to variable X_i is interpreted as the odds ratio for the group in which variable $X_i = 1$ and groups in which the variable assumes a value of 0, with the assumption that variables in products with variable X_i have a value of 0. For example, the odds ratio for a single change of variable X_1 is as follows (Jaccard 2001):

$$\psi(x_1, x_2, \ldots x_k; 1, 0, 0, \ldots, 0) = exp(\beta_1 + \beta_{k+1} x_2 + \beta_{k+2} x_3 + \cdots) = exp(\beta_1) \quad (2)$$

where: Ψ—odds ratio.

In the latter case, $exp(\beta_i)$ may be treated as a quotient of two odds ratios obtained by dividing the odds ratio in the group, where $X_i = 1$ and in the reference group where $X_i = 0$ if the value of moderating variable X_i amounts to 1 and the odds ratio

[1]By shifting to the rank scale, the metric scale has actually been weakened in this research in relation to the natural scale on which the variables presented herein are measured. The values of the rank variable were allocated as per a decrease in the value of original variables by every subsequent 5 percentage points, except for variables *ΔOMI, ΔLC, Δi*, for which the value of the rank variable is allocated as per the increase in these variables by each subsequent 5 percentage points. The values of the *ΔDC* variable are coded as follows: 1—distance up to 700 km, 2—distance from 700 to 1400 km, 3—distance from 1400 to 2100 km, etc.

in the group in which the value of the moderating variable takes 0, with the assumption that all other variables occurring in other products with variable X_i are zeroed (Jaccard 2001).

For the product of exemplary variables X_1 and X_2, where X_1 is the moderating variable, the X_2 expression $exp(\beta_{k+1})$ can be obtained by dividing the relevant odds ratios:

$$\frac{\psi(x_1, x_2 + 1, \ldots x_k; 1, 0, 0, \ldots, 0)}{\psi(x_1, x_2, \ldots x_k; 1, 0, 0, \ldots, 0)} = \frac{exp(\beta_1, \beta_{k+1}(x_2 + 1), \beta_{k+2}x_3 + \cdots)}{exp(\beta_1, \beta_{k+1}x_2, \beta_{k+2}x_3 + \cdots)} = exp(\beta_{k+1}) \quad (3)$$

The parameters of model (1) have been estimated using the method of maximum likelihood. Model estimation has been based on data from the survey conducted in April 2020 among almost 500 enterprises making direct investments in the Visegrad Group countries. Information about the companies participating in the survey was obtained, among others from Orbis and Zephyr databases. The respondents represented all major sectors of the economy. Efforts were made for the industry structure of the surveyed companies to reflect the industry structure of the Visegrad Group countries economies.

4 Results of Empirical Research

Results of the estimation of logit models for three divestment variants (insignificant—up to 20%, moderately pessimistic from 20 to 40%, and pessimistic above 40%) in the Visegrad Group countries are given in Tables 1, 2, 3, 4, 5 and 6. When choosing the final forms of the model, statistical significance of model parameters and model adaptation to empirical data have been considered (based on the McFadden's pseudo R^2 index, the Maximum Likelihood Test (LRT), and the Bayesian information criterion). Selected values of the odds ratio calculated based on the assessment of parameters of individual models have been interpreted.

4.1 Modelling of the Risk of Insignificant Foreign Divestment

Tables 1 and 2, present the results of modelling of the risk of insignificant foreign divestment in Visegrad Group countries.

Tables 1 and 2 show that the IT services sector is most susceptible to divestment in Poland and the Czech Republic (level of up to 20%). Investments in this sector

Table 1 Estimation results of the logit *ΔEx* model describing the inclination of foreign divestment in Poland and the Czech Republic by no more than 20% in comparison with the period before the COVID-19 pandemic

Poland			Czech Republic		
Variable	Coefficient	Odds ratio of divestment	Variable	Coefficient	Odds ratio of divestment
Constant	−12.154***	0.000	Constant	−9.245***	0.000
ΔGDP_c	0.124**	1.132	*ΔGDP_c*	1.108***	3.029
ΔGDP_PL	0.299**	1.349	*ΔGDP_CZ*	−0.063*	0.939
	0.214**	1.239	*ΔEx*	1.046**	2.848
ΔOMI	0.428**	1.534	*ΔOMI*	0.357***	1.428
ΔLC	0.368***	1.445	*ΔLC*	0.883***	2.418
M	−0.187**	0.829	*M*	0.698**	2.009
S	0.425**	1.530	*S*	0.931**	2.537
IT	0.845***	2.328	*IT*	1.491***	4.441
O	−0.245***	0.783	*O*	0.280*	1.324
Δi	0.021*	1.021	*Δi*	−0.209***	0.812
ΔR&D	0.078*	1.081	*ΔR&D*	0.651**	1.917
ΔDC	0.371***	1.449	*ΔDC*	0.206***	1.228
L1	0.754***	2.125	*L1*	0.461*	1.585
L2	0.354*	1.425	*L2*	−0.008**	0.992
L3	−0.265*	0.767	*L3*	0.534**	1.707
*IT*L1*	0.895**	2.447	*M*L1*	1.595***	4.929
*S*L1*	0.745**	2.106	*IT*L2*	1.733**	5.658
*ΔLC*L2*	0.642**	1.900	*ΔLC*L1*	0.209***	1.233
*ΔGDP_PL*M*	0.169***	1.184	*ΔGDP_CZ*S*	0.749***	2.115
*ΔOMI*M*	0.548**	1.730	–	–	–
*M*L2*	0.235**	1.265	–	–	–
McFadden's Pseudo R^2	0.216		McFadden's Pseudo R^2	0.315	
LR	−172.784		LR	−162.089	
LRT	95.2077***		LRT	149.075***	
BIC	482.290		BIC	448.471	

The significance of parameters at the level below 1, 5 and 10% was determined as follows: ***; **; *

lead to increased odds of divestment by approx. 133 and 344% ceteris paribus, respectively. The imposing of minor restrictions on the economy in relation to the epidemic risk is a factor bearing a considerable risk of foreign divestment at a level of up to 20% in Poland and Slovakia. Such restrictions increase the risk of minor divestment by approx. 113 and 395%, respectively. For Hungary, in turn, the greatest risk factor of increased divestment in the variant up to 20% is the reduction of the market openness index (risk increase by 314.7% with a decrease in OMI

Table 2 Estimation results of the logit model describing the inclination of foreign divestment in Slovakia and Hungary by no more than 20% in comparison with the period before the COVID-19 pandemic

Slovakia			Hungary		
Variable	Coefficient	Odds ratio of divestment	Variable	Coefficient	Odds ratio of divestment
Constant	−9.452***	0.000	*Constant*	−13.457**	0.000
ΔGDP_c	0.647***	*1.910*	*ΔGDP_c*	0.677**	1.967
ΔGDP_SK	−0.031*	*0.969*	*ΔGDP_HU*	1.089***	2.972
ΔEx	−0.006***	*0.994*	*ΔEx*	−0.273*	0.761
ΔOMI	0.134**	*1.143*	*ΔOMI*	1.422**	4.147
ΔLC	1.005**	*2.733*	*ΔLC*	1.271***	3.565
M	−0.530***	*0.588*	*M*	−0.261*	0.770
S	1.197**	*3.311*	*S*	0.020***	1.020
IT	0.557***	*1.745*	*IT*	0.375*	1.454
O	0.686***	*1.986*	*O*	−0.568***	0.567
Δi	−0.329*	*0.719*	*Δi*	−0.164**	0.848
ΔR&D	−0.325**	*0.722*	*ΔR&D*	−0.138**	0.872
ΔDC	1.205***	*3.336*	*ΔDC*	1.027***	2.794
L1	1.600*	*4.953*	*L1*	0.532*	1.702
L2	−0.141***	*0.868*	*L2*	1.286***	3.619
L3	−0.556**	*0.574*	*L3*	0.502**	1.652
*M*L1*	1.556***	*4.739*	*M*L1*	1.829***	6.228
*ΔOMI*M*	0.488*	*1.628*	*ΔGDP_HU*M*	1.548***	4.700
*ΔLC*L1*	1.228**	3.415	–	–	–
McFadden's Pseudo R^2	0.416		McFadden's Pseudo R^2	0.230	
LR	−193.589		LR	−182.551	
LRT	275.799***		LRT	109.057***	
BIC	499.042		BIC	470.751	

The significance of parameters at the level below 1, 5 and 10% was determined as follows: ***; **; *

index by another 5 percentage points), followed by the imposing of moderately burdensome restrictions on the Polish economy and its partial freeze due to the spread of the COVID-19 pandemic (increase in divestment risk by 262%). The predicted decrease in the GDP in the investor's country turned out to be a strong risk-bearing factor of divestment reduction at a level of no more than 20% in the Czech Republic (increase of risk by approx. 203% with the OMI index increased by each 5 percentage points). In Slovakia, in turn, the distance between the capital of the host country of FDIs and the capital of the investor's country measured on an ordinal scale turned out to be such a factor (risk increase by approx. 234% with the increase in distance by each 700 km). Considering the interaction of the factors of

Table 3 Estimation results of the logit model describing the inclination of foreign divestment in Poland and the Czech Republic by 20–40% in comparison with the period before the COVID-19 pandemic

Poland			Czech Republic		
Variable	Coefficient	Odds ratio of divestment	Variable	Coefficient	Odds ratio of divestment
Constant	−9.124***	0.000	*Constant*	−6.617***	0.000
ΔGDP_c	0.147*	1.158	*ΔGDP_c*	−0.097***	0.907
ΔGDP_PL	0.328**	1.388	*ΔGDP_CZ*	−0.037***	0.964
ΔEx	0.104***	1.110	*ΔEx*	0.995*	2.705
ΔOMI	0.621***	1.861	*ΔOMI*	1.301**	3.674
ΔLC	0.604***	1.829	*ΔLC*	0.299**	1.348
M	0.378***	1.459	*M*	0.096**	1.101
S	0.220***	1.246	*S*	0.954*	2.596
IT	0.417***	1.517	*IT*	0.153***	1.166
O	−0.192*	0.825	*O*	0.401**	1.494
Δi	−0.216*	0.806	*Δi*	0.637**	1.891
ΔR&D	−0.278**	0.757	*ΔR&D*	−0.664***	0.515
ΔDC	0.217***	1.242	*ΔDC*	−0.101**	0.904
L1	−0.045*	0.956	*L1*	0.456***	1.577
L2	0.642***	1.900	*L2*	0.176***	1.192
L3	0.255**	1.290	*L3*	−0.071**	0.931
*IT*L2*	0.763***	2.145	*IT*L3*	1.617***	5.036
*ΔOMI*L1*	0.632***	1.881	*ΔOMI*L2*	0.504*	1.655
*ΔLC*L3*	0.583***	1.791	*ΔGDP_CZ*M*	0.502**	1.651
*ΔGDP_PL*M*	0.501***	1.650	*S*L2*	0.495***	1.641
*ΔGDP_c*IT*	0.368***	1.445	*ΔBR*S*	1.270**	3.560
*S*L3*	0.314**	1.369	–	–	–
*ΔBR*IT*	−0.354*	0.702	–	–	–
McFadden's Pseudo R^2	0.225		McFadden's Pseudo R^2	0.322	
LR	−159.878		LR	−183.574	
LRT	92.833***		LRT	174.368***	
BIC	462.693		BIC	497.654	

The significance of parameters at the level below 1, 5 and 10% was determined as follows: ***; **; *

divestment to 20% in comparison with the pre-pandemic period, it can be concluded that in Poland the greatest increase in the risk of divestment (by approx. 145%) would result from FDIs in the IT industry combined with minor restrictions on the economy due to the epidemic risk. The combination of FDIs in the IT industry with moderate restrictions on the economy due to the epidemiological risk generates the greatest risk of divestment at a level up to 20% in the Czech Republic.

Table 4 Estimation results of the logit model describing the inclination of foreign divestment in Slovakia and Hungary by 20–40% in comparison with the period before the COVID-19 pandemic

Slovakia			Hungary		
Variable	Coefficient	Odds ratio of divestment	Variable	Coefficient	Odds ratio of divestment
Constant	−7.193***	0.000	*Constant*	−10.209***	0.000
ΔGDP_c	0.849***	2.337	*ΔGDP_c*	0.651*	1.918
ΔGDP_SK	0.094***	1.098	*ΔGDP_HU*	1.112***	3.039
ΔEx	0.723***	2.062	*ΔEx*	1.082**	2.951
ΔOMI	0.477***	1.611	*ΔOMI*	1.307**	3.696
ΔLC	1.358***	3.889	*ΔLC*	0.336***	1.399
M	0.330***	1.391	*M*	−0.073***	0.929
S	1.093***	2.982	*S*	0.129**	1.137
IT	0.192***	1.212	*IT*	0.231*	1.260
O	0.350***	1.420	*O*	0.801**	2.228
Δi	0.676***	1.967	*Δi*	−0.378***	0.685
ΔR&D	−0.314***	0.731	*ΔR&D*	−0.486**	0.615
ΔDC	1.004***	2.730	*ΔDC*	0.141***	1.152
L1	0.717***	2.049	*L1*	−0.350***	0.705
L2	1.391***	4.021	*L2*	0.397**	1.487
L3	0.014***	1.015	*L3*	0.033**	1.034
*S*L2*	1.445***	4.240	*S*L2*	0.387*	1.472
*ΔOMI*L3*	1.600***	4.951	*ΔOMI*L3*	0.313***	1.368
*ΔGDP_SK*S*	0.535***	1.707	*ΔGDP_HU*M*	0.545**	1.725
*ΔBR*IT*	0.451***	1.571	*IT*L2*	1.183***	3.266
–	–	–	*ΔBR*S*	0.351*	1.421
McFadden's Pseudo R^2	0.236		McFadden's Pseudo R^2	0.308	
LR	−193.663		LR	−174.163	
LRT	119.645***		LRT	155.035***	
BIC	511.618		BIC	472.619	

The significance of parameters at the level below 1, 5 and 10% was determined as follows: ***; **; *

In this case, the risk level increases by approx. 466%. When it comes to interactions between factors, the combination of FDIs in the industry with the imposing of minor restrictions on the economy due to the epidemic risk stands out in Slovakia and Hungary; the epidemic increases the risk of divestment at a level of up to 20% by approx. 374% in Slovakia and by 523% in Hungary. A decrease in the GDP of the investor's country turned out to be the factor bearing the greatest risk of minor divestment (by up to 20%) in the Czech Republic and the least risk in Poland. A decrease in the GDP of the host country of FDIs was most conducive to foreign divestment in Hungary, while in the Czech Republic and Slovakia it even reduced

Table 5 Estimation results of the logit model describing the inclination of foreign divestment in Poland and the Czech Republic by more than 40% in comparison with the period before the COVID-19 pandemic

Poland			Czech Republic		
Variable	Coefficient	Odds ratio of divestment	Variable	Coefficient	Odds ratio of divestment
stała	−7.024***	0.000	stała	−5.351***	0.002
ΔGDP_c	−0.098***	0.907	*ΔGDP_c*	−0.467**	0.627
ΔGDP_PL	0.063***	1.065	*ΔGDP_CZ*	1.028***	2.795
ΔEx	−0.058*	0.944	*ΔEx*	0.502***	1.652
ΔOMI	0.897***	2.452	*ΔOMI*	0.729**	2.072
ΔLC	0.458***	1.581	*ΔLC*	0.102*	1.107
M	0.854***	2.349	*M*	0.448**	1.565
S	0.421**	1.523	*S*	0.026***	1.026
IT	0.216*	1.241	*IT*	0.773**	2.167
O	−0.058***	0.944	*O*	−0.255**	0.775
Δi	−0.173**	0.841	*Δi*	0.636***	1.889
ΔR&D	−0.682**	0.506	*ΔR&D*	0.200***	1.221
ΔDC	0.318*	1.374	*ΔDC*	0.830**	2.294
L1	−0.232*	0.793	*L1*	0.495***	1.641
L2	0.109**	1.115	*L2*	0.696***	2.006
L3	0.869***	2.385	*L3*	1.591*	4.908
*IT*L3*	0.762***	2.143	*IT*L3*	0.663***	1.941
*ΔOMI*L3*	0.753***	2.123	*ΔLC*L3*	1.438**	4.213
*ΔLC*L3*	0.583***	1.791	*ΔGDP_CZ*S*	0.557***	1.746
*ΔGDP_PL*M*	0.468**	1.597	*ΔGDP_c*S*	0.196*	1.216
*ΔGDP_c*M*	0.871***	2.389	–	–	–
*IT*L3*	0.276**	1.318	–	–	–
*ΔO*Δi*	−0.316*	0.729	–	–	–
McFadden's Pseudo R^2	0.191		McFadden's Pseudo R^2	0.358	
LR	−154.577		LR	−171.736	
LRT	72.990***		LRT	191.531***	
BIC	452.090		BIC	467.765	

The significance of parameters at the level below 1, 5 and 10% was determined as follows: ***; **; *

the risk of divestment by up to 20%. A higher domestic currency rate in relation to the euro increased the risk of divestment in the Czech Republic the most and reduced the same risk the most in Hungary. An increase in the value of the market openness index of the host country of FDIs had the greatest influence on the increased odds of divestment in Hungary and the lowest in Slovakia. An increase in the cost of labour generated the greatest increase in the risk of foreign divestment in

Table 6 Estimation results of the logit model describing the inclination of foreign divestment in Slovakia and Hungary by more than 40% in comparison with the period before the COVID-19 pandemic

Slovakia			Hungary		
Variable	Coefficient	Odds ratio of divestment	Variable	Coefficient	Odds ratio of divestment
Constant	−9.068***	0.001	*Constant*	−4.093***	0.002
ΔGDP_c	−0.452***	0.636	*ΔGDP_c*	0.501***	1.651
ΔGDP_SK	0.654***	1.923	*ΔGDP_HU*	0.097***	1.102
ΔEx	−0.164***	0.848	*ΔEx*	−0.250**	0.779
ΔOMI	0.565***	1.760	*ΔOMI*	0.746**	2.108
ΔLC	0.437***	1.548	*ΔLC*	1.210**	3.354
M	0.831***	2.296	*M*	0.545*	1.724
S	0.090***	1.095	*S*	1.419***	4.132
IT	1.041***	2.831	*IT*	0.724***	2.063
O	0.766***	2.151	*O*	0.765**	2.148
Δi	−0.357***	0.700	*Δi*	−0.497***	0.609
ΔR&D	−0.974***	0.377	*ΔR&D*	−0.811*	0.444
ΔDC	0.233***	1.262	*ΔDC*	−0.162***	0.850
L1	−0.530***	0.589	*L1*	−0.318**	0.728
L2	0.767***	2.154	*L2*	−0.044***	0.957
L3	0.406***	1.501	*L3*	1.708*	5.517
*IT*L3*	0.759***	2.135	*IT*L2*	0.264***	1.302
*ΔLC*L3*	0.288***	1.334	*ΔLC*L2*	1.424**	4.156
*ΔGDP_SK*S*	1.354***	3.872	*ΔLC*L3*	1.395***	4.036
*ΔGDP_c*IT*	1.018***	2.769	*ΔGDP_c*S*	1.158**	3.184
McFadden's Pseudo R^2	0.198		McFadden's Pseudo R^2	0.271	
LR	−184.771		LR	−164.013	
LRT	91.234***		LRT	121.942***	
BIC	493.834		BIC	452.319	

The significance of parameters at the level below 1, 5 and 10% was determined as follows: ***; **; *

Hungary (by up to 20%) and the lowest in Poland. Only in Poland did an increase in inflation also increase the risk of divestment by up to 20%. In other Visegrad Group countries, it reduced this risk (the most in Slovakia). An increase in expenditures on research and development (R&D) increased the risk of divestment in Poland and the Czech Republic, and reduced it in Slovakia and Hungary. Foreign investment in the processing industry turned out to bear the risk of foreign divestment only in the Czech Republic, while in other countries it reduced the risk of divestment (the most in Slovakia). FDIs in the services sector generate the greatest increase in FDI reduction risk in the Slovakia and the lowest in Hungary. FDIs in the IT industry increase the risk of divestment in the Czech Republic the most and in Hungary the least.

4.2 *Modelling of the Risk of Moderate Foreign Divestment*

Tables 3 and 4 present the results of modelling of the risk of moderate foreign divestment in Visegrad Group countries.

The major factors of increased risk of divestment at a level from 20 to 40% in Poland were moderate restrictions imposed on the economy to prevent the spread of the pandemic and increased value of the market openness index. These restrictions increase the odds of foreign divestment by 90%. Increase of the OMI index by another 5 percentage points causes a higher risk of foreign divestment by approx. 86% ceteris paribus. In the Czech Republic, the major determinants of the increased chances of FDI reduction are restrictions on market openness and the higher exchange rate of the Czech koruna in relation to the euro. An increase in the OMI index by another 5 percentage points causes a greater risk of foreign divestment by approx. 267%, while an increase in currency rate by another 5 percentage points leads to greater chances of reducing FDIs by approx. 171% on average ceteris paribus. In Slovakia, the greatest increase in the risk of restricting FDIs was generated as a result of implementing moderate restrictions on the economy to stop the development of the pandemic and as a result of increased labour costs. Such restrictions increase the chances of foreign divestment by approx. 302%, while the reduction of labour costs by another 5 percentage points causes an increase in the risk of foreign divestment by approx. 289%. In Hungary, increased market openness and reduced GDP are factors which generate the greatest chances of foreign divestment. Reduction of the market openness index by another 5 percentage points causes an increase in the risk of foreign divestment by approx. 270%, while a reduction of Hungary's GDP by another 5 percentage points results in an increase in the odds of divestment by approx. 204% ceteris paribus on average. When analysing divestment factors at a level from 20 to 40% in comparison with the pre-pandemic period, it can be concluded that the greatest increase in the risk of divestment in Poland (approx. 115%) would result from the combination of FDI in the IT sector with the implementation of moderate economic restrictions in relation to the epidemiological risk. In the Czech Republic, in turn, the combination of FDIs in the IT industry with the implementation of strict restrictions on the economy due to the epidemic risk generates the greatest risk of foreign divestment. In this case, the risk increased by approx. 404%. In Slovakia, the combination of increased market openness with the implementation of strict restrictions on the economy due to the epidemic stands out from other factors interactions and increases the risk of divestment at a level of 20–40% by approx. 395% ceteris paribus. The combination of FDIs in IT with moderate economic restrictions to stop the spread of the pandemic in Hungary results in the greatest chances of FDI reduction (by approx. 227%) in comparison with other interactions among variables.

When comparing the Visegrad Group countries in terms of the extent of divestment risk resulting from individual macroeconomic variables, it can be concluded that:

- a decrease in the GDP of the investor's country was a factor bearing the greatest risk of divestment at a level of 20–40% in Slovakia and the lowest risk in the Czech Republic,
- the GDP in the host country of FDIs contributed to the greatest extent of foreign divestment in Hungary, while in the Czech Republic it even decreased the risk of divestment,
- an increase in the level of the exchange rate of the domestic currency in relation to the euro increased the divestment risk in Hungary to the greatest extent and to the least extent in Poland,
- an increase in the value of the market openness index (OMI) in the host country of FDIs increased the chances of divestment in Hungary the most and in Slovakia the least,
- an increase in labour costs generated the greatest increase in the risk of foreign divestment at a level of 20–40% in Slovakia and the lowest in the Czech Republic,
- an increase in inflation increased the risk of divestment in Slovakia the most and reduced the same risk the most in Poland and in Hungary,
- an increase in R&D costs in all compared countries reduced the risk of divestment in the Czech Republic the most and in Poland the least,
- foreign investment in the processing industry intensified the risk of divestment the most in Poland, while in Hungary they even reduced the divestment risk,
- FDIs in the services sector generate the greatest increase in the risk of FDI reduction in Slovakia and the least in Hungary,
- FDIs in the IT industry turn out to increase the risk of divestment the most in Poland and the least in the Czech Republic.

4.3 Modelling of the Risk of Considerable Foreign Divestment

Tables 5 and 6 present the results of modelling of the risk of considerable foreign divestment in Visegrad Group countries.

In Poland, moderate restrictions on the economy to counteract the development of the pandemic and FDIs in the industry sector were the major factors of increased risk of considerable divestment (by more than 40% in comparison with the pre-pandemic era). The restrictions increase the odds of foreign divestment by approx. 139% and FDIs in the industry sector increase these odds by approx. 135%

ceteris paribus, on average. In the Czech Republic, the major determinants of increased chances of FDI reduction included strict restrictions on the economy to counteract the pandemic and a decrease in the Czech GDP. If such restrictions are imposed, this will increase the odds of foreign divestment by 391%, while a decrease in GDP by another 5 percentage points will increase the risk of divestment by approx. 180% ceteris paribus. In Slovakia, the greatest increase in the risk of FDI limitation was generated by imposing moderate restrictions on the economy to suppress the pandemic and by investing in IT. If the discussed restrictions are imposed, it will increase the chances of foreign divestment by approx. 115%. Foreign direct investments in IT cause an increased risk of foreign divestment by approx. 183% ceteris paribus. In Hungary, the following factors generate the greatest chances of foreign divestment: strict restrictions on the economy to counteract the pandemic and FDIs in services. If these restrictions are imposed, this will result in an increased risk of foreign divestment by approx. 452%, whereas FDIs in services increase the risk of divestment by approx. 313% ceteris paribus. Considering the interaction of factors of considerable divestment in comparison with the pre-pandemic period, it can be concluded that the greatest increase in the risk of divestment (approx. 140%) would take place in Poland by the combination of FDIs in the processing industry with the decrease in GDP in the investor's country by another 5 percentage points. In the Czech Republic, in turn, strict restrictions on the economy in relation to the epidemic risk combined with an increase in unit labour costs generate the greatest risk of foreign divestment. In this case, increased risk amounts to approx. 321% with labour costs increased by another 5 percentage costs. In Slovakia, the combination of a decrease in GDP (by every 5 percentage points) with FDIs in services is the most important factor of increased risk of considerable divestment in comparison with other factor interactions; it increases the risk of divestment by approx. 287%. In Hungary, in turn, the combination of strict restrictions on the economy in relation to the epidemic risk with an increase in unit labour costs by every 5 percentage points has the greatest chances of FDI reduction (by approx. 304%) among all different interactions of variables.

When comparing the Visegrad Group countries in terms of the extent of divestment risk above 40% resulting from individual macroeconomic variables, it can be concluded that:

- a decrease in the GDP in the investor's country was a factor bearing a risk of divestment above 40% only with reference to Hungary,
- the GDP in the host country of FDIs contributed to foreign divestment the most in the Czech Republic and the least in Poland,
- increased currency exchange rate in relation to the euro increased the risk of divestment only in the Czech Republic,
- increased value of the market openness index in the host country of FDIs increased the odds of divestment in the Poland the most and in Slovakia the least,

- increased labour costs generated the greatest increase in the risk of foreign divestment above 40% in Hungary and the least in the Czech Republic,
- increased inflation intensified the risk of foreign divestment only in the Czech Republic,
- increased R&D expenditures reduced the risk of divestment in all compared countries, except for the Czech Republic,
- foreign investment in the processing industry increased the risk of divestment in Poland the most, whereas in the Czech Republic an increase in this risk was the least noticeable,
- FDIs in the services sector generate the greatest increase in the risk of reducing FDIs in Hungary and the lowest increase in the Czech Republic,
- FDIs in the IT industry turn out to increase the risk of foreign divestment the most in Slovakia and the least in Poland.

5 Conclusions

In light of the conducted research, it can be concluded that individual macroeconomic variables are associated with different risks depending on the extent of potential FDI reduction. The Visegrad Group countries differ in terms of the risk of divestment associated with macroeconomic variables. It turns out that, in general, FDIs in industry in the compared countries pose greater divestment risk when the potential reduction of FDI increases. Assuming divestment of up to 20% in the Visegrad Group countries, the implementation of minor economic restrictions to stop the spread of the pandemic is, on average, the FDI reduction risk factor associated with the greatest risk. For the expected divestment level of 20–40%, in turn, such factor is the market openness index, while for divestment in excess of 40% the greatest risk factor is the introduction of considerable pandemic-related economic restrictions. When it comes to the compared countries, the analysed variables led to a risk of divestment greater by up to 20% in the Czech Republic than in other countries, while in Slovakia the risk of such divestment turned out to be the lowest. In Slovakia, in turn, these variables were conducive to a higher average increase in the risk of pessimistic divestment (20–40%) than in other countries, while in the Czech Republic the risk of such divestment usually turned out to be the lowest. Furthermore, in the Czech Republic the analysed variables generated a higher risk of considerable divestment (above 40%) than in other countries, and in Slovakia the risk of such divestment turned out to be the lowest. With the introduction of interaction variables to the model, it has been possible to detect feedback between independent variables where the interaction of variables, especially economic restrictions to counteract the pandemic taken into account in logit models, in many cases, increases the risk of divestment multiple times. The estimated logit models may make it easier for decision-makers to choose the direction of export of direct investment. The changing economic environment and the development of the coronavirus pandemic, which is difficult to predict, justify the need to constantly update the results of such research and to repeat them in the future, which will allow for ongoing monitoring of the degree of investment risk of various macroeconomic factors and investment areas.

Acknowledgements Publication was financed from funds allocated to Faculty of Management (Cracow University of Economics) within grants to maintain research capacity.

References

Bergh DD (1997) Predicting divestiture of unrelated acquisitions: an integrative model of ex ante conditions. Strateg Manag J 18(9):715–731

Berry H (2010) Why do firms divest? Organ Sci 21(2):380–396

Berry H (2013) When do firms divest foreign operations? Organ Sci 24(1):246–261

Blake D, Moschieri C (2017) Policy risk, strategic decisions and contagion effects: firm-specific considerations. Strateg Manag J 38(3):732–750. https://doi.org/10.1002/smj.2509

Borga M, Flores PI, Sztajerowska M (2019) Drivers of divestment decisions of multinational enterprises—a crosscountry firm-level perspective. OECD working papers on international investment 3. https://doi.org/10.1787/5a376df4-en

Chatterjee S, Harrison JS, Bergh DD (2003) Failed takeover attempts, corporate governance and refocusing. Strateg Manag J 24(1):87–96

Hamilton RT, Chow YK (1993) Why managers divest—evidence from New Zealand's largest companies. Strateg Manag J 14(6):479–484

Harrell F (2001) Regression modeling strategies with applications to linear models. Logistic regression, and survival analysis. Springer, New York

Harrigan KR (1981) Deterrents to divestiture. Acad Manag J 24(2):306–323. https://doi.org/10.5465/255843

Jaccard J (2001) Interaction effects in logistic regression. Sage University Papers, Series: Quantitative Applications in the Social Sciences Thousand Oaks 07-135

Jovanovic B, MacDonald G (1994) The life cycle of a competitive industry. J Polit Econ 102(2): 322–347

Markides CC (1992) Consequences of corporate refocusing: ex ante evidence. Acad Manag J 35(2):398–412

Martins PS, Esteves LA (2008) Foreign ownership, employment and wages in Brazil: evidence from acquisitions. Divestments and Job Movers, IZA Discussion Papers 3542

Norbäck P-J, Tekin-Koru A, Waldkirchet A (2015) Multinational firms and plant divestiture. Rev Int Econ 23(5):811–845

Pashley MM, Philippatos GC (1990) Voluntary divestitures and corporate life-cycle: some empirical evidence. Appl Econ 22(9):1181–1196. https://doi.org/10.1080/00036849000000038

Sembenelli A, Vannoni D (2000) Why do established firms enter some industries and exit others? Empirical evidence on Italian business groups. Rev Ind Organ 17(4):441–456

Shimizu K, Hitt MA (2005) What constrains or facilitates divestitures of formerly acquired firms? The effects of organizational inertia. J Manag 31(1):50–72. https://doi.org/10.1177/0149206304271381

Shin S (2000) The foreign divestment factors in South Korea: an analysis of the trading sector. Multinational Bus Rev 8(2):98

Analysis of COVID-19 Dynamics in EU Countries Using the Dynamic Time Warping Method and ARIMA Models

Joanna Landmesser

Abstract The aim of the paper is to find the similarities in the evolution of time series for people infected with and died from COVID-19 in different EU countries using dynamic time warping (DTW) as a measure of the distance between time series. Using this method, a joint analysis of the number of infected and deceased will be performed. The DTW distance makes it possible to compare time series of different lengths, which is important when analyzing data for European countries because the virus has not spread to individual countries at the same time. After measuring the similarities between the time series, a hierarchical grouping for countries will be performed, which will allow us to find interesting patterns in the data. Then, ARIMA(p,d,q) models will be used to describe the dynamics of virus distribution in different EU countries. With these models, it is possible to gain knowledge about the mechanisms of pandemic evolution.

Keywords COVID-19 · Dynamic time warping · Hierarchical clustering · ARIMA models

1 Introduction

The COVID-19 outbreak is the biggest public health challenge since the Spanish flu in 1918. First cases of COVID-19 were reported in December 2019 in Wuhan, China. Since then, the disease has been spread to all continents and countries of the world. The first patient detected in Europe was a case from Italy on January 19, 2020. Since March 2020, the number of reported cases increased rapidly worldwide and the World Health Organization categorized COVID-19 as a pandemic (on March 11th, 2020). At the time of writing this post, the number of infected has reached 73,156,745 and the virus has killed 1,627,115 people (Worldometer, status as of December 15, 2020). Many countries have declared a state of emergency

J. Landmesser (✉)
Warsaw University of Life Sciences—SGGW, Warsaw, Poland
e-mail: joanna_landmesser@sggw.edu.pl

K. Jajuga et al. (eds.), *Data Analysis and Classification*, Studies in Classification, Data Analysis, and Knowledge Organization,
https://doi.org/10.1007/978-3-030-75190-6_19

following the surge in COVID-19 infections. All over the world, drastic countermeasures such as ordered school closure, case-based measures, the banning of public events, the encouragement of social distancing, and lockdowns, are being taken to contain the spread of the virus. These countermeasures were aimed at reducing the number of people infected and enabling effective health care activities. Different countries have adopted different restrictions and testing strategies. Disease rate projections allow recommendations on the effective date and the date of withdrawal from government intervention (e.g., Ruktanonchai et al. 2020).

The impact of COVID-19 on our daily lives and economy has led to a great scientific interest in this novel virus. In addition to medicine and biology, the COVID-19 outbreak also attracts attention in the field of mathematics and statistics. In recent months, the evolution of the epidemic in different countries has been analyzed, implementing mathematical models. Traditional predictive models for infectious diseases mainly include differential equation models and models for predicting time series.

The SIR epidemiological models are built as a system of differential equations for Susceptible-Infected-Removed sequences (Kermack and McKendrick 1927; Brauer et al. 2019). SIR models for COVID-19 have been presented in many publications (Vattay 2020; Fanelli and Piazza 2020; Roques et al. 2020; Rojas et al. 2020). For example, Fanelli and Piazza (2020) analyzed the temporal dynamics of the pandemic in mainland China, Italy, and France. Roques et al. (2020) estimated the number of people infected with COVID-19 in France using an approach that combines the SIR model for the actual number of cases and infection mortality rate (IFR), a probabilistic model describing the data collection process, and statistical inference. In Rojas et al. (2020) the SIR model is calculated for the different states of the US. To predict the spread of COVID-19 infections in Japan, a stochastic transmission model by extending the SIR model has been presented (Karako et al. 2020). Many scientists have used the SEIR model (the Susceptible-Exposed-Infectious-Removed model), e.g., Wang et al. (2020a) and Kucharski et al. (2020) for analyzing the dynamics of COVID-19 transmission in China, Kumar et al. (2020b) in Italy, Acuña-Zegarra et al. (2020) in Mexico, Kuniya (2020) in Japan, (Xu et al. 2020) in the USA. The spread of the virus was also analyzed using the logistic and Gompertz models (e.g., Rojas et al. 2020).

To analyze the behavior of pandemic, statistical methods for modeling time series are also used. Publications on forecasting the dynamics of confirmed COVID-19 cases are dominated by the ARIMA models (the Autoregressive Integrated Moving Average models). The ARIMA model is a model widely studied in the context of time series and successfully used in the field of health. Its advantages are simple structure, fast applicability, and ability to explain the data set. For example, the ARIMA model was used to determine the overall prevalence of COVID-19 by (Benvenuto et al. 2020; Ceylan 2020; Perone 2020; Ding et al. 2020; Ribeiro et al. 2020; Dehesh et al. 2020). In Ceylan (2020), the forecast of the total confirmed cases for the next ten days using the ARIMA(0,2,1), ARIMA(1,2,0), and ARIMA(0,2,1) for Italy, Spain and France was presented. In Ribeiro et al. (2020), Brazilian data were analyzed using ARIMA models, the cubistic regression, the random forest, ridge regression, the support vector regression, the stacking-ensemble learning and the number of COVID-19 infected was predicted. Also other statistical models such as SutteARIMA (short-term forecasting method) (Ahmar and del Val 2020), ARIMA-WBF (wavelet-based forecasting) (Chakraborty and Ghosh 2020), LSTM (long short-term memory models) (Chimmula and Zhang 2020) were used to predict COVID epidemic cases. In Fong et al. (2020), 11 methods of machine learning (deep learning) were compared to ARIMA in the forecasting of epidemics. The spatiotemporal dynamics of COVID-19 were modeled by Wang et al. (2020b).

The existing literature on the spread of the coronavirus often considers only country-specific time series and ignores cross-country analyzes. Rare exceptions are the works of Kufel (2020) or Kumar et al. (2020a). Kufel (2020) aimed to assess the usefulness of the ARIMA model for predicting the dynamics of COVID-19 incidence at different stages of the epidemic in 32 European countries.

The main purpose of our paper is to analyze the patterns of COVID-19 evolution in a group of 27 EU countries. Without a doubt, the analysis of the evolution of COVID-19 in Europe is of great importance due to the impact of this pandemic on the economy of the entire European Union. Our article contributes to research in two ways. First, we apply the concept of dynamic time warping (DTW) to identify groups of EU countries affected to varying degrees by the COVID-19 pandemic. We present a measure of similarity based on the measurement of the distance between the time series for the number of infected and deceased people in each country. The main advantage of DTW is that the time sequences are not required to be of the same length (in our case, the compared time series differ in length since the outbreak of the pandemic began differently in each country). The DTW method in the analysis of the spread of the COVID-19 epidemic was previously used by Rojas et al. (2020) (for the different states of the USA) and Stübinger and Schneider (2020) (for the 10 most affected countries in the world). Carrying out the classification is very useful as it will identify similarities and patterns in the evolution of a pandemic between EU countries. Further, within the selected groups of states, we analyze the structure of the time series for infected and deceased COVID-19 patients using ARIMA models.

The remainder of this paper is structured in the following way. The first section covers the basics of the dynamic time warping framework. Then, we carry out data characterization. The next two sections provide the results of the DTW application for the 27 EU countries and the results of the ARIMA modeling. Finally, we summarize our work and conclude.

2 Research Methodology

The evolution of COVID-19 time series in the different countries of the EU presents a different start date, both for the number of confirmed and death cases, and therefore its length is also different. To compare such time series, the dynamic time warping (DTW) distance method is frequently used (Aghabozorgi et al. 2015). This method allows an adjustment of the time axis to find similar but phase-shifted sequences.

Formally, the objective of DTW is to compare two time series X and Y, defined by $X = (x_1, x_2, \ldots, x_N)$ and $Y = (y, y_2, \ldots, y_M)$. The similarity of two series having the same length ($N = M$) can be achieved by calculating the Minkowski or Euclidean distance between points on both time series that happen at the same time. DTW is a method that is able to handle time series of different lengths and can be thought of as an extension of distances mentioned above. DTW calculates an optimal match between two given time series, performing non-linearly in the series (by stretching or shrinking along its time axis). This distortion (called warping) between two time series is used to find corresponding regions and determine the similarity between them.

The DTW method, developed in the 1960s (Bellman and Kalaba 1959), is used in a wide spectrum of different applications, e.g., to compare different speech patterns in automatic speech (Rabiner et al. 1978; Myers and Rabiner 1981; Sakoe and Chiba 1978), in the field of music information retrieval (Müller 2007), gesture recognition (Arici et al. 2014), computer animation (to analyze and align motion data), robotics, signal processing, finance (Stübinger 2019), bioinformatics, and medicine.

In the first step of the method, we define the local cost measure (local distance measure) for two elements of the sequences X and Y as $c(x_i, y_j) = |x_i - y_j|$, $i = 1, 2, \ldots, N$, $j = 1, 2, \ldots, M$. Evaluating this measure for each pair of elements of X and Y, one obtains the local cost matrix (LCM) $C \in \mathbb{R}^{N \times M}$. Then the goal is to find an alignment between X and Y having minimal overall cost.

Following Keogh and Ratanamahatana (2005), the point-to-point alignment between series X and Y can be represented by a time warping path, which is a sequence $p = (p_1, \ldots, p_L)$, with $p_l = (n_l, m_l) \in \{1, \ldots, N\} \times \{1, \ldots, M\}$ for $l \in \{1, \ldots, L\} (L \in \{\max(N, M), \ldots, N + M - 1\})$, satisfying the following three conditions:

(i) $p_1 = (1, 1)$ and $p_L = (N, M)$ (boundary condition, which enforces that the first elements of X and Y as well as the last elements of X and Y are aligned to each other),

(ii) $n_1 \leq n_2 \leq \cdots \leq n_L$ and $m_1 \leq m_2 \leq \cdots \leq m_L$ (monotonicity condition; if an element in X precedes a second one this should also hold for the corresponding elements in Y),

(iii) $p_{l+1} - p_l \in \{(1, 0), (0, 1), (1, 1)\}$ for $i = 1, \ldots, L - 1$ (step size condition, which is a kind of continuity condition: no element in X and Y can be omitted and all index pairs in a path are distinct).

The time warping path $p = (p_1, \ldots, p_L)$ defines an alignment between $X = (x_1, x_2, \ldots, x_N)$ and $Y = (y, y_2, \ldots, y_M)$ by assigning the element x_{n_l} of X to the element y_{m_l} of Y (n_l-th element in time series X is mapping to the m_l-th element in time series Y). Every index from the first time series must be matched with one or more indices from the other time series (and vice versa).

The total cost $c_p(X,Y)$ of a warping path p between X and Y with respect to the local cost measure c is defined as

$$c_p(X,Y) = \sum_{l=1}^{L} c(x_{n_l}, y_{m_l}) \tag{1}$$

and is computed as the sum of absolute differences, for each matched pair of indices, between their values.

The optimal match is denoted by the match that has the minimal total cost:

$$\mathrm{DTW}(X,Y) = c_{p^*}(X,Y) = \min\{c_p(X,Y)|p \in P\} \tag{2}$$

The dynamic time warping algorithm finds the path that minimizes the alignment between X and Y by iteratively stepping through the LCM, starting at $c(x_{n_1}, y_{m_1})$ and finishing at $c(x_{n_L}, y_{m_L})$, and aggregating the cost. The optimal warping path (and the minimal distance) could be found using a dynamic programming algorithm. From a technical point of view, the following recursion scheme is applied:

$$\begin{aligned} D(1,m) &= \sum_{k=1}^{m} c(x_1, y_k) \quad \text{for } m = 1,\ldots,M, \\ D(n,1) &= \sum_{k=1}^{n} c(x_k, y_1) \quad \text{for } n = 1,\ldots,N, \\ D(n,m) &= \min\{D(n-1,m-1), D(n-1,m), D(n,m-1)\} \\ &\quad + c(x_n, y_m) \quad \text{for } 1 < n \le N, 1 < m \le M \end{aligned} \tag{3}$$

where D is called the accumulated cost matrix. It is obvious that the minimal distance between time series X and Y is then defined as $\mathrm{DTW}(X,Y) = D(N,M)$.

In this paper, for each of the EU-country analyzed, both the time series of the number of infected and the time series of deaths are simultaneously taken into account. The following parametric metric is defined:

$$\mathrm{DTW} = 0.5 \cdot \mathrm{DTW}(tc_A, tc_B) + 0.5 \cdot \mathrm{DTW}(td_A, td_B) \tag{4}$$

that measures the similarity in the evolution of the COVID time series for two EU countries (A and B), tc_A and tc_B represent the time series of the number of infected, td_A and td_B represent the time series of the number of deaths for the countries A and B, respectively (compare Rojas et al. 2020). According to the adopted definition, the information of the confirmed patients time series has the same relevance as the time series of deaths for the computation of the DTW distance.

In the next step, after measuring the similarities between the time series for all EU countries (DTW matrix), an attempt is made to group them using an agglomerative hierarchical clustering algorithm. The average linkage cluster analysis with the squared Euclidean distance is used to measure the dissimilarity between each pair of observations. The number of clusters is established following the values of the silhouette, Dunn, and Caliński-Harabasz indices. The results are analyzed and compared across the formulated groups of countries, which allows finding interesting patterns in the data.

In the last step of the analysis, ARIMA dynamic models are estimated within the clusters for all countries. The ARIMA(p,d,q) model is determined by three parameters. The parameters p and q are the lag order in the AR(p) component and the MA(q) component, respectively, while d is the differentiation level (Box et al. 2015). The ARIMA(p,d,q) model has the form:

$$\begin{aligned}(1-u)^d Y_t =& \alpha + \phi_1 Y_{t-1} + \phi_2 Y_{t-2} + \ldots + \phi_p Y_{t-p} + \\ & + \varepsilon_t + \theta_1 \varepsilon_{t-1} + \theta_2 \varepsilon_{t-2} + \ldots + \theta_q \varepsilon_{t-q}\end{aligned} \tag{5}$$

where u is the time-shift operator $u^d Y_t = Y_{t-d}$.

To determine the values of p, d, q, we apply the auto.arima() function in R. The auto.arima() uses a combination of unit root tests, minimization of the AIC and MLE to obtain a target ARIMA model. KPSS test determines the number of differences (d) in the Hyndman-Khandakar algorithm for automatic ARIMA modeling (Hyndman and Khandakar 2008). The algorithm uses a stepwise search to traverse the model space to select the best model with smallest AICc. Then, the obtained lag orders and the differentiation levels of the studied time series are analyzed in the particular groups of countries. As a result, knowledge is gained about the mechanisms of the pandemic evolution and it becomes possible to make forecasts.

3 Empirical Data

Our analysis relies on daily data on the COVID-19 epidemic for 27 EU countries obtained from the Web site https://ourworldindata.org/coronavirus-source-data. The methodology presented above is applied to Coronavirus cases from January 25, 2020, to November 30, 2020.

The following time series were analyzed: *tc*—the number of total confirmed cases per million, *td*—the number of total deaths per million, *ncs*—the number of new cases smoothed per million. The time series for EU countries are of different lengths because the virus has not spread to individual countries at the same time (cf. Table 1).

Table 1 Number of days of the time series *tc* and *td* for EU countries

	n_*tc*	n_*td*		n_*tc*	n_*td*		n_*tc*	n_*td*
FRA	312	290	GRC	279	265	LVA	274	242
DEU	309	267	ROU	279	254	PRT	274	259
FIN	307	255	DNK	278	262	HUN	272	261
ITA	305	284	EST	278	251	POL	272	264
SWE	304	266	NLD	278	270	SVN	271	262
ESP	304	273	LTU	277	255	SVK	270	258
BEL	301	265	IRL	276	265	MLT	269	237
HRV	280	257	LUX	276	262	BGR	268	265
AUT	280	264	CZE	275	254	CYP	267	254

Source Own elaboration

4 The Results of the DTW Method

In the first step, we measured the distance between the time series for the total number of infected and deceased people in each country.

The distance between two time series is computed using the DTW method by stretching or compressing them locally in order to make one resemble the other as much as possible (see Fig. 1 for Sweden and Italy, where we additionally present time series for new cases for illustrative purposes only). The alignments were computed invoking the DTW package for R (Giorgino 2009).

The DTW algorithm finds the path that minimizes the alignment between two time series by iteratively stepping through the local cost matrix and aggregating the cost. Figure 2 gives a visual representation of the optimal warping paths corresponding to Fig. 1 for Sweden and Italy. The shapes of the warping curves provide information about the pairwise correspondences of time points.

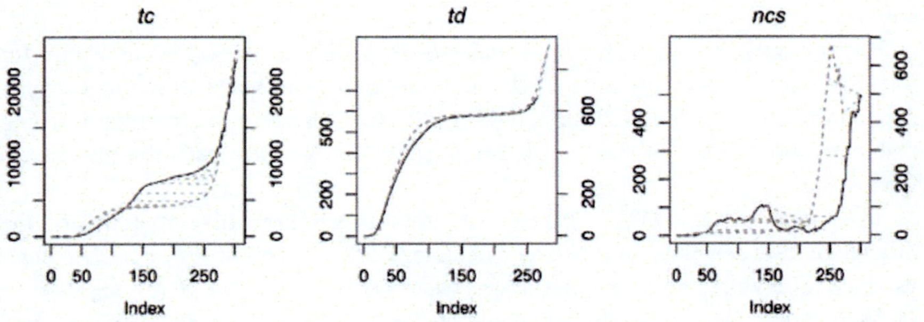

Fig. 1 The alignments performed by the DTW algorithm between time series for total cases per million (*tc*), total deaths per million (*td*), new cases smoothed per million (*ncs*) in Sweden and in Italy (the solid lines—time series for Sweden, the dashed lines—time series for Italy)

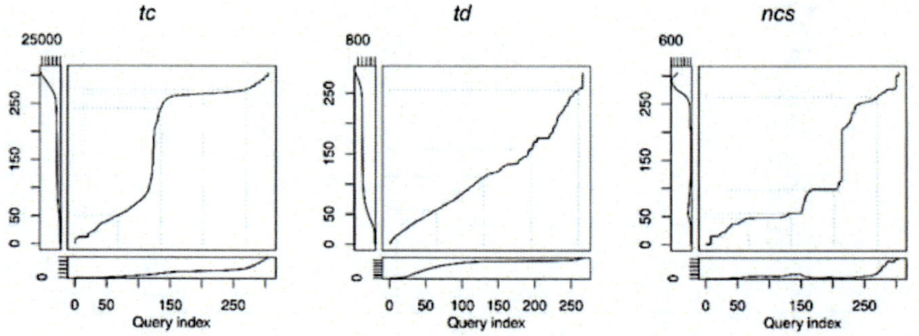

Fig. 2 Three-way plots of the time series alignments: visual representation of the optimal warping paths (time series *tc*, *td* and *ncs* for Sweden and Italy; query indices for Sweden, reference indices for Italy)

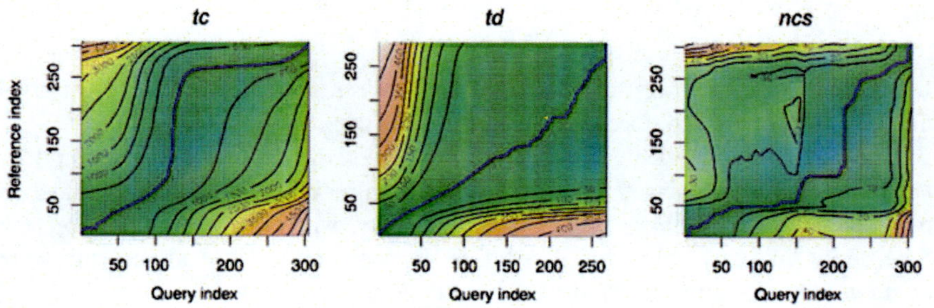

Fig. 3 Local costs and the optimal warping paths (solid lines) for the alignments between time series *tc*, *td* and *ncs* for Sweden and Italy (query indices for Sweden, reference indices for Italy). Regions of low cost (high cost) are presented by dark colors (light colors)

Figure 3 illustrates the local costs and the identified optimal warping paths $p*$ given the adequate time series. Graphically, the sequence of points $p*$ runs along a "valley" of low cost (dark colors) and avoids "mountains" of high cost (light colors).

For comparison purposes, Fig. 4 presents the output of the DTW algorithm for total cases per million (*tc*), total deaths per million (*td*) and new cases smoothed per million (*ncs*) in Sweden and in Poland. In this case, the identified optimal warping paths are above the diagonal, i.e., the time series for Sweden leads the time series for Poland.

The value of the DTW distance, i.e., the stretch-insensitive measure of the difference between time series, was calculated for each country. For further analysis, the following parametric metric was used: $\mathrm{DTW} = 0.5 \cdot \mathrm{DTW}(tc_A, tc_B) + 0.5 \cdot \mathrm{DTW}(td_A, td_B)$ (only the number of infected and deceased people are taken into account for the countries A and B, respectively). The calculated distances for 27 EU countries have a straightforward application in hierarchical clustering and

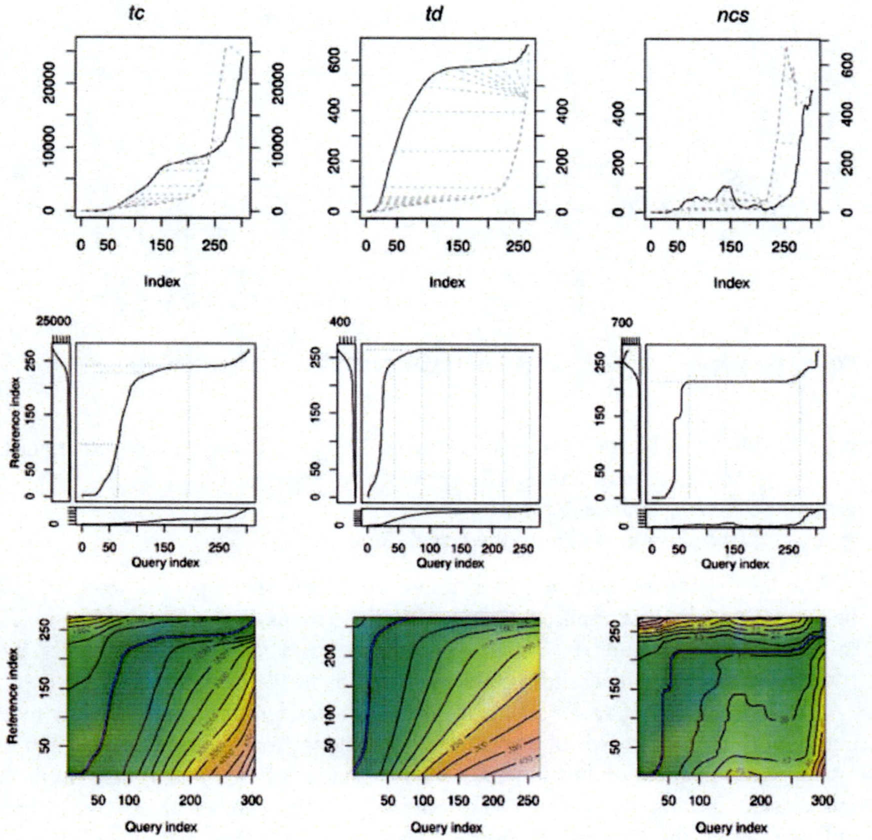

Fig. 4 The alignments performed by the DTW algorithm between time series for *tc*, *td*, *ncs* in Sweden and in Poland (the solid lines and query indices for Sweden, the dashed lines and reference indices for Poland)

classification (Sardá-Espinosa 2019). Therefore, after measuring the similarities between the time series for the EU countries an attempt was made to group them using agglomerative hierarchical clustering algorithm. To carry out the hierarchical cluster tree, average linkage was used with the squared Euclidean distance. The hierarchical cluster tree obtained is presented in Fig. 5.

To analyze the accuracy of the obtained hierarchical clustering, the silhouette, Dunn, and Caliński-Harabasz indices were used. The larger the values of these indices, the better the data partition (the clustering performed). In the problem presented in the paper, the optimal number of clusters was five with the following distribution:

group 1: Finland, Denmark, Germany, Cyprus, Estonia, Greece, Latvia,
group 2: Ireland,

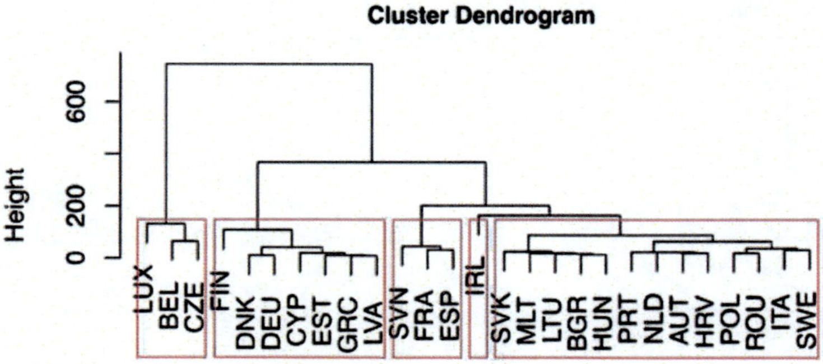

Fig. 5 The hierarchical cluster tree obtained using the DTW as distance metric

group 3: Slovakia, Malta, Lithuania, Bulgaria, Hungary, Portugal, the Netherlands, Austria, Croatia, Poland, Romania, Italy, Sweden,
group 4: Slovenia, France, Spain,
group 5: Luxembourg, Belgium, the Czech Republic.

The time series for total cases per million (*tc*), total deaths per million (*td*), and new cases smoothed per million (*ncs*) in identified groups of countries are examined in Figs. 6, 7, 8, 9 and 10. The results are analyzed and compared across the formulated groups, which allow us to find patterns in the data.

Group 1 consists of seven countries. It is characterized by a relatively low total number of cases up to November, 30 (values ranging from approximately 4000 to 14,000 cases per million inhabitants). The low values of total deaths also occur in this cluster (from about 50 to 230 per million) and the number of new cases (per million) during the first and the second wave of coronavirus were also low (see Fig. 6).

Group 2 is a single-element group (made up of Ireland) and is characterized by the low number of total cases and deaths per million (about 14,000 and 400, respectively). It should be noted that Ireland was marked by a strong first wave of coronavirus and the second wave was also high (Fig. 7).

Group 3, the largest group, includes the set of 13 countries (Slovakia, Malta, Lithuania, Bulgaria, Hungary, Portugal, the Netherlands, Austria, Croatia, Poland,

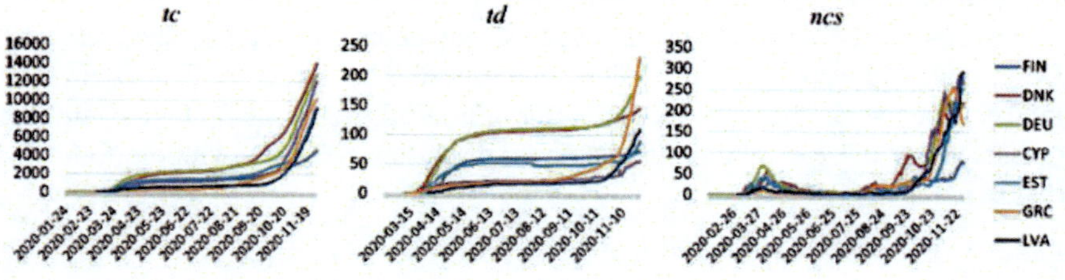

Fig. 6 Total cases (*tc*), total deaths (*td*), and new cases per million (*ncs*) in group 1

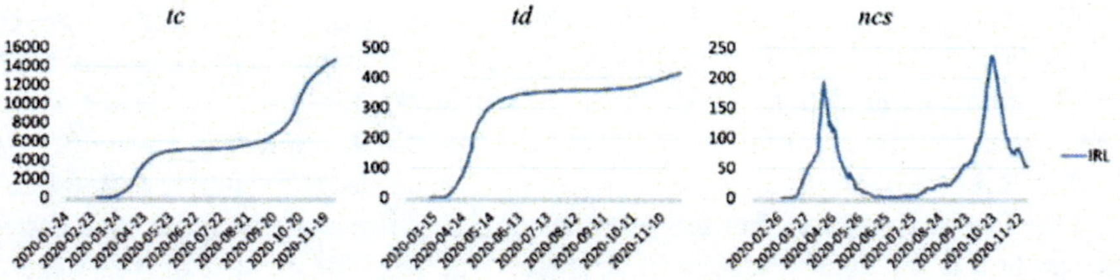

Fig. 7 Total cases (*tc*), total deaths (*td*), and new cases per million (*ncs*) in group 2

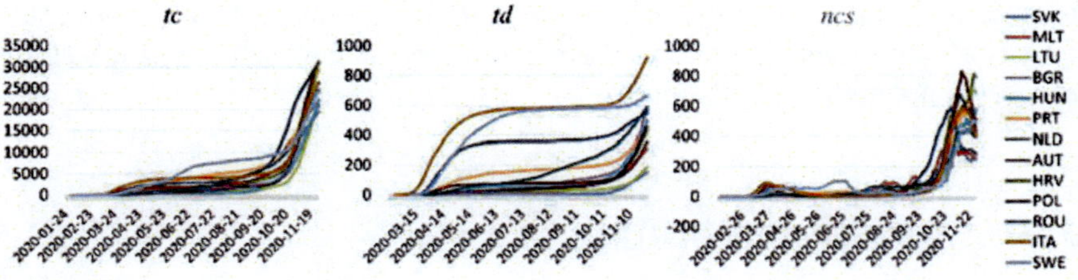

Fig. 8 Total cases (*tc*), total deaths (*td*), and new cases per million (*ncs*) in group 3

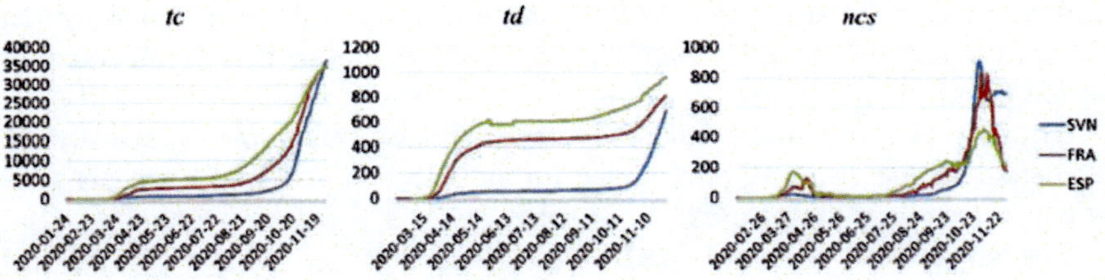

Fig. 9 Total cases (*tc*), total deaths (*td*), and new cases per million (*ncs*) in group 4

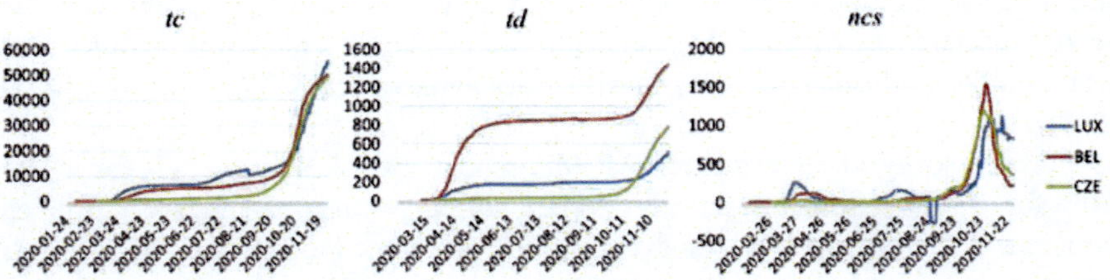

Fig. 10 Total cases (*tc*), total deaths (*td*), and new cases per million (*ncs*) in group 5

Romania, Italy, Sweden). In these countries, the epidemic situation is diverse, although the number of total cases is higher than in the previous groups (about 25,000 per million) and the number of deaths reaches 900 per million (see Fig. 8).

Group 4, with three countries (Slovenia, France, and Spain), presents bigger values of total cases (about 35,000 per million inhabitants) and values of total deaths from about 700 to 1000 per million (Fig. 9). However, in cluster 4 the number of new cases during the second wave was similar to that in cluster 3.

The last group, group 5, consists also of three countries (Luxembourg, Belgium, and the Czech Republic). For the countries in this group, the number of total cases per million is the highest (about 50,000) (see Fig. 10). The number of total deaths per million ranges from 500 to 1400. Luxembourg displays a value of 500, the Czech Republic a value of 800, and Belgium much more—about 1400 deaths per million inhabitants (results for Luxembourg, however, are of questionable quality because errors have been reported in the number of COVID-19 patients in this country). These countries are characterized by a very large second wave (up to 1500 new cases per million per day).

5 The Results of the ARIMA Modeling

Once the hierarchical clustering of the EU countries has been established and performed, it is relevant to model the time series (both infected and dead patients) within the clusters, using the ARIMA models. Table 2 presents the lag order parameters p and q as well as the differentiation level d of models for all countries within the clusters.

The total cases per million (tc) and total deaths per million (td) were analyzed. To determine the values of p, d, q, and for the selection of ARIMA models auto. arima() function in R was used.

For all of the estimated ARIMA(p,d,q) models for p, $q = \{0,1,2,3,4,5\}$ and $d = \{0,1,2\}$, the minimum value of the AIC criterion pointed to the ARIMA (p,2,q) model. When analyzing the time series tc (or td, respectively), the first differences $\Delta tc_t = tc_t - tc_{t-1}$ (or $\Delta td_t = td_t - td_{t-1}$) indicated the daily number of infections per million (or the daily number of deaths per million, respectively). The second difference $\Delta\Delta tc_t = \Delta tc_t - \Delta tc_{t-1} = tc_t - 2tc_{t-1} + tc_{t-2}$ (or $\Delta\Delta td_t$) was statistically significant due to the nonstationary variance for Δtc_t (or Δtd_t) (cf. Kufel 2020).

For a majority of the estimated ARIMA models (for 78% of models for tc and 70% for td) the lag order parameters p were not bigger than parameters q. The lag structure in the MA part was, on average, more developed than in the AR part. It is worth noting that a series displays moving average behavior if it undergoes random "shocks" whose effects are felt in two or more consecutive periods. For time series tc, the parameter $p = 0$ was recorded in 6 models (which means that 22% of all models showed no autoregressive structure), while for the td series, $p = 0$ was noted for 7 models (26% of all models). In contrast, the lack of moving average structure was shown only by 7% of the tc models and 0% of the td models.

Models with extremely high values of parameter p ($p \geq 4$) occurred for countries in group 3 (e.g., Slovakia). On the other hand, models with extremely

Table 2 Lag order parameters *p* and *q* and the differentiation level *d* of ARIMA models for 27 EU countries

Group		*tc*			*td*		
		p	*d*	*q*	*p*	*d*	*q*
1	FIN	0	2	1	2	2	1
	DNK	1	2	2	0	2	3
	DEU	1	2	5	0	2	2
	CYP	3	2	3	2	2	1
	EST	2	2	2	0	2	1
	GRC	3	2	2	1	2	2
	LVA	2	2	2	2	2	2
2	IRL	2	2	0	3	2	2
3	SVK	5	2	1	4	2	1
	MLT	0	2	1	3	2	1
	LTU	5	2	1	2	2	2
	BGR	2	2	3	0	2	4
	HUN	1	2	2	2	2	2
	PRT	0	2	2	1	2	2
	NLD	1	2	1	0	2	2
	AUT	5	2	0	2	2	1
	HRV	1	2	5	1	2	5
	POL	2	2	3	0	2	4
	ROU	0	2	5	2	2	2
	ITA	0	2	5	1	2	3
	SWE	0	2	1	2	2	5
4	SVN	1	2	4	2	2	2
	FRA	3	2	4	2	2	1
	ESP	2	2	5	0	2	1
5	LUX	2	2	1	3	2	4
	BEL	2	2	2	3	2	5
	CZE	2	2	2	2	2	1

Source Own elaboration

high q values ($q = 5$) relate to a large number of infected and deceased patients for countries in groups 3, 4, and 5 (e.g., *tc* in Spain, *td* in Belgium). In these cases, the extinction of the epidemic is still not coming and it will last for a longer time.

For the time series td, it can be observed that the sum of lag orders $p + q$ in ARIMA models increases with the number of classification groups (the higher the group number, the richer the AR and MA structure of the model). Unfortunately, no regularities can be observed for the lag orders p and q in ARIMA models for the time series *tc*, within individual groups of countries. Nevertheless, the smallest average difference between the parameters q and p occurs for group 2 (Ireland) and the largest for group 4 (Slovenia, France, and Spain). At the end of November 2020, the epidemic was slowing down in Ireland, while in Slovenia it escalated.

6 Conclusions

The main goal of this paper was to hierarchically group the EU countries and, within the created groups, to compare the evolution of infected and deceased COVID-19 patients. Grouping time series is a process, with the aim of finding behavioral similarities between the different time series that are analyzed. In this paper, we used the dynamic time warping distance to measure the similarity between time series corresponding to different EU countries, taking into account the behavior of the number of COVID-19 infected and deceased people due to COVID-19.

The optimal number of clusters in which the different countries can be grouped was obtained. A total of five clusters were found. Some of them were groups with a large number of countries (e.g., group 1 or group 3) and one group had only one state (group 2), indicating that his behavior was unique. Within the groups we created, we compared the evolution of the number of infected and deceased COVID-19 patients in each country.

A further goal was to estimate ARIMA models for the time series of infected and dead patients for each country within the created groups. These models have been created by considering the most appropriate AIC values, and the analysis of their parameters was useful to assess the dynamics of the epidemic. ARIMA(p,2, q) models were always chosen (similar to Ceylan 2020). For a majority of the estimated models the lag structure in the MA part was more developed than in the AR part. This confirms that the mechanism of the pandemic evolution is subject to random shocks. Despite this, ARIMA models may enable short-term forecasting of epidemic development (cf. Kufel 2020).

We conclude that the applied methodology is able to detect relationships between EU countries regarding the development of the epidemic, and it is possible to identify some patterns in the underlying data of the EU countries suffering from COVID-19.

References

Acuña-Zegarra M, Santana-Cibrian M, Velasco-Hernandez J (2020) Modeling behavioral change and COVID-19 containment in Mexico: a trade-off between lockdown and compliance. Math Biosci 325(108370). https://doi.org/10.1016/j.mbs.2020.108370

Aghabozorgi S, Shirkhorshidi AS, Wah TY (2015) Time-series clustering—A decade review. Inform Syst 53:16–38. https://doi.org/10.1016/j.is.2015.04.007

Ahmar AS, del Val EB (2020) SutteARIMA: short-term forecasting method, a case: COVID-19 and stock market in Spain. Sci Total Environ 729:138883. https://doi.org/10.1016/j.scitotenv.2020.138883

Arici T, Celebi S, Aydin AS, Temiz TT (2014) Robust gesture recognition using feature pre-processing and weighted dynamic time warping. Multimed Tools Appl 72:3045–3062. https://doi.org/10.1007/s11042-013-1591-9

Bellman R, Kalaba R (1959) On adaptive control processes. IRE T Autom Control 4(2):1–9. https://doi.org/10.1109/TAC.1959.1104847

Benvenuto D, Giovanetti M, Vassallo L, Angeletti S, Ciccozzi M (2020) Application of the ARIMA model on the COVID-2019 epidemic dataset. Data Brief 29: https://doi.org/10.1016/j.dib.2020.105340

Box GE, Jenkins GM, Reinsel GC, Ljung GM (2015) Time series analysis: forecasting and control. John Wiley & Sons, Hoboken

Brauer F, Castillo-Chavez C, Feng Z (2019) Mathematical models in epidemiology. Springer-Verlag, New York. https://doi.org/10.1007/978-1-4939-9828-9

Ceylan Z (2020) Estimation of COVID-19 prevalence in Italy, Spain, and France. Sci Total Environ 729: https://doi.org/10.1016/j.scitotenv.2020.138817

Chakraborty T, Ghosh I (2020) Real-time forecasts and risk assessment of novel coronavirus (COVID-19) cases: a data-driven analysis. Chaos Soliton Fract 135: https://doi.org/10.1016/j.chaos.2020.109850

Chimmula VKR, Zhang L (2020) Time series forecasting of COVID-19 transmission in Canada using LSTM networks. Chaos Soliton Fract 135: https://doi.org/10.1016/j.chaos.2020.109864

Dehesh T, Mardani-Fard HA, Dehesh P (2020) Forecasting of COVID-19 confirmed cases in different countries with ARIMA models. medRxiv preprint. https://doi.org/10.1101/2020.03.13.20035345

Ding G, Li X, Shen Y, Fan J (2020) Brief analysis of the ARIMA model on the COVID-19 in Italy. medRxiv preprint. https://doi.org/10.1101/2020.04.08.20058636

Fanelli D, Piazza F (2020) Analysis and forecast of COVID-19 spreading in China, Italy and France. Chaos Soliton Fract 134: https://doi.org/10.1016/j.chaos.2020.109761

Fong SJ, Li G, Dey N, Crespo RG, Herrera-Viedma E (2020) Composite Monte Carlo decision making under high uncertainty of novel coronavirus epidemic using hybridized deep learning and fuzzy rule induction. Appl Soft Comp 93: https://doi.org/10.1016/j.asoc.2020.106282

Giorgino T (2009) Computing and visualizing dynamic time warping alignments in R: the dtw package. J Stat Softw 31(7):1–24. https://doi.org/10.18637/jss.v031.i07

Hyndman RJ, Khandakar Y (2008) Automatic time series forecasting: the forecast package for R. J Stat Softw 27(3):1–22. https://doi.org/10.18637/jss.v027.i03

Karako K, Song PP, Chen Y, Tang W (2020) Analysis of COVID-19 infection spread in Japan based on stochastic transition model. Biosci Trends 14(2):134–138. https://doi.org/10.5582/bst.2020.01482

Keogh E, Ratanamahatana CA (2005) Exact indexing of dynamic time warping. Knowl Inf Syst 7:358–386. https://doi.org/10.1007/s10115-004-0154-9

Kermack WO, McKendrick AG (1927) A contribution to the mathematical theory of epidemics. Proc Roy Soc A 115:700–721. https://doi.org/10.1098/rspa.1927.0118

Kucharski AJ, Russell TW, Diamond C, Liu Y, Edmunds J, Funk S, Eggo RM (2020) Early dynamics of transmission and control of COVID-19: a mathematical modelling study. Lancet Infect Dis 20(5):553–558. https://doi.org/10.1016/S1473-3099(20)30144-4

Kufel T (2020) ARIMA-based forecasting of the dynamics of confirmed COVID-19 cases for selected European countries. Equilibrium. Quart J Econ Econ Policy 15(2):181–204. https://doi.org/10.24136/eq.2020.009

Kumar P, Kalita H, Patairiya S, Sharma YD, Nanda C, Rani M, Rahmani J, Bhagavathula AS (2020a) Forecasting the dynamics of COVID-19 pandemic in top 15 countries in April 2020: ARIMA model with machine learning approach. medRxiv preprint. https://doi.org/10.1101/2020.03.30.20046227

Kumar S, Sharma S, Kumari N (2020b) Future of COVID-19 in Italy: a mathematical perspective. arXiv preprint arXiv:2004.08588

Kuniya T (2020) Prediction of the epidemic peak of coronavirus disease in Japan. J Clin Med 9 (3):789. https://doi.org/10.3390/jcm9030789

Müller M (2007) Information retrieval for music and motion. Springer-Verlag, Berlin, Heidelberg. https://doi.org/10.1007/978-3-540-74048-3

Myers CS, Rabiner LR (1981) A comparative study of several dynamic time-warping algorithms for connected word recognition. Bell Syst Tech J 60(7):1389–1409. https://doi.org/10.1002/j.1538-7305.1981.tb00272.x

Perone G (2020) An ARIMA model to forecast the spread and the final size of COVID-2019 epidemic in Italy. HEDG—Health econometrics and data group working paper series, University of York. https://doi.org/10.2139/ssrn.3564865

Rabiner L, Rosenberg A, Levinson S (1978) Considerations in dynamic time warping algorithms for discrete word recognition. IEEE Trans Acous Speech Signal Process 26(6):575–582. https://doi.org/10.1109/tassp.1978.1163164

Ribeiro MHDM, da Silva RG, Mariani VC, Coelho L (2020) Short-term forecasting COVID-19 cumulative confirmed cases: perspectives for Brazil. Chaos Soliton Fract 135: https://doi.org/10.1016/j.chaos.2020.109853

Rojas F, Valenzuela O, Rojas I (2020) Estimation of COVID-19 dynamics in the different states of the United States using time-series clustering. medRxiv preprint. https://doi.org/10.1101/2020.06.29.20142364

Roques L, Klein EK, Papaix J, Sar A, Soubeyrand S (2020) Using early data to estimate the actual infection fatality ratio from COVID-19 in France. Biology 9(5):97. https://doi.org/10.3390/biology9050097

Ruktanonchai NW, Floyd JR, Lai1 S, Ruktanonchai1 CW, Sadilek A, Rente-Lourenco P, Ben X, Carioli A, Gwinn J, Steele J E, Prosper O, Schneider A, Oplinger A, Eastham P, Tatem AJ (2020) Assessing the impact of coordinated COVID-19 exit strategies across Europe. Science 369(6510):1465–1470. https://doi.org/10.1126/science.abc5096

Sakoe H, Chiba S (1978) Dynamic programming algorithm optimization for spoken word recognition. IEEE T Acoust Speech 26(1):43–49. https://doi.org/10.1109/tassp.1978.1163055

Sardá-Espinosa A (2019) Time-series clustering in R using the dtwclust package. R J 11(01):22–43. https://doi.org/10.32614/RJ-2019-023

Stübinger J (2019) Statistical arbitrage with optimal causal paths on high-frequency data of the S&P 500. Quant Financ 19:921–935. https://doi.org/10.1080/14697688.2018.1537503

Stübinger J, Schneider L (2020) Epidemiology of coronavirus COVID-19: forecasting the future incidence in different countries. Healthcare 8(2):99. https://doi.org/10.3390/healthcare8020099

Vattay G (2020) Forecasting the outcome and estimating the epidemic model parameters from the fatality time series in COVID-19 outbreaks. Phys Biol 17(6): https://doi.org/10.1088/1478-3975/abac69

Wang H, Wang Z, Dong Y, Chang R, Xu C, Yu X, Zhang S, Tsamlag L, Shang M, Huang J, Wang Y, Xu G, Shen T, Zhang X, Cai Y (2020a) Phase-adjusted estimation of the number of coronavirus disease 2019 cases in Wuhan, China. Cell Discov 6:10. https://doi.org/10.1038/s41421-020-0148-0

Wang L, Wang G, Gao L, Li X, Yu S, Kim M, Wang Y, Gu Z (2020b) Spatiotemporal dynamics, nowcasting and forecasting of COVID-19 in the United States. arXiv preprint arXiv:2004.14103

Worldometer (2020) COVID-19 Coronavirus pandemic. https://www.worldometers.info/coronavirus/. Accessed 15 Dec 2020

Xu C, Yu Y, Chen Y, Lu Z (2020) Forecast analysis of the epidemics trend of COVID-19 in the United States by a generalized fractional-order SEIR model. medRxiv preprint. https://doi.org/10.1101/2020.04.24.20078493